Lecture Notes in Computer Scie

Commenced Publication in 1973
Founding and Former Series Editors:
Gerhard Goos, Juris Hartmanis, and Jan van Leeuwen

Editorial Board

Goichiro Hanaoka Toshihiro Yamauchi (Eds.)

Advances in Information and Computer Security

7th International Workshop on Security, IWSEC 2012
Fukuoka, Japan, November 7-9, 2012
Proceedings

Volume Editors

Goichiro Hanaoka
National Institute of Advanced Industrial Science and Technology (AIST)
Research Institute for Secure Systems (RISEC)
1-1-1 Umezono, 305-8568 Tsukuba, Japan
E-mail: hanaoka-goichiro@aist.go.jp

Toshihiro Yamauchi
Okayama University
Graduate School of Natural Science and Technology
3-1-1 Tsushima-naka, 700-8530 Okayama, Japan
E-mail: yamauchi@cs.okayama-u.ac.jp

ISSN 0302-9743 e-ISSN 1611-3349
ISBN 978-3-642-34116-8 e-ISBN 978-3-642-34117-5
DOI 10.1007/978-3-642-34117-5
Springer Heidelberg Dordrecht London New York

Library of Congress Control Number: 2012948601

CR Subject Classification (1998): E.3, G.2.1, D.4.6, K.6.5, K.4.4, F.2.1, C.2

LNCS Sublibrary: SL 4 – Security and Cryptology

Typesetting: Camera-ready by author, data conversion by Scientific Publishing Services, Chennai, India

Printed on acid-free paper

Springer is part of Springer Science+Business Media (www.springer.com)

Preface

The 7th International Workshop on Security (IWSEC 2012) was held at Nishijin Plaza, Kyushu University, in Fukuoka, Japan, during November 7–9, 2012. The workshop was co-organized by ISEC in ESS of IEICE (Technical Committee on Information Security in Engineering Sciences Society of the Institute of Electronics, Information and Communication Engineers) and CSEC of IPSJ (Special Interest Group on Computer Security of Information Processing Society of Japan).

This year, the workshop received 53 submissions, of which 16 were accepted for presentation. Each submission was anonymously reviewed by at least five reviewers, and these proceedings contain the revised versions of the accepted papers. In addition to the presentations of the papers, the workshop also featured a poster session and four invited talks. The invited talks were given by James Hughes, Matt Bishop, Suguru Yamaguchi, and Katsuyuki Takashima.

The best paper award was given to "Boomerang Distinguishers for Full HAS-160 Compression Function" by Yu Sasaki, Lei Wang, Yasuhiro Takasaki, Kazuo Sakiyama, and Kazuo Ohta, and the best student paper award was given to "Efficient Concurrent Oblivious Transfer in Super-Polynomial-Simulation Security" by Susumu Kiyoshima, Yoshifumi Manabe, and Tatsuaki Okamoto.

A number of people contributed to the success of IWSEC 2012. We would like to thank the authors for submitting their papers to the workshop. The selection of the papers was a challenging and delicate task, and we are deeply grateful to the members of Program Committee and the external reviewers for their in-depth reviews and detailed discussions. We are also grateful to Andrei Voronkov for developing EasyChair, which was used for the paper submission, reviews, discussions, and preparation of these proceedings.

Last but not least, we would like to thank the General Co-chairs, Tsutomu Matsumoto and Kanta Matsuura, for leading the Local Organizing Committee, and we also would like to thank the members of the Local Organizing Committee for their efforts to ensure the smooth running of the workshop.

August 2012

Goichiro Hanaoka
Toshihiro Yamauchi

IWSEC 2012
7th International Workshop on Security

Fukuoka, Japan, November 7–9, 2012
Co-organized by
ISEC in ESS of IEICE
(Technical Committee on Information Security in Engineering Sciences Society of the Institute of Electronics, Information and Communication Engineers)
and
CSEC of IPSJ
(Special Interest Group on Computer Security of Information Processing Society of Japan)

General Co-chairs

Tsutomu Matsumoto	Yokohama National University, Japan
Kanta Matsuura	The University of Tokyo, Japan

Advisory Committee

Hideki Imai	Chuo University, Japan
Kwangjo Kim	Korea Advanced Institute of Science and Technology, Korea
Günter Müller	University of Freiburg, Germany
Yuko Murayama	Iwate Prefectural University, Japan
Koji Nakao	National Institute of Information and Communications Technology, Japan
Eiji Okamoto	University of Tsukuba, Japan
C. Pandu Rangan	Indian Institute of Technology, Madras, India

Program Co-chairs

Goichiro Hanaoka	AIST, Japan
Toshihiro Yamauchi	Okayama University, Japan

Local Organizing Committee

Yoshiaki Hori	Kyushu University, Japan
Takuro Hosoi	The University of Tokyo, Japan
Mitsugu Iwamoto	The University of Electro-Communications, Japan

Takehisa Kato	Toshiba Solutions Corporation, Japan
Akinori Kawachi	Tokyo Institute of Technology, Japan
Fagen Li	Kyushu University, Japan
Kirill Morozov	Kyushu University, Japan
Takashi Nishide	Kyushu University, Japan
Masayuki Numao	The University of Electro-Communications, Japan
Naoyuki Shinohara	National Institute of Information and Communications Technology, Japan
Mio Suzuki	National Institute of Information and Communications Technology, Japan
Tsuyoshi Takagi	Kyushu University, Japan
Kenji Yasunaga	Institute of Systems, Information Technologies and Nanotechnologies, Japan

Program Committee

Rafael Accorsi	University of Freiburg, Germany
Claudio Ardagna	Università degli Studi di Milano, Italy
Nuttapong Attrapadung	AIST, Japan
Andrey Bogdanov	Katholieke Universiteit Leuven, Belgium
Kevin Butler	University of Oregon, USA
Sanjit Chatterjee	Indian Institute of Science, India
Haibo Chen	Shanghai Jiao Tong University, China
Pau-Chen Cheng	IBM Thomas J. Watson Research Center, USA
Koji Chida	NTT, Japan
Bart De Decker	Katholieke Universiteit Leuven, Belgium
Isao Echizen	National Institute of Informatics, Japan
Dario Fiore	New York University, USA
Georg Fuchsbauer	University of Bristol, UK
Eiichiro Fujisaki	NTT, Japan
Steven Furnell	University of Plymouth, UK
David Galindo	University of Malaga, Spain
Dieter Gollmann	Hamburg University of Technology, Germany
Swee-Huay Heng	Multimedia University, Malaysia
Jin Hong	Seoul National University, Korea
Tetsu Iwata	Nagoya University, Japan
Angelos D. Keromytis	Columbia University, USA
Hyung Chan Kim	ETRI, Korea
Takeshi Koshiba	Saitama University, Japan
Noboru Kunihiro	University of Tokyo, Japan
Kwok-Yan Lam	Tsinghua University, China
Benoit Libert	Université Catholique de Louvain, Belgium
Javier Lopez	University of Malaga, Spain
Keith Martin	Royal Holloway, University of London, UK
Masakatsu Nishigaki	Shizuoka University, Japan

Wakaha Ogata	Tokyo Institute of Technology, Japan
Thomas Peyrin	Nanyang Technological University, Singapore
Raphael Phan	Loughborough University, UK
Axel Poschmann	Nanyang Technological University, Singapore
Bart Preneel	Katholieke Universiteit Leuven, Belgium
Kai Rannenberg	Goethe University Frankfurt, Germany
Kazuo Sakiyama	University of Electro-Communications, Japan
Ryoichi Sasaki	Tokyo Denki University, Japan
Joshua Schiffman	Pennsylvania State University, USA
Jae Hong Seo	NICT, Japan
Francesco Sica	University of Waterloo, Canada
Tsuyoshi Takagi	Kyushu University, Japan
Keisuke Takemori	KDDI R&D Laboratories Inc., Japan
Keisuke Tanaka	Tokyo Institute of Technology, Japan
Masayuki Terada	NTT DoCoMo, Japan
Ryuya Uda	Tokyo University of Technology, Japan
Damien Vergnaud	ENS, France
Sabrina De Capitani di Vimercati	University of Milan, Italy
Guilin Wang	University of Wollongong, Australia
Jian Weng	Jinan University, China
Sven Wohlgemuth	Sirrix AG security technologies, Germany
Keita Xagawa	NTT, Japan
Chung-Huang Yang	National Kaohsiung Normal University, Taiwan
Kazuki Yoneyama	NTT, Japan
Katsunari Yoshioka	Yokohama National University, Japan
Hiroshi Yoshiura	University of Electro-Communications, Japan
Rui Zhang	CAS, China
Yunlei Zhao	Fudan University, China

External Reviewers

Elena Andreeva
Lejla Batina
Ji-Jian Chin
Angelo De Caro
Junfeng Fan
Pooya Farshim
Yanfei Guo
Keisuke Hakuta
Takuya Hayashi
Jens Hermans
Johann Heyszl
Takashi Horiyama
Kota Ideguchi
Toshiyuki Isshiki
Malika Izabachene
Charanjit Jutla
Christian Kahl
Chethan Kamath
Akinori Kawachi
Ryo Kikuchi
Ilya Kizhvatov
Ulrich Kuehn
Sebastian Kutzner
Shugo Mikami
Kirill Morozov
Nicky Mouha

Divya Muthukumaran
Phuong Ha Nguyen
Hannah Pruse
Arnab Roy
Ahmad Sabouri
Seog Chung Seo
Kyoji Shibutani
Wook Shin
Ingo Stengel
Mario Strefler
Susan Thomson
Leif Uhsadel
Kerem Varici
Hayawardh Vijayakumar
Lei Wang
Lei Wei
Wun-She Yap
Takanori Yasuda
Kenji Yasunaga
Wei-Chuen Yau
Wei Zhang

Table of Contents

Implementation

Model-Based Conformance Testing for Android 1
Yiming Jing, Gail-Joon Ahn, and Hongxin Hu

Application of Scalar Multiplication of Edwards Curves to Pairing-Based Cryptography 19
Takanori Yasuda, Tsuyoshi Takagi, and Kouichi Sakurai

Standardized Signature Algorithms on Ultra-constrained 4-Bit MCU ... 37
Chien-Ning Chen, Nisha Jacob, Sebastian Kutzner, San Ling, Axel Poschmann, and Sirote Saetang

Very Short Critical Path Implementation of AES with Direct Logic Gates 51
Kenta Nekado, Yasuyuki Nogami, and Kengo Iokibe

Encryption and Key Exchange

One-Round Authenticated Key Exchange with Strong Forward Secrecy in the Standard Model against Constrained Adversary 69
Kazuki Yoneyama

Compact Stateful Encryption Schemes with Ciphertext Verifiability 87
S. Sree Vivek, S. Sharmila Deva Selvi, and C. Pandu Rangan

Structured Encryption for Conceptual Graphs 105
Geong Sen Poh, Moesfa Soeheila Mohamad, and Muhammad Reza Z'aba

Symmetric-Key Encryption Scheme with Multi-ciphertext Non-malleability 123
Akinori Kawachi, Hirotoshi Takebe, and Keisuke Tanaka

Cryptanalysis

Slide Cryptanalysis of Lightweight Stream Cipher RAKAPOSHI 138
Takanori Isobe, Toshihiro Ohigashi, and Masakatu Morii

Boomerang Distinguishers for Full HAS-160 Compression Function 156
Yu Sasaki, Lei Wang, Yasuhiro Takasaki, Kazuo Sakiyama, and Kazuo Ohta

Polynomial-Advantage Cryptanalysis of 3D Cipher and 3D-Based Hash Function 170
Lei Wang, Yu Sasaki, Kazuo Sakiyama, and Kazuo Ohta

Annihilators of Fast Discrete Fourier Spectra Attacks 182
Jingjing Wang, Kefei Chen, and Shixiong Zhu

Meet-in-the-Middle Attack on Reduced Versions of the Camellia Block Cipher 197
Jiqiang Lu, Yongzhuang Wei, Enes Pasalic, and Pierre-Alain Fouque

Secure Protocols

Efficient Concurrent Oblivious Transfer in Super-Polynomial-Simulation Security 216
Susumu Kiyoshima, Yoshifumi Manabe, and Tatsuaki Okamoto

Efficient Secure Primitive for Privacy Preserving Distributed Computations 233
Youwen Zhu, Tsuyoshi Takagi, and Liusheng Huang

Generic Construction of GUC Secure Commitment in the KRK Model 244
Itsuki Suzuki, Maki Yoshida, and Toru Fujiwara

Author Index 261

Model-Based Conformance Testing for Android

Yiming Jing[1], Gail-Joon Ahn[1], and Hongxin Hu[2]

[1] Laboratory of Security Engineering for Future Computing (SEFCOM)
Arizona State University, Tempe, AZ85281, USA
{ymjing,gahn}@asu.edu
[2] Delaware State University, Dover, DE19901, USA
hxhu@asu.edu

Abstract. With the surging computing power and network connectivity of smartphones, more third-party applications and services are deployed on these platforms and enable users to customize their mobile devices. Due to the lack of rigorous security analysis, fast evolving smartphone platforms, however, have suffered from a large number of system vulnerabilities and security flaws. In this paper, we present a model-based conformance testing framework for mobile platforms, focused on Android platform. Our framework systematically generates test cases from the formal specification of the mobile platform and performs conformance testing with the generated test cases. We also demonstrate the feasibility and effectiveness of our framework through case studies on Android Inter-Component Communication module.

1 Introduction

According to a recent report from research firm [5], the worldwide smartphone market ballooned 65.4% year over year in the second quarter of 2011, indicating the total shipments of 100 million units. In addition, with the surging computing power and network connectivity of smartphones, more third-party applications and services are deployed on these platforms and enable users to customize their devices. Many legitimate applications tend to manipulate users' sensitive information such as contact list, locale information, and other credentials [14]. To protect such sensitive attributes, it is necessary to ensure that smartphones are properly configured and rigorously validated.

Fast evolving smartphone platforms, however, have raised considerable security concerns due to the lack of rigorous security analysis. At the same time, a large number of system vulnerabilities and security flaws on smartphone platforms have continuously been reported. For instance, an unprotected component was discovered in the phone application of Android version 1.1 [15]. This flaw allowed any malicious application to make phone calls without the permission it ought to have. Another recent work [10] indicated that the message passing system in Android can be a target for denial-of-service and hijacking if used incorrectly.

Software developers often utilize conformance testing as an indispensable step to check errors and flaws in both developing and maintaining software systems.

G. Hanaoka and T. Yamauchi (Eds.): IWSEC 2012, LNCS 7631, pp. 1–18, 2012.

Conformance testing attempts to bridge the gap between system implementation and design requirements. It compares the expected behaviors described by the system requirements with the observed behaviors of an actual implementation. The observed results reflecting the conformance of implementation strongly depends on the adopted test cases [12]. In addition, test automation [17] has recently become quite common for reducing the cost of software testing procedures. A typical automated testing harness mainly offers automation in managing, executing and evaluating tests. However, such an approach cannot effectively support automated test generation. Manually creating test cases is tedious, error-prone, and often insufficient for proving the conformance of system implementation [19]. Such a problem exists in the widely used test harness for Android, Google's Android testing framework [3] [1]. Android testing framework only adopts hand-crafted test cases for conformance testing and fails to provide a comprehensive set of test cases.

Model-based testing involves developing a data model to generate tests. The model is developed based on the design requirements, and reflects the expected features of the System Under Test (SUT) [7]. Unlike hand-crafted tests, model-based approach helps reuse the generated test cases and improves the efficiency of testing procedures. If any requirement changes, a tester only needs to update the model and get a new suite of test cases, avoiding the tedious work of changing hand-crafted test cases.

In this paper, we present a model-based conformance testing framework for evaluating Android platforms. Our framework automatically generates and executes test cases. Moreover, we demonstrate the feasibility and practicality of our approach through case studies on Android Inter-Component Communication (ICC) module. We chose ICC for several reasons: (1) ICC is one of the core modules of Android as it supports collective interactions of applications; (2) the requirements of ICC are publicly available. To conduct conformance testing in our framework, we first derive the formal models and properties for Android ICC from design requirements. The formal specifications of models and properties are fed into an analysis module to automatically generate test cases, which systematically enable the rigorous conformance testing for the Android platform. MCTF checks whether the SUT's behaviors conform to functional and non-functional requirements. For example, the requirements specify a set of desired behaviors. Therefore, it is necessary to discover invalid and malformed inputs that may violate those requirements and should be caught and handled properly. Having comprehensive conformance testing would ensure the correctness and assurance of ICC in Android.

The remainder of this paper is organized as follows. Section 2 gives an overview of Android ICC. Section 3 discusses our framework and demonstrates how our framework can be applied to examine the conformance of Android ICC. Section 4 presents a tool chain designed with our framework followed by the discussion on performance analysis. Section 5 describes the related work. Section 6 concludes this paper and elaborates the future directions.

2 Overview of Android ICC

Smartphone applications inherently tend to communicate with each other. Android ICC is a sophisticated messaging system designed to support such interactions. In this section, we give a brief overview of Android ICC as described in Android documentation for SDK (SDKD) [2] and Android Compatibility Definition Document (CDD) [1].

2.1 Components

The basic unit in Android application communication is *component*. Each component is a logical building block that could support each other. Four types of components are defined with various requirements.

- *Activities* are components that provide graphic user interface (GUI). The Android GUI is implemented as a stack of activities starting one after another, where each activity is presented as a window on the screen.
- *Services* are components that run in the background to perform long-running operations. Unlike activities, a service does not have any graphic interface. Instead, services provide Remote Procedure Call (RPC) interfaces.
- *Broadcast Receivers* are asynchronous components that receive and reply to system-wide broadcasts from other components.
- *Content Providers* are components that provide public data interfaces to other components. A content provider provides common database commands such as query, insert, update and delete, through which other components can retrieve and store data.

2.2 Intents and Intent Filters

Intents play a leading role in connecting the components of applications. An intent object is a data structure carrying information about its desired recipients and optional data. Applications communicate with each other by sending and receiving intents. All intents are processed and delivered by a centralized "post office", the intent resolver.

Like a post office processing parcels in the real world, the intent resolver finds qualifying recipients by checking the attributes of an intent object.

Primary intent attributes include *action* and *data*:

- *Action* is a string naming the general action to be performed. An intent can contain at most one action.
- *Data* is a tuple consisting of both the URI of the data to be acted on and its MIME media type. This attribute indicates the data to be processed by the action.

Secondary attributes include *component*, *category*, *extras* and *flags*.

- *Component Name* is a string naming the component that should handle the intent.

- *Category* is a string containing additional information about the kind of component that should handle the intent.
- *Extras* is a key-value pair of additional information to be delivered to the recipient component.
- *Flags* is a set of strings that instruct the Android system to launch an activity.

Each component can be bound to one or more *intent filters*, which declare capabilities of the components. An intent filter includes three attributes describing the intents it would accept, including *action*, *category* and *data*. Intents and components are correlated via intent filters. Android maintains a map between public components and intent filters. The intent resolver finds the matching intent filters for a given intent, then delivers the intent to the corresponding components based on the map.

3 Model-Based Conformance Testing Framework (MCTF)

In this section, we present our conformance testing framework, called model-based conformance testing framework (MCTF), which is depicted in Figure 1. Our framework is designed for generating test cases and facilitating rigorous conformance testing with the generated test cases. We divide the framework into four steps as follows:

1. ***System Modeling: Android Modeling.***
   First, all parameters and properties of Android are derived from Android CDD and Android SDKD. Based on the identified parameters and properties, a model is defined. Parameters describe data objects and attributes of the system. Properties lay out rules regulating interactions of parameters. Android parameters and properties are then formally represented.
2. ***Test Case Generation.***
   The most significant recent development in testing is the application of formal reasoning techniques, such as model checking [11], theorem proving [24] and SAT solving [23], to generate test cases from the formal specification. In this step, the formal model is utilized to automatically derive abstract test cases, leveraging a formal reasoning technique.
3. ***Test Case Translation.***
   The generated test cases from the previous step are not suitable for direct execution, since they are generated in an abstraction level. Therefore, it is crucial to bridge the gap between abstract test cases and executable test cases. The translation is performed to extract necessary information from abstract test cases and construct executable test cases.
4. ***Test Case Execution.***
   In this step, executable test packages are generated by compiling executable test cases. With the executable test packages, an Android device or emulator is tested. For each test case, the results are monitored and recorded. Finally,

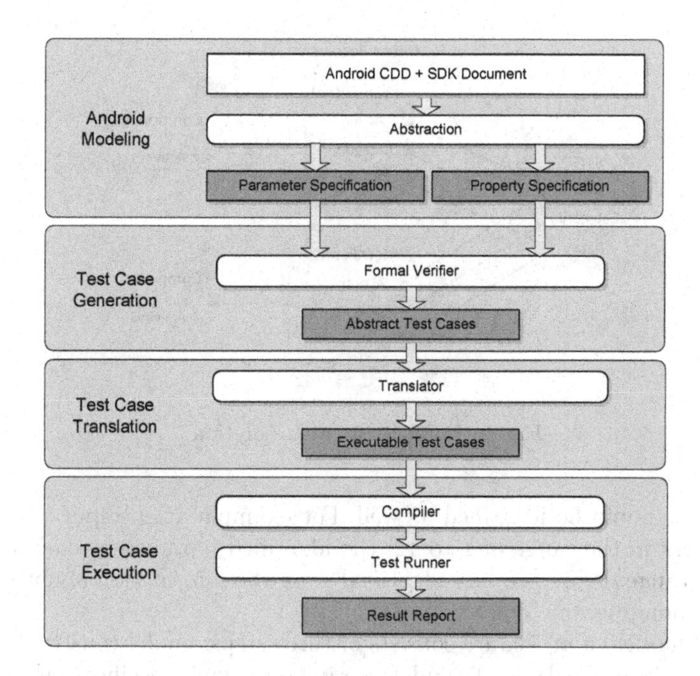

Fig. 1. Model-based Conformance Testing Framework

a human readable report is generated once all the tests are executed. The generated test report may contain supplemental information, such as screenshots, to further examine other functional and non-functional components.

In order to conduct model-based conformance testing, it is crucial to have a well-designed and general purpose language to represent the model. Alloy [20] is a structural modeling language based on first order logic, and has been widely used in the modeling community. The usage of Alloy for the representation of models is an attractive aim. Our framework adopts Alloy to formally represent an Android model. As we discussed earlier, the formal model is in turn utilized by formal reasoning tools such as Alloy Analyzer, to generate abstract test cases, which are then translated into executable test cases.

We now demonstrate how Android ICC can be rigorously tested through the four steps shown in Figure 1, identifying specific mechanisms for each MCTF task.

3.1 System Modeling: Android Modeling

A model for a specific software system is an abstract specification of the system's behaviors. Parameters and properties comprise a typical model for capturing such behaviors. The parameters are attributes or variables that appear in a piece of requirements. After parameters are identified, their types and valid

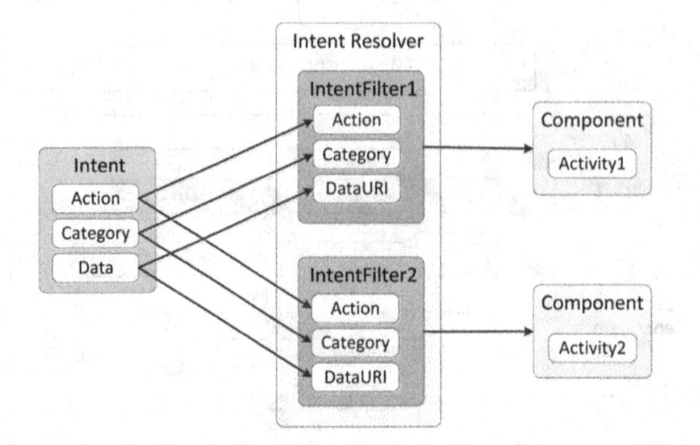

Fig. 2. Implicit Intent Resolution

value ranges should be identified as well. For example, if an input variable accepts integers in the range of 1 to 12, the identified parameters should use the same valid range. Properties are identified from the information about the relationships among parameters.

Android modeling procedure consists of three steps: model construction from requirements, specification of model parameters, and specification of model properties.

Model Construction from Android ICC Requirements. For testing Android systems and applications, testers derive parameters and properties from Android SDKD and Android CDD. Android SDKD defines the requirements of Android system, including objects and logics of Android functions and packages. Android CDD complements Android SDKD by providing additional technical details of various versions of Android platform.

For example, a technical section in Android SDKD says that "there are three Intent characteristics that can be filtered on: actions, data and categories". From this, testers identify three parameters: *action*, *category* and *data*. The definition of these three attributes also shows the data type of each parameter. That is, *action* is any string, *category* is any string set and *data* is a pair (2-tuple) of strings.

Android SDKD and Android CDD describe Android ICC in two categories: *Explicit Intent Resolution* and *Implicit Intent Resolution*, depending on the target attributes for the resolution process. If the component name of an intent is a non-empty set, this intent is an *explicit* intent because the recipient component is given explicitly. The intent resolver delivers explicit intents to the recipients designated by the `ComponentName` attribute, regardless of other attributes in the intent. Such process is called *Explicit Intent Resolution*. Actually, no resolution process is occurred because the recipient is already specified by the sender.

Thus, *intent*, *component* and *intent resolver* are identified as parameters of explicit intent resolution. The attribute *ComponentName* is consulted. The property of explicit intent resolution is trivial, as abstracted below:

- ***Property 1:*** The intent should be delivered to the recipient designated by the component name attribute of the intent.

Implicit intents do not specify any recipient component but wait for the intent resolver to determine which component they should be resolved to, based on the *action*, *data* and *category* attributes specified in the intent. This process is called *Implicit Intent Resolution.*

The parameters of implicit intent resolution include *intent*, *intent filter*, *component*, and *intent resolver*. *Action*, *category* and *data* are attributes that are consulted during the resolution process. Each attribute corresponds to a test, in which the attribute of the intent is matched against that of the intent filter. To be delivered to the component, an implicit intent must pass all the three tests on the intent filters bound with the component. Since a component can be bound with multiple intent filters, an intent that does not pass through one of a component's intent filters may pass another.

In the *action* test, the Android Intent Resolver tests both the action of the intent object and the action set of the intent filter. An intent names a single action while the intent filter specifies one or more actions. To pass the action test, the action specified in the intent object must match at least one of the actions specified in the intent filter. The action set of the intent filter object must not be empty. A special case is an intent without actions, which passes all action tests. The properties of *action* test can be summarized as follows:

- ***Property 2:*** The action specified in the Intent object must match one of the actions listed in the filter.
- ***Property 3:*** An Intent object that does not specify an action automatically passes the test as long as the filter contains at least one action.

The *category* fields in both the intent and intent filter are a set of category strings. To pass the category test, the category set of the intent should be the subset of the category set of the intent filter. The filter can list additional categories, but it cannot omit any in the intent. An intent without category passes all category tests by default. The properties of *category* test can be summarized as follows:

- ***Property 4:*** Every category in the Intent object must match a category in the filter. The filter can list additional categories, but it cannot omit any in the intent.
- ***Property 5:*** An Intent object with no category should always pass this test, regardless of the attributes in the filter.

Data contains URI and type. The URI specifies the location of the data in three sub-attributes: scheme, authority and path. The data type specifies the MIME type of the data. Android also allows wildcards when specifying data subtype in both the intent and intent filter.

- ***Property 6:*** An Intent object that contains neither a URI nor a data type passes the test only if the filter likewise does not specify any URIs or data types.
- ***Property 7:*** An Intent object that contains a URI but no data type passes the test only if its URI matches a URI in the filter and the filter likewise does not specify a type.
- ***Property 8:*** An Intent object that contains a data type but no URI passes the test only if the filter lists the same data types and similarly does not specify a URI.
- ***Property 9:*** An Intent object that contains both a URI and a data type passes the data type part of the test only if its type matches a type listed in the filter.

Figure 2 shows an example of implicit intent resolution. In this example, a public component is bound with two intent filters. An intent resolver attempts to resolve the intent shown on the left. If all of the tests pass for both intent filters, the intent is delivered to the two components on the right.

Specification of Model Parameters. Based on Android SDKD and Android CDD, we formulate the identified parameters. We first define *Component* as follows:

Definition 1. *A component is represented with a* (τ), *where* τ *is a unique name of the component;*

Intent can be defined as follows:

Definition 2. *An intent is represented with a 5-tuple* $(\tau, \alpha, \Gamma, \sigma)$, *where* τ *is the name of the recipient component;* α *is an action string that describes the action to be performed;* Γ *is a set of category strings that represent the type of components which should handle the intent; and* σ *is a 2-tuple* $(uri, type)$ *consisting of data URI and data type.*

Intents can be classified into two categories: *explicit intent* and *implicit intent*, as we discussed earlier. We formally define them as follows:

Definition 3. *Explicit intents designate the target component by its component name field. The set of explicit intents is denoted as EI.* $EI = \{i \mid i \in I \wedge i.\tau \neq null\}$

Definition 4. *Implicit intents do not specify a target. The set of implicit intents is denoted as II.* $II = \{i \mid i \in I \wedge i.\tau = null\}$

Then, the *intent filter* can be defined as:

Definition 5. *An intent filter is represented with a 3-tuple* $(\Lambda, \Gamma, \sigma)$, *where* Λ *is a set of action strings;* Γ *is a set of category strings; and* σ *is is a set of* $(uri, type)$ *tuples consisting of data URI and data type.*

We now formally define the *intent resolver* with sets and relations as:

- C is a set of components, $\{c_1, \cdots, c_p\}$;
- I is a set of intents, $\{i_1, \cdots, i_m\}$;
- F is a set of intent filters, $\{f_1, \cdots, f_q\}$;
- $FC \subseteq F \times C$, a many-to-many filter-to-component assignment relation;
- EIC, a one-to-one explicit intent-to-component assignment relation;
- IIF, a one-to-many implicit intent-to-filter assignment relation;

Based on the above-defined model, we now give the formal specification of identified parameters with Alloy as follows:

```
module android/ICC
abstract sig Str {}
sig actionStr extends Str{}
sig categoryStr extends Str{}
sig uriStr extends Str{}
sig typeStr extends Str{}
sig dataTuple {
    uri: lone uriStr,
    type: lone typeStr }

abstract sig Object {}
sig Component extends Object {
    componentName: lone componentStr }

sig Intent extends Object {
    componentName: lone componentStr,
    action: lone actionStr,
    category: set categoryStr,
    data: lone dataTuple }
sig Filter extends Object {
    action: set actionStr,
    category: set categoryStr,
    data: set dataTuple }
sig Resolver {
    IIF: Intent -> set Filter,
    IIF_A: Intent -> set Filter,
    IIF_C: Intent -> set Filter,
    IIF_D: Intent -> set Filter,
    FC: Filter -> set Component,
    EIC: Intent -> lone Component }
```

The first `sig` statement declares Str, which represents a string that can be assigned to other objects. Then, we define *component*, *intent* and *intent filter* which have all the necessary attributes for intent resolution. We then declare a resolver, which defines several relations which map intents to sets of intent filters. The value ranges of all the parameters are strings.

Specification of Model Properties. Based on Android SDKD and Android CDD, we now formulate and specify properties of Android ICC. A `fact` statement in Alloy puts an explicit constraint on the model. In our cases, we need to represent the identified properties of intent resolution with facts. According to the properties identified from the requirements, we then give their formal specifications.

The formal specification of Property 1, which covers Explicit Intent Resolution, is shown below:

```
fact explicitIntentResolution {
  all r: Resolver, i: Intent, c:Component |
  i.componentName = c.componentName
  <=> i->f in r.EIC }
```

The following shows formal specifications of Property 2-9, which cover Implicit Intent Resolution:

```
fact implicitIntentResolutuion {
 all r: Resolver, i: Intent, f:Filter |
  i->f in r.IIF_A
  and i->f in r.IIF_C
  and i->f in r.IIF_D
<=> i->f in r.IIF }
```

```
fact actionTest {
all r:Resolver| all i:Intent |all f:Filter |
(f.action!=none and i.action!=none
and i.action in f.action)
or (f.action!=none and i.action = none)
<=> i->f in r.IIF_A }
```

```
fact categoryTest {
all r:Resolver| all i:Intent |all f:Filter |
(i.category!=none and i.category in f.category)
or (f.category!=none and i.category = none)
<=> i->f in r.IIF_C }
```

```
fact dataTest {
all r:Resolver| all i:Intent |all f:Filter |
(i.data.uri=none and i.data.type=none
and f.data.uri=none and f.data.type=none)
or (i.data.uri in f.data.uri
and i.data.type = none and f.data.type=none)
or (i.data.type in f.data.type
and i.data.uri = none and i.data.uri=none)
or (i.data.uri in f.data.uri
and i.data.type in f.data.type)
<=> i->f in r.IIF_D }
```

3.2 Test Case Generation

In conformance testing, testers need to generate positive and negative test cases to examine the implementation thoroughly. Positive test cases test whether the system behaves exactly as the specified properties when inputs are valid. Negative test cases test whether the system violates the properties when inputs are invalid. Formal reasoning tools can generate abstract test cases accordingly. They translate the model notations into boolean formulas. Then, the formulas are analyzed to find bindings of the parameters and their values that make the formulas true or false. Such true and false bindings are positive and negative test cases, respectively. To generate abstract test cases, we employ Alloy Analyzer to generate instances that satisfy both `facts` and `predicates`.

Positive test cases for a given property are derived from the formal model representation, in which the property specification serves as a predicate for generating instances that conform to the very property. Similarly, negative test cases are generated from the formal model representation, if we consider it as a predicate to identify counterexamples, which satisfy the negated property. As a model-based testing framework, MCTF can assist test activities at property and behavior levels [13].

Property Testing. We take *Property 2* as an example to demonstrate the process of automated test generation for testing a given property from positive and negative aspects. To simplify the test case generation process, we remove the parameters and properties that are not related with action test. The following predicate is defined to derive the positive test cases for the corresponding facts in the formal property specification.

```
pred P2_pos(r: Resolver, i:Intent) {
  all r: Resolver, i: Intent, f:Filter |
  one i.action and i.action in f.action
      <=> i->f in r.IIF_A}
```

This predicate checks *Property 2* against the model representation of Android ICC, then instances are generated. The generated instances are used to construct positive test cases to ensure that the system should always permit a matched pair of intent object and intent filter object.

The corresponding negative test cases for *negated Property 2* are generated to ensure the system never denies a matching pair or accepts a mismatching pair. In order to derive negative test cases, we specify the negative property with Alloy as follows:

```
pred P2_negDeny(r:Resolver, i:Intent,
                f:Filter)
 {i->f not in r.IIF_A
 and i.action in f.action
 and i.action!=none
}
```

```
pred P2_negAccept(r:Resolver, i:Intent,
                  f:Filter
 ){i->f in r.IIF_A
 and i.action not in f.action
 and i.action!=none
}
```

Alloy Analyzer requires a bounded input domain, specified by the number of intents, intent filters, resolvers, action strings in our example, to generate instances and counterexamples. The size of input domain determines the total number of generated test cases. Then, we come up with the question of choosing an appropriate size for generating test cases that achieve reasonable coverage. Although testers can specify a large input domain and get millions of test cases for a trivial property with respect to the coverage, it is not always the case. The testers need to specify the input size based on practical test requirements[1].

For example, we specify the following input domain to test *Property 2*.

```
run P2_pos for
  exactly 1 Resolver, exactly 2 actionStr,
  exactly 2 Str, exactly 2 Intent,
  exactly 2 Filter
```

```
run P2_negDeny for
  exactly 1 Resolver, exactly 2 actionStr,
  exactly 2 Str, exactly 2 Intent,
  exactly 2 Filter
run P2_negAccept for
  exactly 1 Resolver, exactly 2 actionStr,
  exactly 2 Str, exactly 2 Intent,
  exactly 2 Filter
```

Figure 3 depicts a positive test case generated by Alloy Analyzer for *Property 2*. Both `Intent` and `Filter0` have the same action. Thus, `Resolver` allows the interaction between them. Figure 4 and Figure 5 depict two negative test cases. In Figure 4, `Resolver` unexpectedly denies `Intent` from accessing `Filter1` (marked by `(f)` and `(i)`). In Figure 5, `Resolver` unexpectedly accepts `Intent` and `Filter1` (marked by `(f)` and `(i)`), which have different actions.

Behavior Testing. After each property has been tested independently, we can further check behaviors of the intent resolution module. Here, we give a more complex scenario to test all modeled intent filter properties. Based on the aforementioned properties, we instruct Alloy Analyzer to enumerate all assignments, simulating inter-component communications.

To test if a system always properly delivers the intent to correct recipients, we need positive test cases that are composed of matched pairs of intents and intent filters. In our model, it implies the set of `iif` relation should not be empty. Therefore, we have the following specification:

```
pred Positive(r: Resolver){
    #r.IIF>0 }
```

[1] The testers should balance the coverage and the input size, which are normally obtained from subject matter experts and prior testing results.

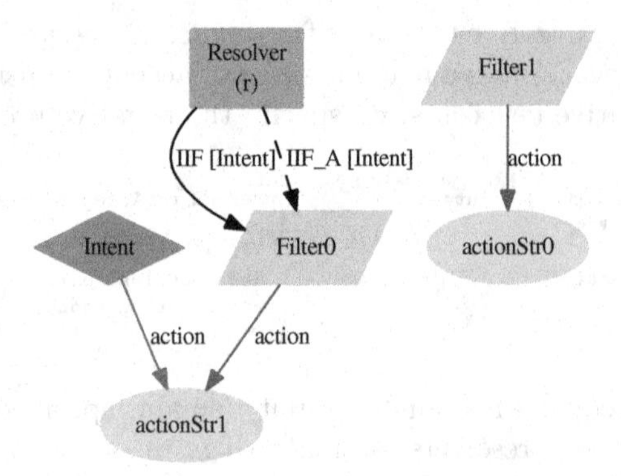

Fig. 3. Abstract Test Cases for Property Testing: Positive

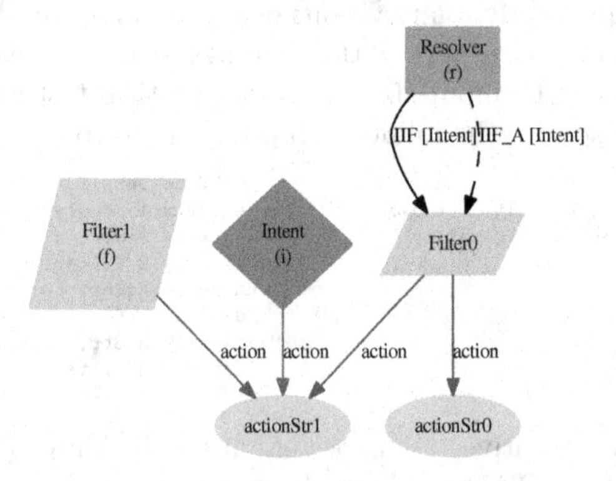

Fig. 4. Abstract Test Cases for Property Testing: Negative Deny

On the contrary, negative test cases are those without paired intents and intent filters. We simply set the size of `iif` to zero.

```
pred Negative(r: Resolver){
    #r.IIF=0 }
```

Figure 6 depicts a positive test case for behavioral testing. In this example, two successful intent deliveries can be identified from the arrows labeled with "`IIF[Intent]`": Intent0→Filter0, Intent1→Filter1.

In addition, the test case generation can be optimized to avoid generating *isomorphic* test cases by adopting the approach proposed in [8]. Finally, each abstract test case is exported to an independent file which contains the test conditions and variables for further processing. Because we are using Alloy Analyzer, one of the available choices is to export test cases to DOT files, which store test cases as hierarchical drawings of direct graphs. This is a perfect choice for visualizing abstract test cases. Another choice is to export test cases into

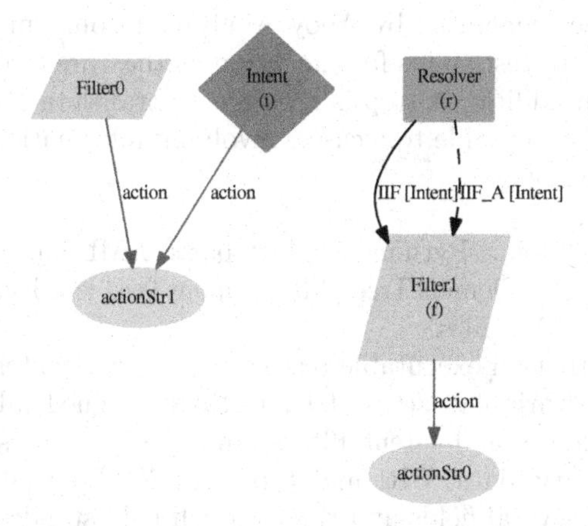

Fig. 5. Abstract Test Cases for Property Testing: Negative Accept

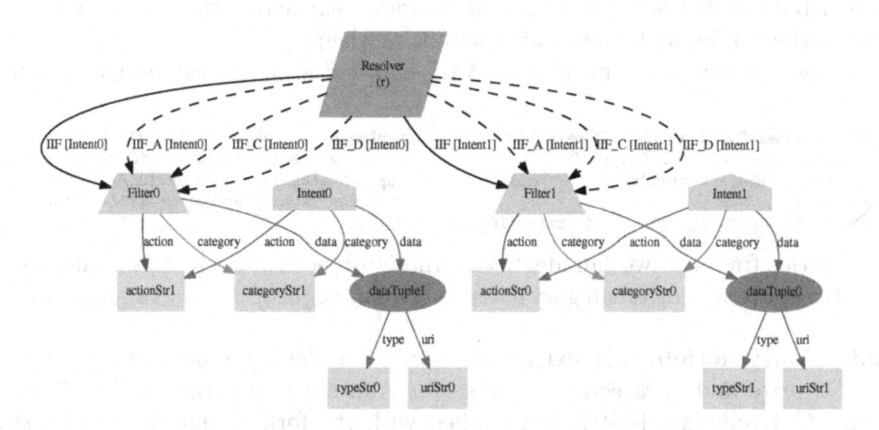

Fig. 6. A Positive Test Case for Behavior Testing

lightweight XML files, which are easy to parse with existing tools. We adopt the latter for generating executable test cases.

3.3 Test Case Translation

Except for requirements, Android SDKD also provides guidelines of Android testing framework and testing Android applications. Android CDD and Compatibility Test Suite (CTS) [1] provides additional guidelines for testing Android. Android test suites are based on JUnit [18] and Android's JUnit extensions. The extensions provide component-specific test classes and helper methods to help creating mock objects and controlling lifecycle of a component. In addition, CTS is shipped with an automated test harness. Testers can choose to use the test harness of Android CTS, use a third-party test harness, or write their own test runner based on the APIs provided by Android testing framework.

Abstract test cases generated by Alloy Analyzer in our approach cannot be directly integrated into test suites for execution as they are at different abstraction levels. Thus, an additional step is required to translate abstract test cases encoded in XML to executable test cases, involving information extraction and source code construction.

Extraction. We employ a Python script to parse XML and regroup essential information fields with cElementTree [4]. cElementTree is a Python package for efficiently managing XML files.

In order to construct an executable test case for testing intent resolution, we need to know all the variables, attributes and their assigned values. In our case, the variables are intents and intent filters, and the attributes are component name, action, category, data, URI and type. An XML-encoded abstract test case is composed of several fields and tuples. Each field stands for an attribute. And each field consists of some tuples, which store a variable and the value of the attribute of that variable. Hence, information extraction can be achieved by enumerating tuples and fields and reorganizing them.

Suppose we have a fragment of an XML-encoded abstract test case as shown below:

```
<field label="action" ID="13" parentID="11">
<tuple> <atom label="Intent$2"/>
    <atom label="actionStr$0"/> </tuple>
</field>
<field label="category" ID="14" parentID="11">
<tuple> <atom label="Intent$2"/>
    <atom label="categoryStr$0"/> </tuple>
<tuple> <atom label="Intent$2"/>
    <atom label="categoryStr$1"/> </tuple>
</field>
```

From this fragment we can identify an Intent object `Intent2`. Its action is assigned to `actionStr0`, its category is assigned to {`categoryStr0`, `categoryStr1`}.

Code Construction. The extracted information fields are utilized for a test case template and Java code fragments for Android Compatibility Test Suite (CTS). Our template is strictly complied with the format and syntax of test cases defined in Android CTS.

The sample code shipped with Android CTS offers practical examples of how to write executable test cases. We give a code template for testing Android ICC.

```
IntentFilter filter = new Match(
    String[] actions, String[] categories,
    String[] dataTypes, String[] uriSchemes,
    String[] uriAuthoroties, String[] uriPorts);
checkMatches(filter, new MatchCondition[] {
    new MatchCondition(
    int expectedResult,
    String action, String[] categories,
    String dataType, String dataURI); }
```

With the extracted information in the template, we get several Java code fragments at the end of this step.

3.4 Test Case Execution

After integrating the code fragments into existing test suites or a new test suite, executable test cases are derived by compiling fragments. Such test suites are run by a test runner that loads the test cases, runs and tears down each test. We use Android's Instrumentation Test Runner [3], which is a set of control methods and hooks in Android platform, to run our generated test cases. For each executable test case, the results are generated accordingly as we discussed in our framework. Finally, a report is presented in an HTML page including test results.

4 Implementation and Evaluation

In this section, we give a brief introduction of our tool set, which constitutes a tool chain for model-based conformance testing. As depicted in Figure 7, our tool chain consists of three tools: Alloy Analyzer, the Translator and the Android Instrumentation Test Runner. The formal representation of models and properties are fed into Alloy Analyzer for automatically generating test cases. Alloy Analyzer exports the generated abstract test cases to intermediate XML files. Then, our translator parses XML and constructs Java code fragments. The output of test case translation is an Android application package containing compiled JUnit test cases. Finally, Android Instrumentation Test Runner executes test suite and generates the test report.

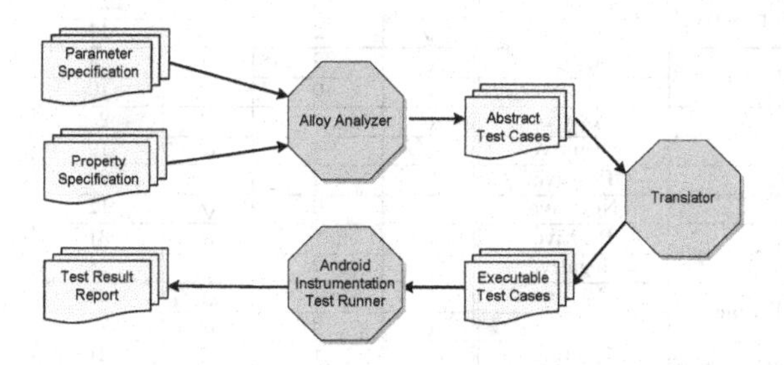

Fig. 7. A tool chain that supports MCTF

We provide a contrastive analysis between Android CTS and our generated test cases to demonstrate effectiveness of our framework in this section. For property testing, every property of the three tests need to be rigorously checked. We identified that Android CTS fails to check some properties from positive or negative aspects. Table 1 shows a comparison between Android CTS and the test cases generated by our approach. The table shows that Android CTS test suites are not offering sufficient test coverage. And our approach could achieve better coverage than that of Android CTS.

To evaluate the efficiency of our approach, we also examined two core processes, test case generation and test case translation, in our implementation.

Figure 8(a) shows that the increase of the total number of generated test cases is proportional to the number of intents and intent filters. Figure 8(b) shows that the processing time taken for test case generation and translation increases linearly with the increase of the number of the test cases, indicating that our approach provides a feasible and promising solution to facilitate and enhance conformance testing for Android platform.

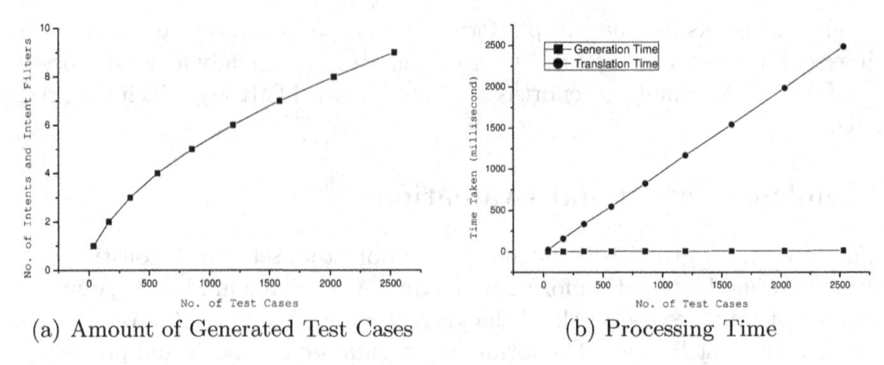

(a) Amount of Generated Test Cases (b) Processing Time

Fig. 8. Performance Evaluation

Table 1. Conformance testing achieved by Android CTS and our approach

Property	Positive/Negative	Android CTS		MCTF	
		Covered	#Test cases	Covered	#Test cases
Property 1	Positive	×	0	√	16
	Negative	×	0	√	18
Property 2	Positive	√	3	√	24
	Negative	√	2	√	14
Property 3	Positive	√	2	√	24
	Negative	×	0	√	10
Property 4	Positive	√	4	√	26
	Negative	√	4	√	10
Property 5	Positive	√	2	√	26
	Negative	×	0	√	12
Property 6	Positive	√	2	√	31
	Negative	√	2	√	18
Property 7	Positive	√	1	√	31
	Negative	√	3	√	20
Property 8	Positive	√	1	√	31
	Negative	×	0	√	20
Property 9	Positive	×	0	√	31
	Negative	√	2	√	26

5 Related Work

Most recent work related to software testing in Android addresses automated GUI testing for Android applications. Amalfitano et al. [6] proposed a crawling-based approach to generate GUI test cases. They designed a tool to simulate events on the user interfaces, generate event transition tree by capturing application responses, and predict future events at runtime. In contrast, our approach is the first attempt to explore rigorous conformance testing for Android. In particular, we adopt a model-based approach to automatically generate test cases.

Model-based approaches have been widely used for testing in various fields. Several researchers proposed automated frameworks for testing Java programs, such as Korat [8] and TestEra [21]. Korat constructs Java predicates and generates all non-isomorphic inputs for which the predicates return true, by searching and enumerating a given bounded input space. TestEra works in a similar way as Korat, but using a first-order relational language and existing SAT solvers.

Both approaches use structural invariants on the input data to automatically generate test cases and then test the output against a set of predicates. However, the generated test cases are abstract and need to perform the translation task to generate the actual code. In our work, we attempt to extend model-based approaches to testing Android platforms. We also demonstrate how test cases can be integrated to perform conformance testing effectively.

Security for mobile devices and applications is a growing concern recently. TaintDroid [14] monitors and controls access to sensitive data by dynamic taint-based information flow tracking. Stowaway [16] identifies vulnerabilities in applications by static analysis on application packages, manifests and bytecodes. Chaudhuri [9] proposed a formal language to describe applications and reason about information flows and the consistency of security specifications.

6 Conclusion

While several automated testing frameworks have been proposed and developed for smartphone platforms, developers still need systematic approaches and corresponding tools to generate test cases for conformance testing efficiently and effectively. To address this issue, we have proposed a novel framework to enable rigorous conformance testing for the Android platform. Our framework adopted a model-based approach which utilizes formal verification techniques to automatically generate test cases. In addition, we have demonstrated the feasibility of our approach with Android ICC.

In our current framework, testers need to manually derive the model from requirements. As part of our future work, we would explore an approach for directly constructing model from the requirements, leveraging the capability of NLP techniques [22]. Moreover, we would apply our approach to other Android modules, such as Activity Manager and Package Manager.

References

1. Android compatibility, `http://source.android.com/compatibility/`
2. Android sdk cocument, `http://developer.android.com/reference/android/package-summary.html/`
3. Android testing fundamentals, `http://developer.android.com/guide/topics/testing/testing_android.html#Instrumentation`
4. The elementtree xml api, `http://docs.python.org/library/xml.etree.elementtree.html`
5. Apple rises to the top as worldwide smartphone market grows 65.4% in the second quarter of 2011, idc finds (August 2011), `http://www.idc.com/getdoc.jsp?containerId=prUS22974611`
6. Amalfitano, D., Fasolino, A.R., Tramontana, P.: A gui crawling-based technique for android mobile application testing. In: IEEE Fourth International Conference on Software Testing, Verification and Validation Workshops (ICSTW), pp. 252–261. IEEE (2011)
7. Beizer, B.: Software testing techniques. Dreamtech Press (2002)

8. Boyapati, C., Khurshid, S., Marinov, D.: Korat: Automated testing based on java predicates. In: Proceedings of the 2002 ACM SIGSOFT International Symposium on Software Testing and Analysis, pp. 123–133. ACM (2002)
9. Chaudhuri, A.: Language-based security on android. In: Proceedings of the ACM SIGPLAN Fourth Workshop on Programming Languages and Analysis for Security, pp. 1–7. ACM (2009)
10. Chin, E., Felt, A.P., Greenwood, K., Wagner, D.: Analyzing inter-application communication in android. In: MobiSys, pp. 239–252 (2011)
11. Clarke, E., Grumberg, O., Peled, D.: Model checking (2000)
12. Constant, C., Jéron, T., Marchand, H., Rusu, V.: Integrating formal verification and conformance testing for reactive systems. IEEE Transactions on Software Engineering 33(8), 558–574 (2007)
13. Dalal, S.R., Jain, A., Karunanithi, N., Leaton, J.M., Lott, C.M., Patton, G.C., Horowitz, B.M.: Model-based testing in practice. In: Proceedings of the 1999 International Conference on Software Engineering, pp. 285–294. IEEE (1999)
14. Enck, W., Gilbert, P., Chun, B.G., Cox, L.P., Jung, J., McDaniel, P., Sheth, A.N.: Taintdroid: An information-flow tracking system for realtime privacy monitoring on smartphones. In: Proceedings of OSDI (2010)
15. Enck, W., Ongtang, M., McDaniel, P.: On lightweight mobile phone application certification. In: Proceedings of the 16th ACM Conference on Computer and Communications Security, pp. 235–245. ACM (2009)
16. Felt, A., Chin, E., Hanna, S., Song, D., Wagner, D.: Android permissions demystified. Technical report (2011)
17. Fewster, M., Graham, D.: Software test automation: effective use of test execution tools. ACM Press/Addison-Wesley Publishing Co. (1999)
18. Gamma, E., Beck, K.: Junit: A cook's tour. Java Report 4(5), 27–38 (1999)
19. Hu, H., Ahn, G.: Enabling verification and conformance testing for access control model. In: Proceedings of the 13th ACM Symposium on Access Control Models and Technologies, pp. 195–204. ACM (2008)
20. Jackson, D.: Alloy: a lightweight object modelling notation. ACM Transactions on Software Engineering and Methodology (TOSEM) 11(2), 256–290 (2002)
21. Khurshid, S., Marinov, D.: Testera: Specification-based testing of java programs using sat. Automated Software Engineering 11(4), 403–434 (2004)
22. Lewis, D.D., Jones, K.S.: Natural language processing for information retrieval. Communications of the ACM 39(1), 92–101 (1996)
23. Mitchell, D.G.: A sat solver primer. Bulletin of the European Association for Theoretical Computer Science 85(112-133), 12 (2005)
24. Robinson, J.A., Voronkov, A.: Handbook of automated reasoning, vol. 1. North-Holland (2001)

Application of Scalar Multiplication of Edwards Curves to Pairing-Based Cryptography

Takanori Yasuda[1], Tsuyoshi Takagi[2], and Kouichi Sakurai[1,3]

[1] Institute of Systems, Information Technologies and Nanotechnologies
[2] Institute of Mathematics for Industry, Kyushu University
[3] Department of Informatics, Kyushu University

Abstract. Edwards curves have efficient scalar multiplication algorithms, and their application to pairing-based cryptography has been studied. In particular, if a pairing-friendly curve used in a pairing-based protocol is isomorphic to an Edwards curve, all the scalar multiplication appearing in the protocol can be computed efficiently. In this paper, we extend this idea to pairing-friendly curves not isomorphic but isogenous to Edwards curves, and add to pairing-friendly curves to which Edwards curves can be applied. Above all, pairing-friendly curves with smaller ρ-values provide more efficient pairing computation. Therefore, we investigate whether pairing-friendly curves with the minimal ρ-values are isogenous to Edwards curves for embedding degree up to 50. Based on the investigation, we present parameters of pairing-friendly curves with 160-bit and 256-bit security level at embedding degree 16 and 24, respectively. These curves have the minimal ρ-values and are not isomorphic but isogenous to Edwards curves, and thus our proposed method is effective for these curves.

Keywords: Pairing-friendly curves, Edwards curves, embedding degree.

1 Introduction

Many pairing-based protocols use not only pairing computations but also scalar multiplications (e.g., [11,40,23,41]). It is known that Edwards curves [18] provide a model of the groups of rational points of elliptic curves that have efficient scalar multiplication algorithms [10]. Therefore, the application of Edwards curves to pairing-based cryptography has been investigated in several studies [1,16,26].

The choice of pairing-friendly curves with efficient arithmetic is an important factor in efficient pairing computation. The parameter ρ-value defined on an elliptic curve is related to the efficiency of arithmetic on the elliptic curve. In general, elliptic curves with small ρ-values are desirable for speeding up arithmetic on the elliptic curves. Moreover, if the curves are isomorphic to an Edwards curve, their scalar multiplication can be computed more efficiently. However, not every curve can be transformed into an Edwards curve. In this paper, we propose how to apply pairing-friendly curves not isomorphic but isogenous to

G. Hanaoka and T. Yamauchi (Eds.): IWSEC 2012, LNCS 7631, pp. 19–36, 2012.

Edwards curves to pairing-based cryptography. In our proposed method, if a pairing-friendly curve E is not isomorphic but isogenous to an Edwards curve, its scalar multiplication is computed on the Edwards curve. On the other hand, its pairing is computed on E if E has a more efficient pairing algorithm than the Edwards curve. Thus, our proposed method changes two curves according to scalar multiplication or pairing. In addition, we investigate whether pairing-friendly curves with the minimal ρ-values are isogenous to Edwards curves. In fact, we list the minimal ρ-values of pairing-friendly curves isogenous to Edwards curves for embedding degree up to 50. Among the constructible pairing-friendly curves, those with minimal ρ-values and embedding degrees less than or equal to 50 have been summarized by Freeman et al. [20]. We compare their results with ours, and we compute the embedding degrees (less than or equal to 50) at which the constructible pairing-friendly curves with minimal ρ-values is isogenous to Edwards curves.

The efficiency of pairing computation on elliptic curves has been improved as a result of numerous studies. (e.g., [5,24,28,8,37,35]). Several approaches to efficient pairing computation on elliptic curves utilize coordinates (affine, projective, Jacobian etc.) in the Weierstrass form. For example, pairing-friendly curves with quartic or sextic twists have efficient pairing computation [24], and this computation requires coordinates in the Weierstrass form. On the other hand, there exist examples of pairing-friendly curves with quartic or sextic twists and with the minimal ρ-values at embedding degree 16 and 24. Our investigation shows that these curves are not isomorphic but isogenous to Edwards curves. the pairing-friendly curves with the minimal ρ-values are isogenous to Edwards curves. For these curves, our proposed method is effective. In Appendix B, we give parameters of these curves with 160-bit and 256-bit security level .

2 Edwards Curves

In this section, we review Edwards curves, their transformation into elliptic curves in the Weierstrass form and their scalar multiplication.

Let $\mathbb{F}_p$ be a finite field of order p, where p is a prime greater than 3. An Edwards curve is a quartic curve over $\mathbb{F}_p$, defined by

$$Ed_d : x^2 + y^2 = 1 + dx^2y^2 \quad (d \in \mathbb{F}_p \backslash \{0, 1\}).$$

Moreover, a twisted Edwards curve over $\mathbb{F}_p$ is defined by the quartic equation

$$Ed_{a,d} : ax^2 + y^2 = 1 + dx^2y^2 \quad (a, d \in \mathbb{F}_p^{\times}),$$

as an extension of an Edwards curve. Hereafter, Ed_d and $Ed_{a,d}$ also represent the sets of $\mathbb{F}$-rational points of Ed_d and $Ed_{a,d}$, respectively. The sum of two points (x_1, y_1) and (x_2, y_2) on the twisted Edwards curve $Ed_{a,d}$ is

$$(x_1, y_1) + (x_2, y_2) = \left(\frac{x_1y_2 + y_1x_2}{1 + dx_1x_2y_1y_2}, \frac{y_1y_2 - ax_1x_2}{1 - dx_1x_2y_1y_1} \right).$$

Addition on the Edwards curve Ed_d is given by that of $Ed_{1,d}$. The point $(0,1)$ is the unit of the addition law. The point $(0,-1)$ has order 2. The points $(1,0)$ and $(-1,0)$ have order 4. The inverse of a point (x,y) on $Ed_{a,d}$ is $(-x,y)$. The addition law is *strongly unified*, i.e., it can also be used to double a point. In [10] (where a $=1$), and later in [9], it was proved that if a is a square and d is a non-square in $\mathbb{F}_p$ then the addition law of $Ed_{a,d}$ is *complete*: it works for all pairs of inputs. The following two propositions show the relation between twisted Edwards curves and elliptic curves.

Proposition 1 ([10,9]).

1. *Over $\mathbb{F}_p$, the twisted Edwards curve $Ed_{a,d}$ is birationally equivalent to a Montgomery curve,*

$$E_{a,d} : \frac{4}{a-d}y^2 = x^3 + \frac{2(a+d)}{a-d}x^2 + x.$$

2. *Moreover, if $Ed_{a,d}$ is complete, then the birational map induces an isomorphism between $Ed_{a,d}$ and $E_{a,d}(\mathbb{F}_p)$ as groups.*

Proposition 2 ([9]).

1. *Let E be an elliptic curve over $\mathbb{F}_p$. The group $E(\mathbb{F}_p)$ has an element of order 4 if and only if E is birationally equivalent to an Edwards curve Ed_d over $\mathbb{F}_p$. If $E(\mathbb{F}_p)$ has an element of order 4, E is defined by a Weierstrass equation,*

$$Y^2 = X^3 + a_2X^2 + a_4X \quad (a_2, a_4 \in \mathbb{F}_p), \tag{1}$$

and if $P = (x_4, y_4)$ is an element in $E(\mathbb{F}_p)$ of order 4, then d is given by $1 - 4x_4^3/y_4^2$.
2. *Moreover, if $Ed_{a,d}$ is complete, then the birational map induces an isomorphism between Ed_d and $E(\mathbb{F}_p)$ as groups.*

Let E_0 be an elliptic curve over $\mathbb{F}_p$ defined by a short Weierstrass equation

$$Y^2 = X^3 + aX + b \quad (a, b \in \mathbb{F}_p). \tag{2}$$

Assume that $E_0(\mathbb{F}_p)$ has an element of order 4; then, (2) is expressed as

$$Y^2 = (X - x_2)(X^2 + CX + D) \tag{3}$$

for some $x_2, C, D \in \mathbb{F}_p$. By changing $X - x_2$ into X in (3), E_0 can be transformed into an elliptic curve E' of the form (1). Let $P = (x_4, y_4)$ be an element in $E'(\mathbb{F}_p)$ of order 4 and $d_0 = 1 - 4x_4^3/y_4^2$. This algorithm is described as follows:

Input: An elliptic curve $E : Y^2 = f(X)$ over $\mathbb{F}_p$ such that $E(\mathbb{F}_p)$ has an element of order 4, and $l = E(\mathbb{F}_p)$.
Output: $d \in K^\times$ such that the Edwards curve E_d: $x^2 + y^2 = 1 + dx^2y^2$ is birationally equivalent to E over $\mathbb{F}_p$.

1. Compute an element $P_2 = (x_2, y_2)$ in $E(\mathbb{F}_p)$ of order 2. (P_2 can be calculated because x_2 satisfies $f(x_2) = 0$, and $y_2 = 0$.)
2. Define a polynomial $f_{x_2}(X) = f(x + x_2)$ and an elliptic curve $E_{x_2} : Y^2 = f_{x_2}(X)$ over $\mathbb{F}_p$.
3. Compute an element $P_4 = (x_4, y_4)$ in $E_{x_2}(\mathbb{F}_p)$ such that $2P_4 = (0,0)$ (see the remark below). If P_4 does not exist, then return to Step 1 and choose another P_2. (P_4 (and its existence) can be calculated because x_4 satisfies $\mathcal{P}(x_4) = 0$ for $\mathcal{P}(X) = X^2 - D$ when $f_{x_0}(X)$ is factorized as $f_{x_0}(X) = X(X^2 + CX + D)$.)
4. $d \leftarrow 1 - 4x_4^3/y_4^2$.

The birational maps between E_0 and Ed_{D_0} can be described explicitly as follows;

- $\mathcal{M} : Ed_{d_0} \to E_0$

$$\mathcal{M}([X, Y, Z]) = [x_4 X(Z + Y) + x_2 X(Z - Y), y_4 Z(Z + Y), X(Z - Y)].$$

- $\mathcal{M}^{-1} : E_0 \to Ed_{d_0}$

$$\mathcal{M}^{-1}([U, V, W]) \\ = [2(U - x_2 W)(U + (x_4 - x_2)W), cV(U - (x_4 + x_2)W), cV(U + (x_4 - x_2)W)],$$

where $c = 2x_4/y_4 \in \mathbb{F}_p$.

Here, the Weierstrass curve and Edwards curve both are expressed by projective coordinates. If Ed_d is complete, $\mathcal{M}$ or $\mathcal{M}^{-1}$ becomes a group isomorphism. However, even though Ed_d is non-complete, under the restriction of the subgroup of elements with odd order of Ed_d, $\mathcal{M}$ and $\mathcal{M}^{-1}$ become group isomorphisms [10,25]. Bernstein et al. compared the efficiency of addition, doubling, etc., of several coordinates of elliptic curves: projective, (modified, Chudnovsky) Jacobi, Doche/Icart/Kohel 2,3, Jacobi quartic, Edwards etc. [10]. Among these, the Edwards curve coordinates recorded top performance in many cases. In general, a twisted Edwards curve with $a = -1$ over $\mathbb{F}_p$ has more efficient scalar multiplication than an Edwards curve over $\mathbb{F}_p$. Hisil et al. introduced the extended Edwards coordinates and proposed efficient scalar multiplication of an Edwards curve by mixing the extended Edwards coordinates and the projective Edwards coordinates [25]. In particular, this scalar multiplication is more efficient than that using a mixture of modified and Chudnovsky Jacobi coordinates.

3 Pairing-Friendly Curves

Pairing-based cryptography uses the following pairing:

$$\omega : \mathbb{G}_1 \times \mathbb{G}_2 \to \mathbb{F}_{p^k}^{\times}.$$

Here, k is the embedding degree of ω, and $\mathbb{G}_1, \mathbb{G}_2$ are subgroups with order r of $E(\mathbb{F}_p)$ and $E(\mathbb{F}_{p^k})$. If E has a pairing with embedding degree k such that $r \geq \sqrt{p}$ and $k < \log_2 r/8$, E is called a *pairing-friendly curve* [20]. It is known that among whole elliptic curves, pairing-friendly curves are very rare [4]. Therefore, it is necessary to construct pairing-friendly curves. In this paper, we treat only ordinary pairing-friendly curves.

3.1 Construction of Pairing-Friendly Curves

Several methods have been proposed for the construction of pairing-friendly curves. The fundamental steps of these methods are similar:

Step 1. Construct a pairing-friendly parameter (t, r, p).
Step 2. From (t, r, p), construct an elliptic curve E (in the Weierstrass form) over $\mathbb{F}_p$ by the CM method of Atkin-Morain [2].

The elliptic curve E constructed by these steps is defined over $\mathbb{F}_p$, the order of its maximal subgroup with prime order of E is r, and t is the Frobenius trace of E. Here, a pairing-friendly parameter is defined as follows.

Definition 1. *The triplet (t, r, p) of integers is called a pairing-friendly parameter of embedding degree k if the following conditions are satisfied:*
1. *r, p are prime,*
2. $r \mid p + 1 - t$,
3. $|t| < 2\sqrt{p}$.
4. $r|p^k - 1$, *and for* $1 \leq i < k$, $r \nmid p^i - 1D$

The known methods for constructing pairing-friendly parameters are as follows [20]: Cocks-Pinch [14], DEM [17], MNT [30], GMV [22], Freeman curve [19], Scott-Barreto [34], Brezing-Weng [13], Barreto-Naehrig curve [7], Kachisa-Schefer-Scott [27], Barreto-Lynn-Scott [6].

Definition 2. *For a pairing-friendly parameter (t, r, p), the ρ-value of (t, r, p) (or the elliptic curve E constructed from (t, r, p) as above) is defined by* $\rho = \rho(t, r, p) := \log(p)/\log(r)$.

From conditions (2), (3) in Definition 1, the minimal ρ-value is almost 1. r is an important parameter related to the security of pairing-based cryptography. When r is constant, a small ρ-value means that $\mathbb{F}_p$ is small. If $\mathbb{F}_p$ is small, the calculation cost of arithmetic on the elliptic curve is low. Therefore, it is necessary to generate pairing-friendly curves with small ρ-values in order to speed up arithmetic on the elliptic curves. Freeman et al. listed the minimal ρ-values of pairing-friendly curves constructed by the methods mentioned above, up to 50 [20].

3.2 Families of Pairing-Friendly Parameters

Several methods for constructing pairing-friendly parameters make use of a triplet $(t(x), r(x), p(x))$ of polynomials over $\mathbb{Q}$, which generates (maybe infinitely) many pairing-friendly parameters by substituting integers.

Definition 3 ([20]).

1. *Let k be a positive integer, and D, a positive square-free integer. We say that a triplet $(t(x), r(x), p(x))$ of polynomials with rational coefficients is a (pairing-friendly) family with embedding degree k and discriminant D if the following conditions are satisfied:*
   (a) *$p(x)$ represents primes, i.e.,*
      - *$p(x)$ is non-constant,*
      - *$p(x)$ has a positive leading coefficient,*
      - *$p(x)$ is irreducible,*
      - *$p(a) \in \mathbb{Z}$ for some $a \in \mathbb{Z}$,*
      - *$gcd(\{p(x) : x, f(x) \in Z\}) = 1$.*

   (b) *$r(x)$ is non-constant, irreducible, integer-valued, and has a positive leading coefficient.*
   (c) *$r(x)$ divides $p(x) + 1 - t(x)$.*
   (d) *$r(x)$ divides $\Phi_k(t(x) - 1)$, where Φ_k is the k-th cyclotomic polynomial.*
   (e) *The equation $Dy^2 = 4p(x) - t(x)^2$ has infinitely many integer solutions (x, y).*
2. *The ρ-value of a family $(t(x), r(x), p(x))$ is defined by $\deg p(x)/\deg r(x)$.*

MNT [30], GMV [22], Freeman curve [19], Scott-Barreto [34], Brezing-Weng [13], Barreto-Naehrig curve [7], and Kachisa-Schefer-Scott [27] output pairing-friendly families. The ρ-value of a family $(t(x), r(x), p(x))$ coincides with the limit of $\log p(a)/\log r(a)$, the ρ-value of the elliptic curve by the CM method of Atkin-Morain from $(t(a), r(a), p(a))$, as $a \to \infty$. Therefore, the definition of the ρ-value of a family is natural.

4 Pairing-Friendly Edwards Curves

In order to utilize an Edwards curve in pairing-based protocols, we have to construct a pairing-friendly Edwards curve. However, from Proposition 2 not every pairing-friendly curve can be transformed into an Edwards curve. In this section, we investigate the following: (1) methods for constructing a pairing-friendly Edwards curve, and (2) transformability of constructible pairing-friendly curves with minimal ρ-values listed in [20]. With regard to (1), we explain how to modify any method for constructing general pairing-friendly curves using pairing-friendly parameters so as to output pairing-friendly Edwards curves. We obtain the list of minimal ρ-values of constructible pairing-friendly Edwards curves using (1). By comparing this list with that of minimal ρ-values of constructible pairing-friendly curves in [20], we can investigate (2).

4.1 Constructing Pairing-Friendly Edwards Curves

From Proposition 2, an elliptic curve E is birationally equivalent to an Edwards curve over $\mathbb{F}_p$ if and only if $E(\mathbb{F}_p)$ has an element of order 4. If $E(\mathbb{F}_p)$ has an element of order 4 then $\sharp E(\mathbb{F}_p)$ is divisible by 4. The opposite is not always true; however, if $\sharp E(\mathbb{F}_p)$ is divisible by 8, then $E(\mathbb{F}_p)$ has an element of order 4 because the number of 2-torsion points of $E(\mathbb{F}_p)$ is less than or equal to 3. Therefore, the following procedure constructs a pairing-friendly Edwards curve:

Step 1. Construct a pairing-friendly parameter (t, r, p) such that $8 \mid p + 1 - t$.
Step 2. From (t, r, p), construct an elliptic curve (in the Weierstrass form) over $\mathbb{F}_p$ by the CM method of Atkin-Morain [2].
Step 3. Transform the elliptic curve in Step 2 into an Edwards curve.

The algorithm for Step 3 has been described in §2. There are several methods for constructing pairing-friendly parameter, as explained in §3.1.

By using 2-isogeny, the above procedure can be modified. We will explain this modification in the following subsection.

4.2 Constructing Pairing-Friendly Complete Edwards Curves

Morain showed the following fact.

Proposition 3 ([31]). *Assume that a prime p is expressed as $p = \frac{1}{4}(t^2 + Dy^2)$ for some positive integer D and integers t, y. Let E be an elliptic curve over $\mathbb{F}_p$ with the trace of Frobenius t, which is constructed by the CM method of Atkin-Morain. Moreover, assume that either of the following is satisfied.*
(1) D is odd,
(2) D, y both are even.
Then, E is not birationally equivalent to any complete Edwards curve over $\mathbb{F}_p$.

This proposition implies that in many cases, Edwards curves constructed by the CM method are not complete. The method for constructing complete Edwards curves using 2-isogenies has been discussed by Aréne et al. [1]. (A 2-isogeny means an isogeny whose kernel consists of 2 elements.) The algorithm is described as follows:

Step 1. Construct a pairing-friendly parameter (t, r, p) such that $4 \mid p + 1 - t$.
Step 2. From (t, r, p), construct an elliptic curve E (in the Weierstrass form) over $\mathbb{F}_p$ by the CM method of Atkin-Morain [2].
Step 3. Find an elliptic curve E' that can be transformed into a complete Edwards curve by compositions of 2-isogenies from E.
Step 4. Transform the elliptic curve E' in Step 2 into a complete Edwards curve.

This algorithm is an improved version of the algorithm in §4.1 because the condition of (t, r, p) in Step 1 becomes weaker and the output Edwards curve is always complete.

We need to explain Step 3. 2-isogenies of elliptic curves can be described explicitly using Vélu's classical formula in finite fields [39].

Proposition 4 ([31] Prop. 6). *Assume that $E : Y^2 = X^3 + a_2X^2 + a_4X + a_6$ has a rational point of order 2, denoted by $P = (x_2, 0)$. Put $s = 3x_2^2 + 2a_2x_2 + a_4$ and $w = x_2s$. Then, E is 2-isogenous to the elliptic curve $E_1 : Y_1^2 = X_1^3 + A_2X_1^2 + A_4X_1 + A_6$ where $A_2 = a_2, A_4 = a_4 - 5s, A_6 = a_6 - 4a_2s - 7w$. Moreover, the 2-isogeny $\psi : E \to E_1$ whose kernel is generated by P sends $[X; Y; Z]$ to*

$$[X_1; Y_1, Z_1] = [(X - x_2)^2X + (X - x_2)s; Y((X - x_2)^2 - s); (X - x_2)^2].$$

Here, the points of the elliptic curves are described by projective coordinates.

We write 2-Isog(E, P) for E_1 in the above proposition. If $E(\mathbb{F}_p)$ in Step 2 has an element of order 4, Step 3 can be omitted from Proposition 2. Assume that $E(\mathbb{F}_p)$ does not have an element of order 4. Since $4 \mid p + 1 - t$, $E(\mathbb{F}_p)$ must have three 2-torsion points. The following is an algorithm for Step 3. It is essentially a special case of FINDDESCENDINGPATH of [21] when $l = 2$, which can input an elliptic curve of j-invariant 0 or 1728.

Input An elliptic curve E that has three rational 2-torsion points.
Output An elliptic curve E' transformable into a complete Edwards curve.
1. $F \leftarrow \{$2-Isog$(E, P_i) \mid i = 1, 2, 3\}$ where P_1, P_2, P_3 are the three rational 2-torsion points of E.
2. For $i = 1$ to 3, do
 (a) $G[i] \leftarrow E$; $G'[i] \leftarrow F[i]$.
 (b) If $G'[i]$ has a unique rational 2-torsion point, then $i_0 \leftarrow i$ and goto **5**, else $S[i] \leftarrow \{$2-Isog$(G'[i], P_i) \mid i = 1, 2, 3\}$ where P_1, P_2, P_3 are the three rational 2-torsion points of $G'[i]$.
3. $i_0 = -1$.
4. while $i_0 = -1$ do
 For $i = 1$ to 3, do
  If $S[i] = \emptyset$, then use next i,
  else
  (a) If $(j(S[i][1])) = j(G[i])$, then $G[i] \leftarrow G'[i]$; $G'[i] \leftarrow S[i][2]$, else $G[i] \leftarrow G'[i]$; $G'[i] \leftarrow S[i][1]$.
  (b) If $G'[i]$ has unique rational 2-torsion point, then $i_0 \leftarrow i$, else $S[i] \leftarrow \{$2-Isog$(G'[i], P_i) \mid i = 1, 2, 3\}$ where P_1, P_2, P_3 are the three rational 2-torsion points of $G'[i]$.
5. Return $G'[i_0]$.

From the volcano theory of isogenies [29,21], it is known that the number of loops of **4** is bounded.

4.3 Families of Pairing-Friendly Edwards Parameters

In this subsection, we investigate pairing-friendly families that yield pairing-friendly Edwards curves by the algorithm for constructing pairing-friendly Edwards curves in the last subsection.

Let a triplet $(t(x), r(x), p(x))$ of polynomials be a pairing-friendly family constructed by the Brezing-Weng method, the MNT method, etc. We can determine whether the family $(t(x), r(x), p(x))$ yields (infinitely) many pairing-friendly parameters (t, r, p) such that $4 \mid p + 1 - t$. In general, the following algorithm determines the condition of integers x_0 satisfying $4 \mid p(x_0) + 1 - t(x_0)$.

Input: A pairing-friendly family $(t(x), r(x), p(x))$.
Output A set of integers modulo $4m$, where m is the common denominator of the coefficients of $p(x)$ and $t(x)$.
1. $S \leftarrow \{\}$.
2. For $i = 0$ to $4m - 1$, do
(a) If $p(i) + 1 - t(i)$ is an integer and $4 \mid p(i) + 1 - t(i)$, then $S \leftarrow S \cup \{i\}$.
3. Return S.

For an integer x_0, $p(x_0) + 1 - t(x_0)$ is an integer and $4 \mid p(x_0) + 1 - t(x_0)$ if and only if $x_0 \bmod 4m$ belongs to the output set S of the above algorithm. In particular, if S is empty, the family $(t(x), r(x), p(x))$ yields no pairing-friendly parameter (t, r, p) such that $4 \mid p + 1 - t$, or no pairing-friendly Edwards curve by the algorithm for constructing pairing-friendly Edwards curves in the last subsection. We say that $(t(x), r(x), p(x))$ is a *(pairing-friendly) Edwards family* if S is not empty. There exist families that are not Edwards families. For example, the family of Barreto-Naehrig curves [7] is not an Edwards family.

4.4 Minimal ρ-values of Pairing-Friendly Edwards Curves

We compute the minimal ρ-values of pairing-friendly Edwards curves at embedding degrees up to 50, which are constructible using the algorithm described in §4.2. In order to construct pairing-friendly parameters, we consider Cocks-Pinch [14], DEM [17], MNT [30], GMV [22], Freeman curve [19], Scott-Barreto [34], Brezing-Weng [13], Barreto-Naehrig curve [7], Kachisa-Schefer-Scott [27], Barreto-Lynn-Scott [6], and the method in [20].

The ρ-values of (t, r, p) constructed by Cocks-Pinch and DEM method are almost 2. The remaining methods output pairing-friendly families $(t(x), r(x), p(x))$ whose ρ-values are less than 2. From these and the observation in §4.3, it is sufficient to investigate only pairing-friendly Edwards families in order to obtain the minimal ρ-values of constructible pairing-friendly Edwards curves. For any method, except for the Brezing-Weng method, the number of output families $(t(x), r(x), p(x))$ is finite. On the other hand, if the degree of $r(x)$ is bounded, the number of Brezing-Weng families of fixed embedding degree is finite. By using an argument similar to that in §8 in [20], if degree of $r(x)$ is more than 100, the expected number of pairing-friendly parameters with a security level less than 1000 bits generated by $(t(x), r(x), p(x))$ is less than 0.03. Therefore, we impose the assumption that the degree of $r(x)$ is less than 100. The column "Edwards PF curve" in Table 1 lists the minimal ρ-values of families $(t(x), r(x), p(x))$ with embedding degree up to 50, which construct pairing-friendly Edwards curves, under the condition that the degree of $r(x)$ is less than 100. Table 8.2 in [20] lists the minimal ρ-values of general constructible pairing-friendly curves. For comparison, we list the result in the column "General PF curve". The embedding degrees in boldface implies that the minimal ρ-values of general constructible pairing-friendly curves and constructible pairing-friendly Edwards curves coincide at the embedding degree.

Table 1. Comparison of ρ-values of Constructible Pairing-friendly Curves and Constructible Pairing-friendly Edwards Curves

k	General PF curve			Edwards PF curve		
	ρ	D	type	ρ	D	type
3	1.000	some	MNT	1.000	some	GMV
4	1.000	some	MNT	1.000	some	GMV
5	1.500	3	BW	1.833	7 (19)	BW
6	1.000	some	MNT	1.000	some	GMV
7	1.333	3 mod 4	FST	1.667	7 (3)	BW
8	1.250	3	BW	1.500	1	FST
9	1.333	3	BW	1.333	3	BW
10	1.000	some	F	1.500	1	BW
11	1.200	3 mod 4	FST	1.400	3	BW
12	1.000	3	BN	1.167	3	FST
13	1.167	3	BW	1.333	3	BW
14	1.333	3	BW	1.500	1	BW
15	1.500	3	BW	1.500	3	BW
16	1.250	1	KSS	1.250	1	KSS
17	1.125	3	BW	1.250	3	BW
18	1.333	3	KSS	1.583	2	BW
19	1.111	3	BW	1.222	3	BW
20	1.375	3	BW	1.500	1	BW
21	1.333	3	BW	1.333	3	BW
22	1.300	1	BW	1.300	1	BW
23	1.091	3 mod 4	FST	1.182	3	BW
24	1.250	3	BW	1.250	3	BW
25	1.300	3	BW	1.400	3	BW
26	1.167	3	BW	1.250	1	BW

k	General PF curve			Edwards PF curve		
	ρ	D	type	ρ	D	type
27	1.111	3	BW	1.111	3	BW
28	1.333	1	BW	1.333	1	BW
29	1.071	3	BW	1.143	3	BW
30	1.500	3	BW	1.500	3	BW
31	1.067	3 mod 4	FST	1.133	3	BW
32	1.063	3	BW	1.125	1	KSS
33	1.200	3	BW	1.200	3	BW
34	1.188	1	BW	1.188	1	BW
35	1.500	3 mod 4	FST	1.583	3	BW
36	1.167	3	KSS	1.167	3	KSS
37	1.056	3	BW	1.111	3	BW
38	1.111	3	BW	1.167	1	BW
39	1.167	3	BW	1.167	3	BW
40	1.375	1	KSS	1.375	1	KSS
41	1.050	3	BW	1.100	3	BW
42	1.333	3	BW	1.333	3	BW
43	1.048	3 mod 4	FST	1.095	3	BW
44	1.150	3	BW	1.200	1	BW
45	1.333	3	BW	1.333	3	BW
46	1.136	1	BW	1.136	1	BW
47	1.043	3	BW	1.087	3	BW
48	1.125	3	BW	1.125	3	BW
49	1.190	3	BW	1.238	3	BW
50	1.300	3	BW	1.350	1	BW

BW: Brezing-Weng, MNT: MNT, GMV: GMV, F: Freeman curveC BN: Barreto-Naehrig curve, KSS: Kachisa-Schefer-Scott, FST: [20].

4.5 An Example of Brezing-Weng families

Let k be a positive integer divisible by 3, but not divisible by 18. Polynomials $t_1(x), r_1(x), p_1(x)$ over $\mathbb{Q}$ are defined as follows:

(1)
$$k \equiv 3 \bmod 6,$$
$$t_1(x) = x + 1,$$
$$r_1(x) = \Phi_k(x),$$
$$p_1(x) = \tfrac{1}{3}(x-1)^2(x^{2k/3}+x^{k/3}+1)+x,$$

(2)
$$k \equiv 0 \bmod 6 \text{ ([20] Construction 6.6)},$$
$$t_1(x) = x + 1,$$
$$r_1(x) = \Phi_k(x),$$
$$p_1(x) = \tfrac{1}{3}(x-1)^2(x^{k/3}-x^{k/6}+1)+x.$$

The above family has embedding degree k and discriminant $D = 3$. The ρ-value is equal to $(2k/3+2)/\phi(k)$ if $k \equiv 3 \bmod 6$, and $(k/3+2)/\phi(k)$ if $k \equiv 0 \bmod 6$.

Remark 1. If k is divisible by 18, $p_1(x)$ has a factor $x^2 + x + 1$; therefore, it is not irreducible.

Table 2. ρ-value of $(t_1(x), r_1(x), p_1(x))$ and minimal ρ-value of constructible pairing-friendly curves

emb. deg.	3	6	9	12	15	21	24
ρ-value	2.000	2.000	**1.333**	1.500	**1.500**	**1.333**	**1.250**
minimal ρ	1.000	1.000	**1.333**	1.000	**1.500**	**1.333**	**1.250**

emb. deg.	27	30	33	39	42	45	48
ρ-value	**1.111**	**1.500**	**1.200**	**1.167**	**1.333**	**1.333**	**1.125**
minimal ρ	**1.111**	**1.500**	**1.200**	**1.167**	**1.333**	**1.333**	**1.125**

Lemma 1. *If $x \equiv 1 \bmod 6$, then $t_1(x), r_1(x), p_1(x)$ all represent integers, and $p_1(x)+1-t_1(x)$ is divisible by 4. Moreover, if $x \equiv 1 \bmod 12$, then $p_1(x)+1-t_1(x)$ is divisible by 16.*

Proof. If $x \equiv 1 \bmod 6$, $p_1(x)$ represents an integer because $x-1$ is divisible by 3. For both cases (1) and (2), $p_1(x)+1-t_1(x)$ is divisible by $(x-1)^2/3$. Since $x-1$ is divisible by 2, $p_1(x)+1-t_1(x)$ is divisible by 4. If $x \equiv 1 \bmod 12$, $x-1$ is divisible by 4; thus, $p_1(x)+1-t_1(x)$ is divisible by 16.

Table 2 shows that for many embedding degrees, the ρ-value of $(t_1(x), r_1(x), p_1(x))$ is minimal among those of the constructible pairing-friendly curves.

5 Application of Pairing-Friendly Edwards Curves

In this section, we propose how to apply the construction algorithm of complete Edwards curve in § 4.2 and the list in § 4.3 to pairing-based cryptography.

For embedding degree k written in boldface in Table 1, there is a pairing-friendly curve E in the Weierstrass form over $\mathbb{F}_p$ with a minimal ρ-value listed in Table 8.2 in [20], a complete Edwards curve Ed_d over $\mathbb{F}_p$ and a birational map $\phi : Ed_d \dashrightarrow E$ whose restriction of the subgroup of rational points with order r becomes a group isomorphism. (We remark that ϕ need not induce a group isomorphism between Ed_d and $E(\mathbb{F}_p)$.) In fact, these factors all are obtained in the algorithm constructing a complete Edwards curve in § 4.2. The pairing-friendly curve E is constructed by Step 1 and 2 , and the birational map ϕ is obtained by the composite of the 2-isogenies in Step 3 and the transformation in Step 4. (The 2-isogenies and the transformation are described concretely by Proposition 1, 2 and 4.) Then we have the Edwards curve Ed_d as the output of the algorithm. In this situation, we assume that E has an efficient pairing $\omega : \mathbb{G}_1 \times \mathbb{G}_2 \to \mathbb{F}_{p^k}^{\times}$, where $\mathbb{G}_1, \mathbb{G}_2$ are subgroups with order r of $E(\mathbb{F}_p)$ and $E(\mathbb{F}_{p^k})$. Then we propose that in a pairing-based protocol, scalar multiplication and pairing are computed as follows:

1. Scalar multiplication.
   All scalar multiplications are calculated on the Edwards curve Ed_d. These scalar multiplications are more efficient than those on the curve $E(\mathbb{F}_p)$ in the Weierstrass form.

2. Pairing computation.
   $\omega' = \omega \circ (\phi \times \phi)$ defines a pairing on $\phi^{-1}(\mathbb{G}_1) \times \phi^{-1}(\mathbb{G}_2)$. (We remark that ϕ defines a group isomorphism from the subgroup of elements with odd order of Ed_d to that of $E(\mathbb{F}_{p^k})$ cf. [25, Th. 1].) We use ω' as a pairing in the protocol. In fact, the pairing $\omega'(P,Q)$ for $P \in \phi^{-1}(\mathbb{G}_1)$, $Q \in \phi^{-1}(\mathbb{G}_2)$ is calculated by $\omega'(P,Q) = \omega(\phi(P), \phi(Q))$.

One advantage of our proposal is that we can use the most efficient pairing implemented on an elliptic curve in the Weierstrass form because the ρ-value of the elliptic curve is minimal among constructible pairing-friendly curves by the assumption. Since the scalar multiplication described above is faster than that on the elliptic curve in the Weierstrass form, our proposal is faster than the protocol implemented on the elliptic curve in the Weierstrass form.

One achievement of our proposal is that we need not choose ϕ such that it induces a group isomorphism between Ed_d and $E(\mathbb{F}_p)$. This implies that the pairing-friendly elliptic curve used in a protocol need not to be transformed into an Edwards curve. Therefore, Edwards curves can be applied for more pairing-friendly curves. For example, let E be a pairing-friendly curve with order divisible by 4, but not by 8, and with sextic or quartic twists. Since E has sextic or quartic twists, E has an efficient pairing ω [24]. On the other hand, E can not be transformed into an Edwards curve because E has no element of order 4 and by Proposition 2. However, there is an Edwards curve Ed_d birational to E by the algorithm constructing a complete Edwards curve in § 4.2. In our proposal, we can use both the efficient pairing ω on E and the scalar multiplication on Ed_d, although Ed_d and $E(\mathbb{F}_p)$ are not isomorphic.

Example 2 and 3 in Appendix A are examples of pairing-friendly curves with order divisible by 4, but not by 8, and with sextic or quartic twists at embedding degrees 16 and 24. These curves have the security level recommended in [36]. Therefore, our proposal is effective for these curves. In these cases, the overhead of each transformation between pairing-friendly curves and Edwards curves is less than or equal to 10 field multiplications.

6 Conclusion

We investigate pairing-friendly curves isogenous to Edwards curves. Accordingly, we listed the minimal ρ-values of pairing-friendly curves isogenous to Edwards curves which are constructed by GMV, Brezing-Weng, Kachisa-Schefer-Scott method, etc., up to embedding degree 50. We compared these and the minimal ρ-values of known constructible pairing-friendly curves, and we determined the embedding degree (less than or equal to 50) such that these two types of minimal ρ-values coincide. For these embedding degrees, the scalar multiplication of pairing-friendly curves with the minimal ρ-values can be computed on Edwards curves efficiently. In fact, we propose a method to make use of the scalar multiplication on Edwards curves which is not isomorphic but isogenous to the pairing-friendly curves. We also present examples of pairing-friendly curves to which our method is applicable at embedding degree $16, 24$.

Acknowledgements. This work was partially supported by the Japan Science and Technology Agency (JST) Strategic Japanese-Indian Cooperative Programme for Multidisciplinary Research Fields, which aims to combine Information and Communications Technology with Other Fields.

References

1. Aréne, C., Lange, T., Naehrig, M., Ritzenthaler, C.: Faster Pairing Computation of the Tate Pairing. Journal of Number Theory 131, 842–847 (2011)
2. Atkin, A.O.L., Morain, F.: Elliptic Curves and Primarity Proving. Math. Comp. 61(203), 29–68 (1993)
3. Bach, E., Shallit, J.: Algorithmic number theory. Efficient algorithms. Foundations of Computing Series, vol. 1. MIT Press, Cambridge (1996)
4. Balasubramanian, R., Koblitz, N.: The Improbability that an Elliptic Curve has Subexponential Discrete Log Problem under the Menezes-Okamoto-Vanstone Algorithm. J. Cryptology 11(2), 141–145 (1998)
5. Barreto, P.S.L.M., Galbraith, S., O'hEigeartaigh, C., Scott, M.: Efficient Pairing Computation on Supersingular Abelian Varieties. Designs, Codes and Cryptography, 239–271 (2004)
6. Barreto, P.S.L.M., Lynn, B., Scott, M.: Constructing Elliptic Curves with Prescribed Embedding Degrees. In: Cimato, S., Galdi, C., Persiano, G. (eds.) SCN 2002. LNCS, vol. 2576, pp. 257–267. Springer, Heidelberg (2003)
7. Barreto, P.S.L.M., Naehrig, M.: Pairing-Friendly Elliptic Curves of Prime Order. In: Preneel, B., Tavares, S. (eds.) SAC 2005. LNCS, vol. 3897, pp. 319–331. Springer, Heidelberg (2006)
8. Benger, N., Scott, M.: Constructing Tower Extensions of Finite Fields for Implementation of Pairing-Based Cryptography. In: Hasan, M.A., Helleseth, T. (eds.) WAIFI 2010. LNCS, vol. 6087, pp. 180–195. Springer, Heidelberg (2010)
9. Bernstein, D.J., Birkner, P., Joye, M., Lange, T., Peters, C.: Twisted Edwards Curves. In: Vaudenay, S. (ed.) AFRICACRYPT 2008. LNCS, vol. 5023, pp. 389–405. Springer, Heidelberg (2008)
10. Bernstein, D.J., Lange, T.: Faster Addition and Doubling on Elliptic Curves. In: Kurosawa, K. (ed.) ASIACRYPT 2007. LNCS, vol. 4833, pp. 29–50. Springer, Heidelberg (2007)
11. Boneh, D., Boyen, X.: Efficient Selective-ID Secure Identity-Based Encryption Without Random Oracles. In: Cachin, C., Camenisch, J.L. (eds.) EUROCRYPT 2004. LNCS, vol. 3027, pp. 223–238. Springer, Heidelberg (2004)
12. Boneh, D., Franklin, M.: Identity-Based Encryption from the Weil Pairing. In: Kilian, J. (ed.) CRYPTO 2001. LNCS, vol. 2139, pp. 213–229. Springer, Heidelberg (2001)
13. Brezing, F., Weng, A.: Elliptic Curves Suitable for Pairing based Cryptography. Designs, Codes and Cryptography 37, 133–141 (2005)
14. Cocks, C., Pinch, R.G.E.: Identity-based Cryptosystems based on the Weil pairing. Unpublished manuscript (2001)
15. Cohen, H., Miyaji, A., Ono, T.: Efficient Elliptic Curve Exponentiation Using Mixed Coordinates. In: Ohta, K., Pei, D. (eds.) ASIACRYPT 1998. LNCS, vol. 1514, pp. 51–65. Springer, Heidelberg (1998)
16. Das, M.P.L., Sarkar, P.: Pairing Computation on Twisted Edwards Form Elliptic Curves. In: Galbraith, S.D., Paterson, K.G. (eds.) Pairing 2008. LNCS, vol. 5209, pp. 192–210. Springer, Heidelberg (2008)

17. Dupont, P., Enge, A., Morain, F.: Building Curves with Arbitrary Small MOV Degree over Finite Prime Fields. Journal of Cryptology 18, 79–89 (2005)
18. Edwards, H.M.: A Normal Form for Elliptic Curves. Bulletin of the American Mathematical Society 44, 393–422 (2007)
19. Freeman, D.: Constructing Pairing-Friendly Elliptic Curves with Embedding Degree 10. In: Hess, F., Pauli, S., Pohst, M. (eds.) ANTS 2006. LNCS, vol. 4076, pp. 452–465. Springer, Heidelberg (2006)
20. Freeman, D., Scott, M., Teske, E.: A Taxonomy of Pairing-Friendly Elliptic Curves. Journal of Cryptology 23(2), 224–280 (2010)
21. Fouquet, M., Morain, F.: Isogeny Volcanoes and the SEA Algorithm. In: Fieker, C., Kohel, D.R. (eds.) ANTS 2002. LNCS, vol. 2369, pp. 276–291. Springer, Heidelberg (2002)
22. Galbraith, S.D., McKee, J., Valença, P.: Ordinary Abelian Varieties Having Small Embedding Degree. Finite Fields and Their Applications 13, 800–814 (2007)
23. Gentry, C.: Practical Identity-Based Encryption Without Random Oracles. In: Vaudenay, S. (ed.) EUROCRYPT 2006. LNCS, vol. 4004, pp. 445–464. Springer, Heidelberg (2006)
24. Hess, F., Smart, N., Vercauteren, F., Berlin, T.U.: The Eta Pairing Revisited. IEEE Transactions on Information Theory 52, 4595–4602 (2006)
25. Hisil, H., Wong, K.K.-H., Carter, G., Dawson, E.: Twisted Edwards Curves Revisited. In: Pieprzyk, J. (ed.) ASIACRYPT 2008. LNCS, vol. 5350, pp. 326–343. Springer, Heidelberg (2008)
26. Ionica, S., Joux, A.: Another Approach to Pairing Computation in Edwards Coordinates. In: Chowdhury, D.R., Rijmen, V., Das, A. (eds.) INDOCRYPT 2008. LNCS, vol. 5365, pp. 400–413. Springer, Heidelberg (2008)
27. Kachisa, E.J., Schaefer, E.F., Scott, M.: Constructing Brezing-Weng Pairing-Friendly Elliptic Curves Using Elements in the Cyclotomic Field. In: Galbraith, S.D., Paterson, K.G. (eds.) Pairing 2008. LNCS, vol. 5209, pp. 126–135. Springer, Heidelberg (2008)
28. Koblitz, N., Menezes, A.: Pairing-Based Cryptography at High Security Levels. In: Smart, N.P. (ed.) Cryptography and Coding 2005. LNCS, vol. 3796, pp. 13–36. Springer, Heidelberg (2005)
29. Kohel, D.: Endomorphism Rings of Elliptic Curves over Finite Fields. PhD thesis, University of California at Berkeley (1996)
30. Miyaji, A., Nakabayashi, M., Takano, S.: New Explicit Conditions of Elliptic Curve traces for FR-reduction. IEICE Transactions on Fundamentals E84-A(5), 1234–1243 (2001)
31. Morain, F.: Edwards Curves and CM Curves (2009), `http://arxiv.org/PS_cache/arxiv/pdf/0904/0904.2243v1.pdf`
32. Sahai, A., Waters, B.: Fuzzy Identity-Based Encryption. In: Cramer, R. (ed.) EUROCRYPT 2005. LNCS, vol. 3494, pp. 457–473. Springer, Heidelberg (2005)
33. Sakai, R., Ohgishi, K., Kasahara, M.: Cryptosystems based on Pairing. In: SCIS 2000 (2000)
34. Scott, M., Barreto, P.S.L.M.: Generating more MNT Elliptic Curves. Designs, Codes and Cryptography 38, 209–217 (2006)
35. Scott, M.: Computing the Tate Pairing. In: Menezes, A. (ed.) CT-RSA 2005. LNCS, vol. 3376, pp. 293–304. Springer, Heidelberg (2005)

36. Scott, M.: On the Efficient Implementation of Pairing-Based Protocols. In: Chen, L. (ed.) IMACC 2011. LNCS, vol. 7089, pp. 296–308. Springer, Heidelberg (2011)
37. Scott, M., Benger, N., Charlemagne, M., Dominguez Perez, L.J., Kachisa, E.J.: On the Final Exponentiation for Calculating Pairings on Ordinary Elliptic Curves. In: Shacham, H., Waters, B. (eds.) Pairing 2009. LNCS, vol. 5671, pp. 78–88. Springer, Heidelberg (2009)
38. Tanaka, S., Nakamula, K.: Constructing Pairing-Friendly Elliptic Curves Using Factorization of Cyclotomic Polynomials. In: Galbraith, S.D., Paterson, K.G. (eds.) Pairing 2008. LNCS, vol. 5209, pp. 136–145. Springer, Heidelberg (2008)
39. Vélu, J.: Isogenies entre courbes elliptiques. Comptes Rendus De L'Academie Des Sciences Paris, Serie I-Mathematique, Serie A 273, 238–241 (1971)
40. Waters, B.: Efficient Identity-Based Encryption Without Random Oracles. In: Cramer, R. (ed.) EUROCRYPT 2005. LNCS, vol. 3494, pp. 114–127. Springer, Heidelberg (2005)
41. Waters, B.: Ciphertext-Policy Attribute-Based Encryption: An Expressive, Efficient, and Provably Secure Realization. In: Catalano, D., Fazio, N., Gennaro, R., Nicolosi, A. (eds.) PKC 2011. LNCS, vol. 6571, pp. 53–70. Springer, Heidelberg (2011)

A Concrete Parameters of Pairing-Friendly Curves

In this section, we present some parameters of pairing-friendly curves with embedding degree 6, 16, 24, which can be transformed into Edwards curves and achieve the minimal ρ-values among the constructible pairing-friendly curves. We also present the parameters of Edwards curves associated with the pairing-friendly curves and their birational maps.

Example 1 (embedding degree 6). We present concrete parameters of a pairing-friendly curve of minimal ρ-value by the GMV method [22] for $k = 6$ and discriminant $D = 128083$. The GMV method uses a family $(t(x), r(x), p(x))$ with ρ-value 1, where $t(x)$ is the trace of Frobenius, $r(x)$ is the prime order of the maximal subgroup, and $p(x)$ is the prime of the base field. When $k = 6$, we can choose parameters of 80-bit security level [20,36], and thus, $r(x)$ must be larger than 160 bits. The following pairing-friendly parameter is obtained from a prime $r(x)$ of $r(x) \geq 2^{159}$ and the corresponding $t(x)$ and $p(x)$.

$$\begin{aligned} t &= -512443546777372184617955 2, \\ r &= 2019987604875175648454545408192574280537212419951 \text{ (161-bit)}, \\ p &= 8079950419500702593818176508334829348427003500251 \text{ (163-bit)}. \end{aligned}$$

From this (t, r, p), using the algorithm in §4.2, we obtain the following elliptic curve,

$$\begin{aligned} E : y^2 =& x^3 + 1998898220505475498985523800218737638994117359471x \\ &+ 1485504696264522858267183942852126541062537134957, \end{aligned}$$

over $\mathbb{F}_p$. The order of $E(\mathbb{F}_p)$ is $4r$ and thus the ρ-value of this curve is $\log(p)/\log(r) = 1.012$. From the algorithm in §2, E is birationally equivalent to the complete Edwards curve,

$$Ed_d : x^2 + y^2 = 1 + dx^2y^2,$$
$$d = 5447142112983792947243789310208468523057475861758$$

over $\mathbb{F}_p$. The birational map is given by $\mathcal{M}$ (or $\mathcal{M}^{-1}$) in §2 for

$$x_2 = 3657207110027107395510706995842511528822735475634,$$
$$x_4 = 1585104739241067019245770351764733165593020279937,$$
$$y_4 = 7654609595387770473489409319104778065105857892185.$$

$\mathcal{M}$ (or $\mathcal{M}^{-1}$) induces a group isomorphism between $E(\mathbb{F}_p)$ and Ed_d.

Example 2 (embedding degree 16). We present concrete parameters of a pairing-friendly curve for $k = 16$ and $D = 1$ by using a family of Kachisa-Schefer-Scott in [27] Example 4.3:

$$t(x) = \frac{1}{35}(2x^5 + 41x + 35),$$
$$r(x) = x^8 + 48x^4 + 625,$$
$$p(x) = \frac{1}{980}(x^{10} + 2x^9 + 5x^8 + 48x^6 + 152x^5 + 240x^4 + 625x^2 + 2398x + 3125).$$

The ρ-value of this family is 1.250, which is minimal. We choose parameters of 160-bit security level, and thus, $r(x)$ is $321 (\fallingdotseq 320)$ bits. We have a parameter (t, r, p) from this family:

$$t = 9421491671814145509134223576122784471820154601489389274892 7714,$$
$$r = 229269484538237404769845466018193408635494162139970701161378 0070//$$
$$6978278427133652906255410672441 13 \text{ (321-bit)},$$
$$p = 27738907913157391241888841689555045766744968140788405824011445844//$$
$$78218136859751438675039143383901154081701677760538397710213 \text{ (411-bit)}.$$

From the CM method [2], we obtain the pairing-friendly elliptic curve E over $\mathbb{F}_p$,

$$E : y^2 = x^3 - 4x.$$

The ρ-value of E is 1.276. E has quartic twists, but it cannot be transformed into an Edwards curve from Proposition 3. E is 2-isogenous to

$$E' : y^2 = x^3 - 44x + 112.$$

The 2-isogeny $E' \to E$ is given by ψ in Proposition 4 for

$$x_2 = 4, \; s = 4.$$

E' can be transformed into the complete Edwards curve

$$Ed_d : x^2 + y^2 = 1 + 1/2x^2y^2.$$

The birational map is given by $\mathcal{M}$ (or $\mathcal{M}^{-1}$) in §2 for

$$x_2 = 4, \; x_4 = 2, \; y_4 = -8.$$

For a pairing on $E(\mathbb{F}_p)$, we can apply a technique using quartic twists. There is a homomorphism $\phi : Ed_d \to E(\mathbb{F}_p)$, which is an isomorphism on the restriction of the subgroup of order r. Therefore, scalar multiplication on the subgroup of order r of $E(\mathbb{F}_p)$ can be calculated on the Edwards curve Ed_d.

Example 3 (embedding degree 24). We present concrete parameters of a pairing-friendly curve for $k = 24$ and $D = 3$ by using a Brezing-Weng family $(t_1(x), r_1(x), p_1(x))$ of Example 1 in §4.5. The ρ-value of this family is 1.250, which is minimal. When $k = 24$, we can choose parameters of 256-bit security level [20,36], and thus, $r(x)$ must be larger than 512 bits. Substituting $x = -(2^{64} + 2^{24} + 2^{22} + 2^{10} + 1)$ for $(t_1(x), r_1(x), p_1(x))$, we obtain

$t =-$ 18446744073730524160,
$r =$13407807930064546362398767933349089388959380713288989663936765586//
88701320601174041435552889229362679203517575956740970726902979324//
618447307198723438045184 1 (513-bit),
$p =$15208135392246888996279084342090403162542447988198528268807465285 6//
61057031869257038249315452379353541837370964091953841205854377127 0//
42202576296800182377516690985145880243642726548595169600529 07 (639-bit).

From the CM method [2], we obtain the pairing-friendly elliptic curve E over $\mathbb{F}_p$,

$$E : y^2 = x^3 + 1.$$

The ρ-value of E is 1.247. E has sextic twists, but it cannot be transformed into an Edwards curve from Proposition 3. E is 2-isogenous to

$$E' : y^2 = x^3 + ax + b,$$

$a =$15208135392246888996279084342090403162542447988198528268807465285 66//
10570318692570382493154523793535418373709640919538412058543771270 42//
20257629680018237751669098514588024364272654859516960052892,
$b = 22$.

The 2-isogeny $E' \to E$ is given by ψ in Proposition 4 for

$$
\begin{aligned}
x_2 &= 2, \\
s &= 152081353922468889962790843420904031625424479881985282688074652 85// \\
&\quad 66105703186925703824931545237935354183737096409195384120585437712// \\
&\quad 7042202576296800182377516690985145880243642726548595169600529 04.
\end{aligned}
$$

E' can be transformed into the complete Edwards curve,

$$Ed_d : x^2 + y^2 = 1 + dx^2y^2,$$

$$
\begin{aligned}
d &= 24733040147563529679719615563444037470778085395385057049701717401// \\
&\quad 39186163689714776227485799390309958095455324067311822030840188850// \\
&\quad 9798737532617133046338622012819150225181 1849.
\end{aligned}
$$

The birational map is given by $\mathcal{M}$ (or $\mathcal{M}^{-1}$) in §2 for

$$
\begin{aligned}
x_2 &= 2, \\
x_4 &= 15208135392246888991332476312577697226598524875509720774651848 20// \\
&\quad 65840456219289135579709431249999239893823993653113339250149437289// \\
&\quad 924185585146164224804177691844617192709759183239847651245642 9210, \\
y_4 &= 143957992950429139853707429155021014239683027301057818257580 7827// \\
&\quad 90663845733639244785551826824274608314781592647947400376411849623// \\
&\quad 806199111835164687730217267331893589926119562461788385850077 7743.
\end{aligned}
$$

For a pairing on $E(\mathbb{F}_p)$, we can apply a technique using sextic twists. There is a homomorphism $\phi : Ed_d \to E(\mathbb{F}_p)$, which is an isomorphism on the restriction of the subgroup of order r. Therefore, scalar multiplication on the subgroup of order r of $E(\mathbb{F}_p)$ can be calculated on the Edwards curve Ed_d.

Standardized Signature Algorithms on Ultra-constrained 4-Bit MCU*

Chien-Ning Chen[1], Nisha Jacob[2], Sebastian Kutzner[1], San Ling[2], Axel Poschmann[1,2], and Sirote Saetang[2]

[1] Physical Analysis & Cryptographic Engineering (PACE)
Nanyang Technological University, Singapore
{chienning,skutzner}@ntu.edu.sg

[2] School of Physical and Mathematical Sciences
Nanyang Technological University, Singapore
{njacob,lingsan,aposchmann,sirote.tang}@ntu.edu.sg

Abstract. In this work, we implement all three digital signature schemes specified in Digital Signature Standard (FIPS 186-3), including DSA and RSA (based on modular exponentiation) as well as ECDSA (based on elliptic curve point multiplication), on an ultra-constrained 4-bit MCU of the EPSON S1C63 family. Myriads of 4-bit MCUs are widely deployed in legacy devices, and some in security applications due to their ultra low-power consumption. However, public-key cryptography, especially digital signature, on 4-bit MCU is usually neglected and even regarded as infeasible. Our highly energy-efficient implementation can give rise to a variety of security functionalities for these ultra-constrained devices.

Keywords: 4-bit MCU, DSA, ECDSA, Elliptic Curve Cryptography, Lightweight Cryptography, RSA, SHA-1.

1 Introduction

In recent years, the area footprint of hardware implementations of standardized algorithms has been continuously brought down to a level, where it is hard to yield any further gain, e.g. for AES [29] from 5,400 GE [34] down to 2400 GE [28]. In the meantime, a great deal of research work has been spent on the design of new lightweight cryptographic primitives. Notably examples for block ciphers and hash functions include KLEIN [11], KATAN [5], LED [13], PICCOLO [40] and PRESENT [4] for the former, and QUARK [1], PHOTON[12] and SPONGENT[3] for the latter, amongst many others. A major optimization goal for those lightweight algorithms is to reduce the area footprint in silicon in order to reduce the cost and the power consumption. The recent adoption of PRESENT as an ISO standard [16] shows the maturity of the field, and, hence, it is no wonder that state-of-the-art lightweight algorithms require close to the theoretical optimal area [12].

* The authors were supported in part by the Singapore National Research Foundation under Research Grant NRF-CRP2-2007-03.

G. Hanaoka and T. Yamauchi (Eds.): IWSEC 2012, LNCS 7631, pp. 37–50, 2012.

At the same time, there is a surprising lack of improvements on the software side. 8-bit microcontrollers (MCUs) have been long used as the platform of choice to evaluate the efficiency of cryptographic algorithms in embedded devices. However, one of the simplest, cheapest and most-abundant computing platforms is 4-bit MCUs that are embedded in a wide variety of everyday items. Applications range from watches and toys to security sensitive applications such as remote access and control systems, car immobilizers, one-time password generators, and all sorts of sensors. The ultra low power consumption of a few micro ampere [37] makes it a fitting choice for passive RFID-tags and a reasonable choice for active RFID-tags as well.

Previous works on 4-bit MCUs are mostly on symmetric crypto, i.e. block cipher implementations using a legacy device from ATMEL. `PRESENT` is reported in [41], `HUMMINGBIRD` [8] in [9], and `AES` in [17,20]. [17] also reports the first implementations of the hash function, `SHA-1` [31], and the public key primitive, `ECC`. In this work we partially build on the results of [17] and combine for the first time `SHA-1` and `ECC` to `ECDSA` on a 4-bit MCU. We also present the first implementations of `DSA`, `RSA`, and Rabin cryptosystem on a 4-bit MCU and compare the results. Our implementations provide functionalities of digital signature on 4-bit MCU for applications that are not timing critical, e.g., legally binding sensor/meter readings and secure firmware updates.

The remainder of this work is organized as follows. In Section 2 the target platform and the design flow are briefly introduced. Section 3 discusses modular exponentiation and in particular the Montgomery multiplication. Subsequently, `DSA` is treated in Section 4, before Section 5 describes our `ECDSA` implementation. Finally, we conclude this paper in Section 6.

2 Target Platform and Design Flow

The Epson S1C63 family of MCUs was introduced in 2011 and is one of the most recent 4-bit low-power architectures. All members of the S1C63 family have a 4-bit core along with ROM, RAM, LCD drivers, and I/O ports. It also has a two-stage pipeline (fetch and execute) and a maximum of 15 and 63 hardware and software interrupt vectors respectively, depending on the model being used. The MCUs differ mainly in the memory size and on-board components, such as UART or hardware multiplier [36]. In this work, due to the extensive space requirement for public-key cryptography, we use S1C63016, which has 26kB (16k*13 bits) of code ROM, 1kB (2k*4 bits) of RAM and 2kB (4k*4 bits) of data ROM as well as an integer multiplier/divider communicated through memory I/O.

The S1C63 MCU core supports a wide instruction set with a linear addressing space without pages. It has two 4-bit data registers `A` and `B`; a 4-bit flag register `F` consisting of extension `E`, interrupt `I`, carry `C` and zero flag `Z`; two 16-bit index registers `X` and `Y` supporting post increment instructions; two stack pointers, SP1 for address and SP2 for data. Table 1 gives a list of some frequently used instructions and their instruction cycles. One instruction cycle is equal to 2 clock cycles[1].

Table 1. Frequently used instruction list [36] of S1C63 family MCU

Mnemonic*				Cycles
LD [%ir]+,%r	LD %r,[%ir]+	ADC %r,[%ir]	ADC %r,[%ir]+	1
CMP %r,%[ir]+	CMP [%ir]+,%r	AND %r,imm4	OR %r,imm4	1
JR sign8	JRNC sign8	CALR imm8	RET	1
LDB %EXT,%BA	LDB %rr,imm8	ADD %ir,%BA	ADD %ir,sign8	1
LD [%ir]+,[%ir]+	LDB [%X]+,%BA	ADC [%ir]+,%r		2
INC [addr6]	DEC [addr6]	XOR [%ir]+,%r	EX %r,[%ir]+	2

*ir = index register (X or Y); r = data register (A or B); rr = XL, XH, YL, YH; imm4 = 4-bit immediate data; imm8 = 8-bit immediate data; sign8 = signed 8-bit digit; addr6 = 6-bit absolute data address.

Details of the design flow of this MCU can be found in [39]. For debugging we use a software simulator on PC (Fig. 1(a)) and a FPGA-based hardware emulation board, called In-Circuit Emulator (ICE) (Fig. 1(b)) [38]. The code will be tested first on the software simulator or on the ICE and then burned on the target board (Fig. 1(c)). The advantage of using the ICE over the software simulator is to ensure the proper operation of the system before burning it on the target board. The software simulator is also used to get the cycle count and code size of our implementations, which are the two most common performance metrics for embedded platforms. Furthermore, energy consumptions are estimated based on datasheets.

3 Modular Exponentiation

Modular exponentiation is widely used in public-key cryptosystems, like RSA [33] and DSA [30]. It is the most time consuming operation in these cryptosystems and determines their performance. This section presents our implementation of 512-bit and 1024-bit modular exponentiation on the EPSON 4-bit MCU, S1C63016.

The computation of modular exponentiation can be divided into two parts: modular multiplication of multi-precision integers at the bottom and exponentiation evaluation on the top. In our implementation, modular multiplication is realized by using the Montgomery multiplication [27] to avoid expensive modular operations, and exponentiation is evaluated by the binary left-to-right exponentiation algorithm. The implementation details of these two parts are provided in Sec. 3.1 and Sec. 3.2.

[1] In the remainder of the paper we refer to instruction cycles as cycles.

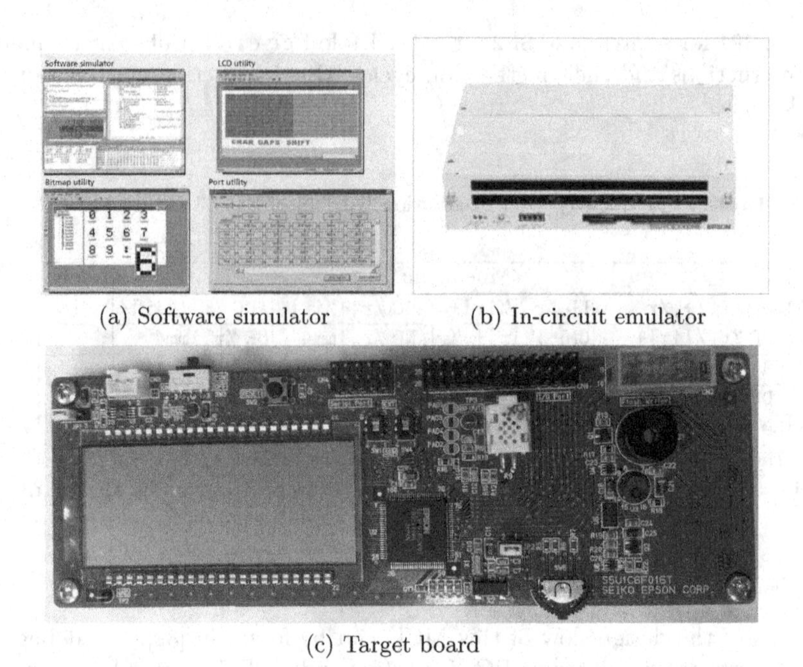

(a) Software simulator (b) In-circuit emulator

(c) Target board

Fig. 1. S1C63 family development tools [38]

3.1 Montgomery Multiplication

Montgomery multiplication introduced by Peter Montgomery is commonly used in modular arithmetic. It computes ($\mathtt{A} \times \mathtt{B} \times 2^{-nt}$ mod $\mathtt{M}$) instead of ($\mathtt{A} \times \mathtt{B}$ mod $\mathtt{M}$) to avoid expensive modular operations (divisions). Figure 2 provides the typical Montgomery multiplication, where A and B are the two operands, M is the modulus, A[], B[], M[] are their (2^t)-ary representation, $m' = (-\mathtt{M[0]})^{-1} \bmod 2^t$, and $0 \leq \mathtt{A,B,M} < 2^{nt}$ as well as $0 \leq \mathtt{A[}i\mathtt{],B[}i\mathtt{],M[}i\mathtt{]} \leq 2^t - 1$ for all $0 \leq i \leq n-1$.

```
Input: A[], B[], M[]
Output: R[] = MontMul(A[],B[])
01 T[] = 0
02 for i = 0 to n-1
03   T[] = T[] + A[i]B[]
04   u = (T[0] × m') mod 2^t
05   T[] = (T[] + u×M[])/2^t
06 output R[] = T[] or R[] = T[] - M[]
```

Fig. 2. Montgomery multiplication algorithm

In a naïve implementation, the inputs and the result of the Montgomery multiplication will satisfy $0 \le \mathtt{A}, \mathtt{B}, \mathtt{R} < \mathtt{M}$. It needs to check if $\mathtt{T} \ge \mathtt{M}$ and optionally performs a subtraction $\mathtt{T} - \mathtt{M}$ before outputting the result. The check and the optional subtraction will cause the execution time to depend on the operands. In addition, the downward scanning in the check is less efficient in both computational time and code size because the EPSON 4-bit MCU only supports post-increment instructions.

C. Walter [42] as well as G. Hachez and J.-J. Quisquater [15] proposed some techniques to eliminate the check and the subtraction, in order to have a constant run-time. In their methods, the parameters will satisfy $\mathtt{A}, \mathtt{B}, \mathtt{R} < 2\mathtt{M}$ as well as $2\mathtt{M} < 2^{(n'-1)t}$ or $\mathtt{M} < 2^{(n'-1)t}$. However, in order to satisfy the extra condition for the modulus, we will have $n' = n + 1$ or $n + 2$ for an nt-bit modulus, which will cause a large overhead on ultra-constrained devices. When n is replaced by $n' = n + 1$, the Montgomery multiplication will require $2(n + 1)^2 + 1$ t-bit multiplications instead of $2n^2 + 1$ multiplications. In addition, since n is usually of 2's power, replacing n by $n' = n + 1$ might also cause some extra costs in memory management.

In order to avoid either the slow check or the extra cost of extending $\mathtt{M}$ to $2\mathtt{M}$, our implementation only keeps the inputs and the result within $0 \le \mathtt{A}, \mathtt{B}, \mathtt{R} < 2^{nt}$ (i.e., $\mathtt{A}$, $\mathtt{B}$, and $\mathtt{R}$ might be greater than $\mathtt{M}$). The temporary result after each iteration (lines 02–05 in Fig.2) will satisfy $\mathtt{T} \le 2^{nt} + \mathtt{M} - 1$. After the whole n iterations, we only check if $\mathtt{T} \ge 2^{nt}$, which is much easier than checking $\mathtt{T} \ge \mathtt{M}$, and a final subtraction $\mathtt{T} - \mathtt{M}$ is required when $\mathtt{T} \ge 2^{nt}$. To achieve a constant time implementation, the optional final subtraction can be evaluated by

$$\mathtt{T}[i] = \mathtt{T}[i] - (\mathtt{mask\ AND\ M}[i]) - \mathtt{c}$$

from $i = 0$ to $n - 1$, where $\mathtt{c}$ is the carry (borrow) flag and $\mathtt{mask} = (-\mathtt{T}[n] \bmod 2^t) = 0$ or $2^t - 1$.

Our implementation achieves the constant execution time of 242,916 and 960,944 cycles for a 512-bit and 1024-bit Montgomery multiplication, respectively. Detailed results are provided in Table 2.

3.2 Exponentiation Computation

We implement the binary left-to-right exponentiation algorithm. In order to use the Montgomery multiplication, some additional computations are required before and after the exponentiation. When computing $\mathtt{X}^{\mathtt{E}} \bmod \mathtt{M}$, the base number $\mathtt{X}$ will be converted to $\mathtt{X}' = (\mathtt{X} \times 2^{nt} \bmod \mathtt{M})$ before exponentiation. After exponentiation, one extra Montgomery multiplication $\mathtt{R} = \mathtt{MontMul}(\mathtt{R}', 1) = \mathtt{R}' \times 2^{-nt} \bmod \mathtt{M}$ is required to get the final result. Although the Montgomery multiplication in our implementation only ensures its output being smaller than 2^{nt} (i.e., might be greater than the modulus $\mathtt{M}$), the output of $\mathtt{MontMul}(\mathtt{R}', 1)$ will always be smaller than $\mathtt{M}$ when $\mathtt{R}' \ne \mathtt{M}$ and $\mathtt{M} > 2^{nt-1}$ (i.e., $\mathtt{M}$ is nt-bit).

Table 2 provides the implementation results including the code size[2] and the execution time. The execution time of the exponentiation with a full-length

[2] Each instruction takes 13 bits, and we provide the code size in byte (8 bits).

exponent is the average value by assuming the Hamming weight of exponent is equal to half of its bit length.

Table 2. Implementation results of Montgomery multiplication and exponentiation.

Operation	Code Size [bytes]	Cycles [million]		Energy [mJ] @3V	
		512-bit	1024-bit	512-bit	1024-bit
Montgomery multiplication	260	0.243	0.961	0.0801	0.317
Exponentiation (full exponent)	499	187.1	1,476	61.74	487.08
Exponentiation (exponent $= 2^{16}+1$)	463	5.156	19.15	1.70	6.31

It is clear that exponentiation with a full length exponent (e.g, RSA signature generation) is impracticality for this ultra-constrained MCU. However, exponentiation with a short exponent (e.g., RSA signature verification with public key $e = 2^{16} + 1$) might still be practical. For Rabin cryptosystem [32], only one modular squaring is required for signature verification. The computation can be further reduced to one Montgomery multiplication (i.e., without pre- and post-computation) by using a modified signature $\mathtt{S}' = \mathtt{S} \times 2^{nt/2} \bmod \mathtt{M}$.

4 Digital Signature Algorithm

The digital signature standard was announced by the US National Institute of Standards and Technology (NIST) in 1991, of which the latest specification can be found in FIPS 186-3 [30]. It includes the secure hash algorithm (SHA) specified in FIPS 180-4 [31], and the digital signature algorithm (DSA). In this section, we combine the SHA-1 implementation in [17] and the modular exponentiation in the previous section and then implement DSA with domain parameters $L = 1024$ bits and $N = 160$ bits (i.e., 1024-bit modulus and 160-bit exponent in modular exponentiation).

4.1 SHA-1 Implementation

SHA-1 is a secure hash standard published by NIST in 1995. It processes arbitrary messages up to a length of 2^{64} bits and produces a 160-bit message digest. There are two stages of SHA-1 computation, *preprocessing* and *hash computation*.

Preprocessing stage of SHA-1 consists of the following three steps.

1. *Padding*: The message is padded by a bit '1' followed by the necessary number ($0 \sim 511$) of bits '0', and then the bit length of the original message (a 64-bit integer) is appended. The length of the message after padding will be a multiple of 512 bits.
2. *Parsing the padded message*: This step divides the padded message into blocks of 512 bits.

3. *Initialize hash value*: The 160-bit starting value is initialized by the five 32-bit words: A = 0x67452301, B = 0xEFCDAB89, C = 0x98BADCFE, D = 0x10325476 and E = 0xC3D2E1F0 in big-endian.

Hash computation: SHA-1 consists of 80 rounds for each block (512 bits) of the message. A block of message will be divided into 16 32-bit words, $\mathsf{M}_0 \sim \mathsf{M}_{15}$. In each round, $\mathsf{W}_t = \mathsf{M}_t$ for $0 \geq t \geq 15$, or $\mathsf{W}_t = \text{ROTL}^1(\mathsf{W}_{t-3} \oplus \mathsf{W}_{t-8} \oplus \mathsf{W}_{t-14} \oplus \mathsf{W}_{t-16})$ for $16 \geq t \geq 79$, where $\text{ROTL}^n()$ is the n-bit rotate left (circular left shift) operation. The round function $f_t(\mathsf{B}, \mathsf{C}, \mathsf{D})$, constants K_t, and the round computations are described in Table 3 and Fig. 3.

Table 3. SHA-1 function $f_t(\mathsf{B}, \mathsf{C}, \mathsf{D})$ and constants K_t.

Round (t)	$f_t(\mathsf{B}, \mathsf{C}, \mathsf{D})$	K_t
0 to 19	$(\mathsf{B} \wedge \mathsf{C}) \oplus (\neg \mathsf{B} \wedge \mathsf{D})$	0x5A827999
20 to 39	$\mathsf{B} \oplus \mathsf{C} \oplus \mathsf{D}$	0x6ED9EBA1
40 to 59	$(\mathsf{B} \wedge \mathsf{C}) \oplus (\mathsf{B} \wedge \mathsf{D}) \oplus (\mathsf{C} \wedge \mathsf{D})$	0x8F1BBCDC
60 to 79	$\mathsf{B} \oplus \mathsf{C} \oplus \mathsf{D}$	0xCA62C1D6

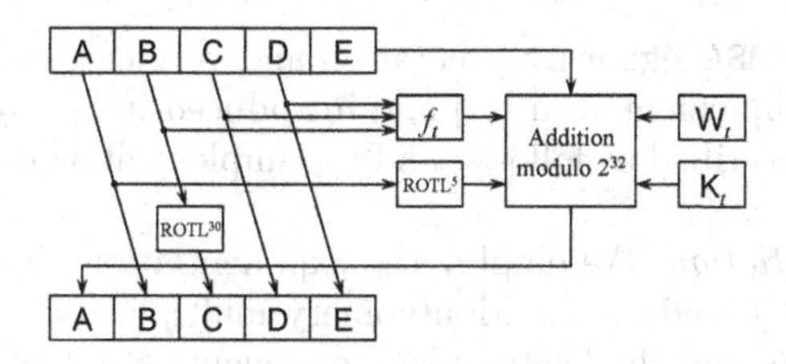

Fig. 3. One round of SHA-1 computation

Details of our SHA-1 implementation can be found in [17]. Table 4 summaries the results for space (code size) and speed optimization.

Table 4. Implementation results of SHA-1

Optimization	Code Size [bytes]	Cycles	Energy Consumption [μJ] @3V
Space	2,038	108,666	35.85
Speed	2,324	87,788	28.97

4.2 DSA Implementation

The digital signature algorithm provides the capability of generation and verification of a digital signature. The system parameters include two prime numbers

p and q, satisfying $2^{1023} < p < 2^{1024}$, $2^{159} < q < 2^{160}$, and q divides $(p-1)$, as well as a base number $g \in \mathbb{Z}_p^*$ of the order q. Signer's private key is x, satisfying $0 < x < q$, and the public key is $y = g^x \bmod p$. The signature generation and verification algorithms are given below:

Signature generation

1. Compute $h(m)$ by using SHA-1.
2. Generate a random ephemeral key k satisfying $0 < k < q$.
3. Compute $k^{-1} \pmod{q}$.
4. Compute $r = (g^k \bmod p) \bmod q$.
5. Compute $s = \left(k^{-1}(h(m) + x \times r)\right) \bmod q$.

Signature verification

1. Verify the signature (r', s') satisfying $0 < r' < q$ and $0 < s' < q$.
2. Compute $w = (s')^{-1} \bmod q$.
3. Compute $u_1 = (h(m) \times w) \bmod q$.
4. Compute $u_2 = (r' \times w) \bmod q$.
5. Compute $v = (g^{u_1} \times y^{u_2} \bmod p) \bmod q$.
6. If $v = r'$, the signature is valid.

We implement both DSA signature generation and verification. The SHA-1 hash function has been implemented in [17] and introduced in Sec. 4.1, other atomic computations are described as follows, and the implementation results are summarized in Table 5.

Modular exponentiation: We employ the exponentiation algorithm described in Sec. 3.2, which is based on the Montgomery multiplication. The modulus is the 1024-bit prime p, and the length of the exponents k, u_1 and u_2 are 160-bit. We also implement Shamir's double-exponentiation algorithm [10, section V.B] for signature verification.

Multiplication and reduction: Except the modular exponentiation, other multiplications modulo q are achieved by using row-wise multiplication and Barrett reduction [2]. When reducing a 1024-bit integer by the 160-bit modulus q, seven reductions are required, starting from MSB of the 1024-bit integer.

Inversion: We employ the binary extended GCD algorithm [21, Ch 4.5.2] implemented in [17].

5 Elliptic Curve Digital Signature Algorithm

Elliptic curve cryptography (ECC), introduced independently by Neil Koblitz [22] and Victor Miller [25], is an alternative of public-key cryptography. Similar to the discrete logarithm problem on modular exponentiation, ECC can be employed in a variety of applications, like key exchange (e.g., ECDH [35]), digital signature (e.g., ECDSA [18]). The main advantage of ECC is the small key size. ECC with much smaller key size can provide the same level of security as RSA or DLP-based cryptography, e.g., ECC with 160-bit key is as secure as RSA with 1024-bit

Table 5. Implementation results of DSA

Operation	Code Size [bytes]	Cycles [million]	Energy Consumption [mJ] @ 3V
Exponentiation (160-bit exponent)	499	232.68	76.78
Double-exp (160-bit exponents)	655	274.69	90.65
Barrett reduction	425	0.036	0.011
Inversion	703	0.13	0.043
Signature generation	3,951	239.18	78.92
Signature verification	4,154	290.78	95.96

key [14]. The small key size reduces the cost of communication, storage, and even computation, and makes it particularly suitable for constrained devices.

ECC relies upon group operations in an elliptic curve group, and a group $\mathbb{E}$ over field $\mathbb{F}_p$ can be defined by the points (x, y) satisfying the short Weierstrass form:

$$\mathbb{E} : y^2 = x^3 + ax + b, \text{ where } 4a^3 + 27b^2 \neq 0.$$

In "Standards for Efficient Cryptography 2" (SEC2) [6], an elliptic curve over a prime field is specified by a sextuple: (p, a, b, G, n, h), where p is the prime, a and b are the curve parameters, G is a base point with order n, and h is the cofactor.

In this section, we combine the SHA-1 and the SEC2 curve secp160r1 implementation in [17] and then provide the first implementation of the standardized elliptic curve digital signature, ECDSA, on the 4-bit MCU, S1C63016.

5.1 ECDSA Implementation

ECDSA is one of the standardized digital signature schemes. The system parameters of ECDSA include the specification of the underlying curve (secp160r1 for our implementation), signer's private key d ($0 < d < n$) and public key $Q = dG$. The signature generation and verification algorithms are described as follows.

ECDSA *signature generation*

1. Compute $h(m)$ by using SHA-1.
2. Generate a random number k satisfying $0 < k < n$.
3. Compute $k^{-1} \pmod{n}$.
4. Compute $r = x \bmod n$, where $(x, y) = kG$.
5. Compute $s = (k^{-1}(h(m) + d \times r)) \bmod n$.

ECDSA *signature verification*

1. Verify the signature (r', s') satisfying $0 < r' < n$ and $0 < s' < n$.
2. Compute $w = (s')^{-1} \bmod n$.
3. Compute $u_1 = (h(m) \times w) \bmod n$.

4. Compute $u_2 = (r' \times w) \bmod n$.
5. $P = (x, y) = u_1 G + u_2 Q$.
6. If $r' = x \bmod n$, the signature is valid.

The computations in ECDSA can be divided into three parts: prime field arithmetic, point arithmetic, and protocol layer. There are two different types of prime field arithmetic in ECDSA, where the moduli are the pseudo Mersenne prime p in the point arithmetic and the curve order n in the protocol layer, respectively.

Prime Field Arithmetic. The curve secp160r1 in SEC2 employs the pseudo Mersenne prime, $p = 2^{160} - 2^{31} - 1$, which makes the modulo operations much more efficient compared to the Barrett reduction [2] and the Montgomery multiplication [27]. The computations modulo the 164-bit curve order n in the protocol layer are performed by using Barrett reduction. We have implemented the following operations for both moduli p and n, and the results are summarized in Table 6.

Modular Addition and Subtraction: We further optimize the implementation in [17] and generalize it for both 160-bit and 164-bit moduli by using two entry points. The functions for the two moduli share most of the code, but some additional code is required for processing the 164-bit modulus.

Multiplication (M) *and Squaring* (S): According to the implementation result in [17], the row-wise multiplication is more efficient for this MCU than the column-wise or hybrid multiplication. We separately implement the modular multiplication for the modulus n because a generalized implementation for both moduli will cause 10% increase in execution time for the 160-bit multiplication and this significantly slows down the overall run time (as about 90% run time of a point multiplication is spent on underlying field multiplications).

Bisection is required only for point arithmetic. For an even number, division by 2 is a right shift of its binary representation. For an odd number, an extra addition of the odd prime p is required before the right shift.

Inversion (I) is achieved by the binary extended GCD algorithm [21, Ch 4.5.2], which requires multi-precision addition/subtraction and bisection. The implementation in [17] only supports the 160-bit modulus p, and we extend it to support both 160-bit and 164-bit moduli.

Reduction: Since the prime number of the curve *secp160r1* is a *pseudo mersenne prime*, the reduction modulo p can be implemented efficiently by using only shifts and additions. However, the reduction modulo the curve order n requires the Barrett reduction (BR). Each Barrett reduction requires 2M and some additions/subtractions. It also requires some space to store the pre-computed values.

Point Arithmetic. The major computation in ECC is the point multiplication nP which can be evaluated through the combination of point doubling $2P$ and point addition $P_1 + P_2$. Instead of representing points in affine coordinates ($\mathcal{A}$), we employ Jacobian projective coordinates ($\mathcal{J}$) to implement point doubling and addition. A point (x, y) in $\mathcal{A}$ can be represented by $(x, y, 1)$ in $\mathcal{J}$, and a point

Table 6. Implementation results of prime field arithmetic.

Operation	modulo 160-bit p		modulo 164-bit n	
	Code Size [bytes]	Cycles	Code Size [bytes]	Cycles
Modular add/sub	292	340	302	344
Multiplication	318	16,226	333	17,836
Bisection	208	207	299	212
Fast reduction	624	679	-	-
Barrett reduction	-	-	425	36,194

(X, Y, Z) in $\mathcal{J}$ is identical to the point $(X/Z^2, Y/Z^3)$ in $\mathcal{A}$. The following are the three point operations, and detailed results are summarized in Table 7 and Table 8.

Point Doubling (D): We implement point doubling in Jacobian coordinates ($2\mathcal{J} \rightarrow \mathcal{J}$). It requires 4M and 4S as well as some minor field operations, or alternatively 3M and 5S by using the trick $\alpha\beta = \frac{1}{2}\left((\alpha+\beta)^2 - \alpha^2 - \beta^2\right)$ [23].

Point Addition/Subtraction (A): Point addition in mixed coordinates ($\mathcal{J}+\mathcal{A} \rightarrow \mathcal{J}$) takes 8M and 3S, or 7M and 4S by using the trick described above. Point subtraction is similar to point addition but has an extra subtraction to calculate the y-coordinate.

Point Multiplication (PM): As shown in [17], we have implemented various scalar multiplication algorithms, including the basic binary left-to-right method, the left-to-right NAF recoding [19], and some side-channel countermeasures[3] [7,24]. For signature verification, we also need double point multiplication (d-PM). Employing NAF recoding in either Shamir's double-exponentiation algorithm [10] or Möller's interleaving algorithm [26] can achieve the average complexity of $1.55 \log_2 n$ or $1.66 \log_2 n$, respectively. However, recoding both scalars into NAF will cause huge overhead in code size on this MCU. We only implement Shamir's method with binary scalars which achieves the average complexity of $1.75 \log_2 n$.

Table 7. Implementation results of ECC point arithmetic

Operation	Description	Code Size [bytes]	Cycles
Point doubling	4M + 4S	900	128,453
$2\mathcal{J} \rightarrow \mathcal{J}$	3M + 5S	940	123,781
Point addition	8M + 3S	1,700	178,956
$\mathcal{J} + \mathcal{A} \rightarrow \mathcal{J}$	7M + 4S	1,748	176,601

Protocol Layer. We implement the ECDSA signature generation and verification. Besides hash computation, for signature generation, we need $2\mathtt{M}_{164}+\mathtt{I}+2\mathtt{BR}$ (modulo 164-bit n) and one PM. For signature verification, we need $2\mathtt{M}_{164}+\mathtt{I}+2\mathtt{BR}$ and one d-PM. Table 9 shows the implementation results using a 160-bit message m (including one SHA-1 computation).

[3] Please refer to [17, Sec. 5.2] for the security-efficiency trade-off on 4-bit MCU.

Table 8. Implementation results of ECC point multiplication

Point Multiplication Algorithm	Code Size [bytes]	Cycles [millions]	Side-channel Immunity
Left-to-right PM	5,724	34.37	-
Left-to-right PM with left-to-right NAF	8,127	29	-
Double-and-add-always	6,562	48.04	SPA
BRIP	7,681	49	SPA,RPA,ZPA,DPA
Randomization of scalar (20-bit)	8,215	32.21	DPA
Randomization of scalar (64-bit)	8,342	42.04	DPA
Randomized projective coordinates	8,093	30.50	DPA
Randomization of scalar (20-bit) & Randomized projective coordinates	8,312	32.52	DPA

Table 9. Implementation results of ECDSA

Operation*	Code Size [Bytes]	Cycles [Million]	Energy Consumption [mJ] @ 3V
Signature generation	8,546	35.28	11.64
Signature verification	8,611	41.9	13.87

* Using Left-to-Right PM or d-PM

6 Conclusion

In this work, we implement the three standardized signature schemes, RSA (512- and 1024-bit), DSA (1024-bit) and ECDSA (160-bit) on a 4-bit MCU. Our implementation results show that ECDSA is the most practical signature scheme for ultra-constrained devices, and when only signature verification is required, RSA with small public key is also practical. Through this work, we show that public-key cryptography is possible on constrained devices. Future work includes the investigation of side-channel immunity of our implementations.

References

1. Aumasson, J.-P., Henzen, L., Meier, W., Naya-Plasencia, M.: QUARK: A Lightweight Hash. In: Mangard, S., Standaert, F.-X. (eds.) CHES 2010. LNCS, vol. 6225, pp. 1–15. Springer, Heidelberg (2010), http://131002.net/quark/
2. Barrett, P.: Implementing the Rivest Shamir and Adleman Public Key Encryption Algorithm on a Standard Digital Signal Processor. In: Odlyzko, A.M. (ed.) CRYPTO 1986. LNCS, vol. 263, pp. 311–323. Springer, Heidelberg (1987)
3. Bogdanov, A., Knežević, M., Leander, G., Toz, D., Varıcı, K., Verbauwhede, I.: SPONGENT: A Lightweight Hash Function. In: Preneel, B., Takagi, T. (eds.) CHES 2011. LNCS, vol. 6917, pp. 312–325. Springer, Heidelberg (2011)
4. Bogdanov, A., Knudsen, L.R., Leander, G., Paar, C., Poschmann, A., Robshaw, M.J.B., Seurin, Y., Vikkelsoe, C.: PRESENT: An Ultra-Lightweight Block Cipher. In: Paillier, P., Verbauwhede, I. (eds.) CHES 2007. LNCS, vol. 4727, pp. 450–466. Springer, Heidelberg (2007), http://lightweightcrypto.org/present/

5. De Cannière, C., Dunkelman, O., Knežević, M.: KATAN and KTANTAN — A Family of Small and Efficient Hardware-Oriented Block Ciphers. In: Clavier, C., Gaj, K. (eds.) CHES 2009. LNCS, vol. 5747, pp. 272–288. Springer, Heidelberg (2009)
6. Certicom Research. Standards for efficient cryptography, SEC 2: Recommended elliptic curve domain parameters (2000)
7. Coron, J.-S.: Resistance against Differential Power Analysis for Elliptic Curve Cryptosystems. In: Koç, Ç.K., Paar, C. (eds.) CHES 1999. LNCS, vol. 1717, pp. 292–302. Springer, Heidelberg (1999)
8. Engels, D., Fan, X., Gong, G., Hu, H., Smith, E.M.: Ultra-lightweight cryptography for low-cost RFID tags: Hummingbird algorithm and protocol. Technical report, Centre for Applied Cryptographic Research, CACR (2009), `http://cacr.uwaterloo.ca/techreports/2009/cacr2009-29.pdf`
9. Fan, X., Hu, H., Gong, G., Smith, E.M., Engels, D.: Lightweight implementation of Hummingbird cryptographic algorithm on 4-bit microcontrollers. In: International Conference for Internet Technology and Secured Transactions, pp. 1–5 (2009)
10. Gamal, T.E.: A public key cryptosystem and a signature scheme based on discrete logarithms. IEEE Transactions on Information Theory 31(4), 469–472 (1985)
11. Gong, Z., Nikova, S., Law, Y.W.: KLEIN: A New Family of Lightweight Block Ciphers. In: Juels, A., Paar, C. (eds.) RFIDSec 2011. LNCS, vol. 7055, pp. 1–18. Springer, Heidelberg (2012)
12. Guo, J., Peyrin, T., Poschmann, A.: The PHOTON Family of Lightweight Hash Functions. In: Rogaway, P. (ed.) CRYPTO 2011. LNCS, vol. 6841, pp. 222–239. Springer, Heidelberg (2011)
13. Guo, J., Peyrin, T., Poschmann, A., Robshaw, M.J.B.: The LED Block Cipher. In: Preneel, B., Takagi, T. (eds.) CHES 2011. LNCS, vol. 6917, pp. 326–341. Springer, Heidelberg (2011)
14. Gura, N., Patel, A., Wander, A., Eberle, H., Shantz, S.C.: Comparing Elliptic Curve Cryptography and RSA on 8-bit CPUs. In: Joye, M., Quisquater, J.-J. (eds.) CHES 2004. LNCS, vol. 3156, pp. 119–132. Springer, Heidelberg (2004)
15. Hachez, G., Quisquater, J.-J.: Montgomery Exponentiation with no Final Subtractions: Improved Results. In: Koç, Ç.K., Paar, C. (eds.) CHES 2000. LNCS, vol. 1965, pp. 293–301. Springer, Heidelberg (2000)
16. ISO/IEC. 29192-2: Information technology – security techniques – lightweight cryptography – part 2: Block ciphers, `http://www.iso.org/iso/iso_catalogue/catalogue_tc/catalogue_detail.htm?csnumber=56552`
17. Jacob, N., Saetang, S., Chen, C.-N., Kutzner, S., Ling, S., Poschmann, A.: Feasibility and practicability of standardized cryptography on 4-bit micro controllers. To appear in SAC (2012)
18. Johnson, D., Menezes, A., Vanstone, S.: The elliptic curve digital signature algorithm (ECDSA). International Journal of Information Security 1(1), 36–63 (2001)
19. Joye, M., Yen, S.-M.: Optimal left-to-right binary signed-digit recoding. IEEE Trans. Computers 49(7), 740–748 (2000)
20. Kaufmann, T., Poschmann, A.: Enabling standardized cryptography on ultra-constrained 4-bit microcontrollers. In: IEEE International Conference on RFID, Orlando, USA, pp. 32–39 (April 2012)
21. Knuth, D.E.: The Art of Computer Programming, vol. II: Seminumerical Algorithms, 3rd edn. Addison-Wesley (1997)
22. Koblitz, N.: Elliptic curve cryptosystems. Mathematics of Computation 48(177), 203–209 (1987)

23. Longa, P., Miri, A.: Fast and flexible elliptic curve point arithmetic over prime fields. IEEE Trans. Computers 57(3), 289–302 (2008)
24. Mamiya, H., Miyaji, A., Morimoto, H.: Efficient Countermeasures against RPA, DPA, and SPA. In: Joye, M., Quisquater, J.-J. (eds.) CHES 2004. LNCS, vol. 3156, pp. 343–356. Springer, Heidelberg (2004)
25. Miller, V.S.: Use of Elliptic Curves in Cryptography. In: Williams, H.C. (ed.) CRYPTO 1985. LNCS, vol. 218, pp. 417–426. Springer, Heidelberg (1986)
26. Möller, B.: Algorithms for Multi-exponentiation. In: Vaudenay, S., Youssef, A.M. (eds.) SAC 2001. LNCS, vol. 2259, pp. 165–180. Springer, Heidelberg (2001)
27. Montgomery, P.L.: Modular multiplication without trial division. Mathematics of Computation 44(170), 519–521 (1985)
28. Moradi, A., Poschmann, A., Ling, S., Paar, C., Wang, H.: Pushing the Limits: A Very Compact and a Threshold Implementation of AES. In: Paterson, K.G. (ed.) EUROCRYPT 2011. LNCS, vol. 6632, pp. 69–88. Springer, Heidelberg (2011)
29. National Institute of Standards and Technology. FIPS 197: Announcing the advanced encryption standard (AES) (November 2001), `http://csrc.nist.gov/publications/PubsFIPS.html`
30. National Institute of Standards and Technology. FIPS 186-3: Digital signature standard (DSS) (June 2009), `http://csrc.nist.gov/publications/PubsFIPS.html`
31. National Institute of Standards and Technology. FIPS 180-4: Secure hash standard (SHS) (March 2012), `http://csrc.nist.gov/publications/PubsFIPS.html`
32. Rabin, M.O.: Digitalized signatures and public key functions as intractable as factorization (1979), `http://publications.csail.mit.edu/lcs/pubs/pdf/MIT-LCS-TR-212.pdf`
33. Rivest, R.L., Shamir, A., Adleman, L.M.: A method for obtaining digital signatures and public-key cryptosystems. Commun. ACM 21(2), 120–126 (1978)
34. Satoh, A., Morioka, S., Takano, K., Munetoh, S.: A Compact Rijndael Hardware Architecture with S-Box Optimization. In: Boyd, C. (ed.) ASIACRYPT 2001. LNCS, vol. 2248, pp. 239–254. Springer, Heidelberg (2001)
35. Schroeppel, R., Orman, H., O'Malley, S., Spatscheck, O.: Fast Key Exchange with Elliptic Curve Systems. In: Coppersmith, D. (ed.) CRYPTO 1995. LNCS, vol. 963, pp. 43–56. Springer, Heidelberg (1995)
36. Seiko Epson Corporation. CMOS 4-bit single chip microcomputer S1C63000 core CPU manual (2011), `http://www.epson.jp/device/semicon_e/product/index_mcu.htm`
37. Seiko Epson Corporation. CMOS 4-bit single chip microcontroller S1C63003/004/008/016 technical manual (2011), `http://www.epson.jp/device/semicon_e/product/index_mcu.htm`
38. Seiko Epson Corporation. Microcontrollers 2011 (2011), `http://www.epsondevice.com/webapp/docs_ic/DownloadServlet?id=ID000463`
39. Seiko Epson Corporation. Program development process (2011), `http://www.epson.jp/device/semicon_e/product/mcu/development/tool.htm`
40. Shibutani, K., Isobe, T., Hiwatari, H., Mitsuda, A., Akishita, T., Shirai, T.: *Piccolo*: An Ultra-Lightweight Blockcipher. In: Preneel, B., Takagi, T. (eds.) CHES 2011. LNCS, vol. 6917, pp. 342–357. Springer, Heidelberg (2011)
41. Vogt, M., Poschmann, A., Paar, C.: Cryptography is feasible on 4-bit microcontrollers - a proof of concept. In: IEEE International Conference on RFID, Orlando, USA, pp. 267–274 (2009)
42. Walter, C.D.: Montgomery's Multiplication Technique: How to Make It Smaller and Faster. In: Koç, Ç.K., Paar, C. (eds.) CHES 1999. LNCS, vol. 1717, pp. 80–93. Springer, Heidelberg (1999)

Very Short Critical Path Implementation of AES with Direct Logic Gates

Kenta Nekado, Yasuyuki Nogami, and Kengo Iokibe

Graduate School of Natural Science and Technology, Okayama University, 700–8530 Tsushima–naka, Kita–ward, Okayama–city, Okayama–pref., Japan
{nekado,nogami}@trans.cne.okayama-u.ac.jp, iokibe@cne.okayama-u.ac.jp

Abstract. A lot of improvements and optimizations for the hardware implementation of AES algorithm have been reported. These reports often use, instead of arithmetic operations in the AES original $\mathbb{F}_{2^8}$, those in its isomorphic tower field $\mathbb{F}_{((2^2)^2)^2}$ and $\mathbb{F}_{(2^4)^2}$. This paper focuses on $\mathbb{F}_{(2^4)^2}$ which provides higher–speed arithmetic operations than $\mathbb{F}_{((2^2)^2)^2}$. In the case of adopting $\mathbb{F}_{(2^4)^2}$, not only high–speed arithmetic operations in $\mathbb{F}_{(2^4)^2}$ but also high–speed basis conversion matrices from the $\mathbb{F}_{2^8}$ to $\mathbb{F}_{(2^4)^2}$ should be used. Thus, this paper improves arithmetic operations in $\mathbb{F}_{(2^4)^2}$ with *Redundantly Represented Basis* (RRB), and provides basis conversion matrices with *More Miscellaneously Mixed Bases* (MMMB).

Keywords: AES, SubBytes, MixColumns, type–I optimal normal basis, mixed bases.

1 Introduction

Since NIST published Advanced Encryption Standard (AES), namely a special class of Rijndael [1], many hardware implementations of AES algorithm have been reported [5,6,7,8,9,10,11]. Thus, this paper also proposes approaches for more *efficient* hardware implementaions, where the "*efficient*" is, in this paper, meant as primarily "*high–speed*", and secondly "*compact*".

In the encryption procedure of AES algorithm, 4 steps such as SubBytes, ShiftRows, MixColumns and AddRoundKey [2] are iterated in sequence. On the other hand, in the decryption procedure of AES algorithm, 4 steps such as InvSubBytes, InvShiftRows, InvMixColumns, AddRoundKey [2] are iterated in sequence. For software implementations, SubBytes and InvSubBytes are often implemented with the lookup–table [1]. On the other hand, for hardware implementations, SubBytes and InvSubBytes are often implemented with some arithmetic operation circuits in octic binary extension field (Galois field) $\mathbb{F}_{2^8}$. In SubBytes and InvSubBytes, an inversion in $\mathbb{F}_{2^8}$ is carried out, and it plays a important role to prevent *linear cryptanalysis* [3]. Additionally, it is the most complex among the arithmetic operations. On the other hand, in the case of hardware implementations, not only SubBytes and InvSubBytes but also MixColumns and InvMixColumns should be efficient. In MixColumns and InvMixColumns, some multiplications in $\mathbb{F}_{2^8}$ are carried out. Thus, this paper first considers to implement

G. Hanaoka and T. Yamauchi (Eds.): IWSEC 2012, LNCS 7631, pp. 51–68, 2012.

more efficient arithmetic operation circuits in $\mathbb{F}_{2^8}$ by using only some logic gates such as AND, XOR, and XNOR gates.

In the case of the original AES algorithm [1], an element in $\mathbb{F}_{2^8}$ is represented by the polynomial basis, whose modular polynomial is the octic irreducible polynomial $t^8+t^4+t^3+t+1$ over $\mathbb{F}_2$. Therefore, originally, SubBytes and InvSubBytes implementations require inversion circuits in the $\mathbb{F}_{2^8}$. However, by adopting inversion circuits in towering fields (composite fields [4]) isomorphic to the $\mathbb{F}_{2^8}$, some researchers have been provided faster and more compact SubBytes and InvSubBytes circuits. At the beginning, Rudra et al. have shown such implementation with a certain $\mathbb{F}_{(2^4)^2}$ as the isomorphic towering field [5]. On the other hand, Satoh and Morioka et al. have shown that with a certain $\mathbb{F}_{((2^2)^2)^2}$ [6,7]. After those, some implementations with the other $\mathbb{F}_{(2^4)^2}$ and $\mathbb{F}_{((2^2)^2)^2}$ have been reported [8,9,10,11]. To the authors' knowledge, the implementations with $\mathbb{F}_{(2^4)^2}$ [5,11] can provide faster inversion circuits than those with $\mathbb{F}_{((2^2)^2)^2}$ [6,7,8,9,10]. Thus, this paper focuses on $\mathbb{F}_{(2^4)^2}$, and proposes *Redundantly Represented Basis* (RRB) which can provide faster inversion circuits in $\mathbb{F}_{(2^4)^2}$ than the bases adopted by [5,11]. Then, this paper also considers multiplication circuits in the $\mathbb{F}_{(2^4)^2}$ with RRB. By adopting RRB, an inversion in $\mathbb{F}_{(2^4)^2}$ can be carried out in $4T_{\mathrm{AND}} + 7T_{\mathrm{XOR}}$, where T_{AND} and T_{XOR} respectively denote the critical path delays of AND and XOR gates.

On the other hand, in the case that arithmetic operations in towering field isomorphic to the $\mathbb{F}_{2^8}$ are adopted for the encryption and decryption procedures of AES algorithm, not only arithmetic operations in an isomorphic towering field but also basis conversion from the $\mathbb{F}_{2^8}$ to the isomorphic towering field should be efficient. However, when many kinds of basis conversion matrices can not be prepared, it is quite difficult to select some efficient conversion matrices. In order to prepare more kinds of basis conversion matrices, Nogami et al. have proposed *Mixed Bases* (MB) technique [10]; however, when using RRB, MB is not enough to provide efficient matrices. Thus, this paper proposes *More Miscellaneously Mixed Bases* (MMMB), and then shows how to find efficient conversion matrices.

This paper has the following proposals.

PR1: To make arithmetic operations in $\mathbb{F}_{(2^4)^2}$ more efficient
PR2: To find more efficient basis conversion matrices

As described above, the former proposal is achieved by RRB, and the latter proposal is achieved by MMMB. With RRB and MMMB, this paper theoretically shows that the encryption and decryption circuits of AES can be provided by the critical path delay $4T_{\mathrm{AND}} + 13T_{\mathrm{XOR}}$.

2 AES Algorithm Applied Basis Conversion

In encryption and decryption procedures of AES algorithm, a plaintext is split into 128–bit blocks. Every block is described as the following 4×4 matrix, whose each element is dealt with as an element in the $\mathbb{F}_{2^8}$.

$$\begin{bmatrix} H_{0,0} & H_{0,0} & H_{0,2} & H_{0,3} \\ H_{1,0} & H_{1,1} & H_{1,2} & H_{1,3} \\ H_{2,0} & H_{2,1} & H_{2,2} & H_{2,3} \\ H_{3,0} & H_{3,1} & H_{3,2} & H_{3,3} \end{bmatrix} \quad (H_{j,l} \in \mathbb{F}_{2^8}). \tag{1}$$

The original AES algorithm [1] represents an element in $\mathbb{F}_{2^8}$ with the polynomial basis $\{1, \alpha, \alpha^2, \ldots, \alpha^6, \alpha^7\}$, where α is a zero of the irreducible polynomial $f_0(t) = t^8 + t^4 + t^2 + t + 1$ over $\mathbb{F}_2$. Let H denote an element in the $\mathbb{F}_{2^8}$, then this paper arbitrarily represents H as **Table** 1.

This section introduces the encryption and decryption procedures of AES algorithm applied *basis conversion* from the $\mathbb{F}_{2^8}$ to its isomorphic towering field. Although the paper fundamentally follows the approach in [5], some parts of the procedures are improved. In what follows, the improved parts are clarified.

Table 1. Representation styles of an element in the $\mathbb{F}_{2^8}$

Style	Representation $(h_j \in \{0, 1\})$
basis in $\mathbb{F}_{2^8}$	$h_0 + h_1\alpha + h_2\alpha^2 + \cdots + h_6\alpha^6 + h_7\alpha^7$
vector	$[h_0\ h_1\ h_2\ \cdots\ h_6\ h_7]$
integer	'h' $(h = h_0 + h_1 2 + h_2 2^2 + \cdots + h_6 2^6 + h_7 2^7)$

2.1 Encryption Procedure Applied Basis Conversion

0–th Round: Only AddRoundKey is carried out. Then, each element of the 4×4 matrix is processed as

$$C_{0,j,l} = \left(H_{j,l} + K_{0,j,l}\right)\mathbf{B} \quad (0 \leq j,\, l < 4), \tag{2}$$

where $K_{0,j,l}$ is the j–th row and l–th column element of the 0–th round key (4×4 matrix), and $\mathbf{B}$ denotes a basis conversion matrix from the $\mathbb{F}_{2^8}$ to its isomorphic towering field. $C_{0,j,l}$ in Eq. (2) becomes an element in the isomorphic towering field. From there to the last round, each element of the 4×4 matrix is dealt with as an element in the isomorphic towering field.

From 1–st to 2–nd Last Round: First, SubBytes is carried out. Then, each element of the 4×4 matrix is processed as

$$G_{r,j,l} = \left(C_{r-1,j,l}\right)^{-1}\bar{\mathbf{B}}\mathbf{A}\mathbf{B} \quad (0 \leq j,\, l < 4), \tag{3}$$

where r is the ordinal number of the round, $\bar{\mathbf{B}}$ denotes the inverse matrix of $\mathbf{B}$, and $\mathbf{A}$ denotes the Affine transformation matrix [2]. $\bar{\mathbf{B}}\mathbf{A}\mathbf{B}$ in Eq. (3) can be preliminarily calculated. Additionally, $(C_{r-1,j,l})^{-1}$ in Eq. (3) is the inverse element in the isomorphic towering field, and it should be efficiently calculated.

Next, ShiftRows, MixColumns, and AddRoundKey are carried out. In order to perform these steps faster, this paper applies a new approach different from that in [5]. Actually, each element of the 4×4 matrix can be processed as Eq. (4a) or (4b).

$$C_{r,j,l} = \left(\left(G_{r,\langle j+1\rangle,\langle l+j\rangle} + G_{r,\langle j+2\rangle,\langle l+j\rangle}\right) + \left(G_{r,\langle j+3\rangle,\langle l+j\rangle} + (K_{r,j,l} + L)\mathbf{B}\right)\right)$$
$$+ \left((\text{`2'}\mathbf{B})\left(G_{r,j,\langle l+j\rangle} + G_{r,\langle j+1\rangle,\langle l+j\rangle}\right)\right) \quad (0 \leq j,\, l < 4), \quad \text{(4a)}$$

$$C_{r,j,l} = \left(\left(G_{r,j,\langle l+j\rangle} + G_{r,\langle j+2\rangle,\langle l+j\rangle}\right) + \left(G_{r,\langle j+3\rangle,\langle l+j\rangle} + (K_{r,j,l} + L)\mathbf{B}\right)\right)$$
$$+ \left((\text{`3'}\mathbf{B})\left(G_{r,j,\langle l+j\rangle} + G_{r,\langle j+1\rangle,\langle l+j\rangle}\right)\right) \quad (0 \leq j,\, l < 4), \quad \text{(4b)}$$

where $\langle j\rangle$ means "$j \mod 4$", $K_{r,j,l}$ is the j–th row and l–th column element of the r–th round key (4×4 matrix), and L denotes the Affine transformation vector [2]. In Eq. (4), '02'**B** and '03'**B** can be preliminarily calculated, and $(K_{r,j,l}+L)\mathbf{B}$ can be calculated when the round key is generated.

Last Round: First, SubBytes is carried out. Then, each element of the 4×4 matrix is processed as Eq. (3).

Next, ShiftRows and AddRoundKey are carried out. Then, each element of the 4×4 matrix is processed as

$$\tilde{C}_{j,l} = G_{r,j,\langle l+j\rangle}\bar{\mathbf{B}} + (K_{r,j,l} + L) \quad (0 \leq j,\, l < 4). \tag{5}$$

$K_{r,j,l} + L$ in Eq. (5) can be calculated when the round key is generated. $\tilde{C}_{j,l}$ in Eq. (5) is dealt with in the same way as $H_{j,l}$, namely as an element in the $\mathbb{F}_{2^8}$. The 4×4 matrix which consists of $\tilde{C}_{j,l}$ in Eq. (5) forms a 128–bit block of the cipher text. This 128–bit block is the same of that not applied basis conversion, namely that in the original AES algorithm.

2.2 Decryption Procedure Applied Basis Conversion

0–th Round: Only AddRoundKey is carried out. Then, each element of the 4×4 matrix is processed as

$$C_{r-1,j,l} = \left(\tilde{C}_{j,l} + (K_{r,j,l} + L)\right)\mathbf{B} \quad (0 \leq j,\, l < 4). \tag{6}$$

$K_{r,j,l} + L$ in Eq. (6) can be calculated when the round key is generated. $C_{r-1,j,l}$ in Eq. (6) is an element in the isomorphic towering field. From there to the last round, each element of the 4×4 matrix is dealt with as an element in the isomorphic towering field.

From 1–st to 2–nd Last Round: First, InvShiftRows and InvSubBytes are carried out. Then, each element of the 4×4 matrix is processed as

$$G_{r,j,l} = \left(C_{r,j,\langle l-j\rangle}\bar{\mathbf{B}}\bar{\mathbf{A}}\mathbf{B}\right)^{-1} \quad (0 \leq j,\, l < 4), \tag{7}$$

where $\bar{\mathbf{A}}$ denotes the inverse Affine transformation matrix [2]. $\bar{\mathbf{B}}\bar{\mathbf{A}}\mathbf{B}$ in Eq. (7) is preliminarily calculated. Additionally, $(C_{r,j,l}\bar{\mathbf{B}}\bar{\mathbf{A}}\mathbf{B})^{-1}$ in Eq. (7) is the inverse element in the isomorphic towering field, and it should be efficiently calculated.

Next, AddRoundKey and InvMixColumns are carried out. In order to perform these steps faster, this paper applies a new approach different from that in [5]. For example, each element of the 4×4 matrix can be processed as

$$C_{r-1,j,l} = \Bigg(('14'\mathbf{B})G_{r,j,l} + ('11'\mathbf{B})G_{r,\langle j+1\rangle,l}\Bigg) + \Bigg(('13')\mathbf{B}G_{r,\langle j+2\rangle,l} + \Big(('9'\mathbf{B})G_{r,\langle j+3\rangle,l} + J_{r,j,l}\Big)\Bigg), \tag{8a}$$

$$J_{r,j,l} = ('14'K_{r,j,l} + '11'K_{r,\langle j+1\rangle,l} + '13'K_{r,\langle j+2\rangle,l} + '9'K_{r,\langle j+3\rangle,l} + L)\mathbf{B}, \tag{8b}$$

where '14'$\mathbf{B}$, '11'$\mathbf{B}$, '13'$\mathbf{B}$, and '9'$\mathbf{B}$ can be preliminarily calculated, and $J_{r,j,l}$ can be calculated when the round key is generated.

Last Round: First, InvShiftRows and InvSubBytes are carried out. Then, each element of the 4×4 matrix is processed as Eq. (7).

Next, AddRoundKey is carried out. Then, each element of the 4×4 matrix is processed as

$$H_{j,l} = G_{1,j,l}\bar{\mathbf{B}} + K_{0,j,l} \quad (0 \leq j,\, l < 4). \tag{9}$$

3 Arithmetic Operations in Towering Field $\mathbb{F}_{(2^4)^2}$

In the AES algorithm applied basis conversion from the $\mathbb{F}_{2^8}$ to $\mathbb{F}_{(2^4)^2}$, inversions and multiplications in $\mathbb{F}_{(2^4)^2}$ are required as described in Eqs. (3), (4), (7) and (8). Thus, this section introduces how to prepare $\mathbb{F}_{(2^4)^2}$ and its subfield $\mathbb{F}_{2^4}$, and efficient arithmetic operations in these extension fields.

In the case of $\mathbb{F}_{(2^4)^2}$, first construct $\mathbb{F}_{2^4}$, then 2–nd tower over the $\mathbb{F}_{2^4}$. Most of researchers [5,6,7,8,9,10,11] use normal bases and polynomial bases to prepare extension fields and towering fields. This paper also adopts normal bases to achieve 2–nd towering over $\mathbb{F}_{2^4}$. On the other hand, this paper adopts an innovative basis to construct $\mathbb{F}_{2^4}$. This section introduces the detail of the adopted bases and the arithmetic operations.

3.1 Quartic Extension Field $\mathbb{F}_{2^4}$

Irreducible Polynomial and an Innovative Basis: There exist 3 kinds of quartic irreducible polynomials over $\mathbb{F}_2$ as

$$f_1(t) = t^4+t+1, \quad f_2(t) = t^4+t^3+1, \quad f_3(t) = t^4+t^3+t^2+t+1. \tag{10}$$

Normal bases and polynomial bases in $\mathbb{F}_{2^4}$ can be distinguished from a zero of these polynomials. For a zero β of $f_1(t)$, the set $\{\beta, \beta^2, \beta^{2^2}, \beta^{2^3}\}$ does not form normal bases; however, $\{1, \beta, \beta^2, \beta^3\}$ forms a polynomial basis. Rudra et al. [5] and Joen et al. [11] have shown that the polynomial basis efficiently carries out arithmetic operations, especially inversion, in $\mathbb{F}_{2^4}$.

On the other hand, for a zero β of $f_3(t)$, the sets $\{\beta, \beta^2, \beta^{2^2}, \beta^{2^3}\}$ and $\{1, \beta, \beta^2, \beta^3\}$ respectively form a normal basis and a polynomial basis. The normal basis is especially called type–I optimal normal basis (ONB) [12], and it carries out arithmetic operations in $\mathbb{F}_{2^4}$ as efficiently as in Rudra et al.'s and Jeon et al.'s implementations. However, this paper adopts an innovative basis instead of type–I ONB and the polynomial basis. The basis is the union $\{\beta, \beta^2, \beta^{2^2}, \beta^{2^3}, 1\}$ of type–I ONB $\{\beta, \beta^2, \beta^{2^2}, \beta^{2^3}\}$ and $\{1\}$, and it can provide faster arithmetic operations than the type–I ONB and the polynomial basis. This paper especially calls it *Redundantly Represented Basis* (RRB). In what follows, the properties of RRB is described.

β which is a zero of $f_3(t)$ has the following relations.

$$f_3(\beta) = \beta^4 + \beta^3 + \beta^2 + \beta + 1 = 0, \Leftrightarrow f_3(\beta) = \beta + \beta^2 + \beta^{2^2} + \beta^{2^3} + 1 = 0, \quad (11a)$$

$$\because \ (\beta + 1) f_3(\beta) + 1 = \beta^5 = 1. \quad (11b)$$

According to Eq. (11b), type–I ONB $\{\beta, \beta^2, \beta^{2^2}, \beta^{2^3}\}$ is described as

$$\{\beta, \beta^2, \beta^{2^2}, \beta^{2^3}\} = \{\beta, \beta^2, \beta^3, \beta^4\}. \quad (12)$$

Because β, β^2, β^{2^2}, β^{2^3} are conjugate zeros of $f_3(t)$, 4 kinds of polynomial bases are considered according to Eq. (11b)$'$ as

$$\{1, \beta, \beta^2, \beta^3\} = \{1, \beta, \beta^2, \beta^3 \quad\}, \quad (13a)$$

$$\{1, \beta^2, (\beta^2)^2, (\beta^2)^3\} = \{1, \beta, \beta^2, \quad \beta^4\}, \quad (13b)$$

$$\{1, \beta^{2^2}, (\beta^{2^2})^2, (\beta^{2^2})^3\} = \{1, \quad \beta^2, \beta^3, \beta^4\}, \quad (13c)$$

$$\{1, \beta^{2^3}, (\beta^{2^3})^2, (\beta^{2^3})^3\} = \{1, \beta, \quad \beta^3, \beta^4\}. \quad (13d)$$

According to Eqs. (12), (13), a basis is obtained by removing some one element from the set $\{1, \beta, \beta^2, \beta^3, \beta^4\}$. On the other hand, RRB $\{\beta, \beta^2, \beta^{2^2}, \beta^{2^3}, 1\} = \{1, \beta, \beta^2, \beta^3, \beta^4\}$ uses all. Thus, the conversion from RRB to the bases in Eqs. (12), (13) can be easily achieved from Eq. (11a).

Let D denote an element in $\mathbb{F}_{2^4}$, then D is represented with RRB as Eq. (14a).

$$D = d_0\beta + d_1\beta^2 + d_2\beta^{2^2} + d_3\beta^{2^3} + d_4 \quad (d_j \in \mathbb{F}_2). \quad (14a)$$

$$= (d_0 + d_4)\beta + (d_1 + d_4)\beta^2 + (d_2 + d_4)\beta^{2^2} + (d_3 + d_4)\beta^{2^3} \quad (14b)$$

$$= (d_4 + d_2) + (d_0 + d_2)\beta + (d_1 + d_2)\beta^2 + (d_3 + d_2)\beta^3. \quad (14c)$$

As described above, according to Eq. (11a), D represented with RRB can be easily converted to that represented with type–I ONB and the polynomial bases in Eqs. (12), (13a) as Eqs. (14b), (14c).

In principle, RRB in $\mathbb{F}_{2^4}$ can not uniquely represent an element in $\mathbb{F}_{2^4}$. For example, $D = \beta + \beta^2$ is also described as $D = \beta^{2^2} + \beta^{2^3} + 1$ according to Eq. (11a). However, D is uniquely represented when the Hamming weight of D is restricted to be equal to or less than 2. On the other hand, the Hamming weight of D can be easily reduced to be equal to or less than 2 according to Eq. (11a) when it is more than 2.

Arithmetic Operations: Let E denote an element in $\mathbb{F}_{2^4}$, then E is represented with RRB as

$$E = e_0\beta + e_1\beta^2 + e_2\beta^{2^2} + e_3\beta^{2^3} + e_4 \quad (e_j \in \mathbb{F}_2). \tag{15}$$

A multiplication $M = D \times E$ is given as follows. Note that it is derived from type–I *Cyclic Vector Multiplication Algorithm* (CVMA) [13] and Eq. (14).

$$\begin{aligned} M &= m_0\beta + m_1\beta^2 + m_2\beta^{2^2} + m_3\beta^{2^3} + m_4 \quad (m_j \in \mathbb{F}_2) \\ &= (d_4e_0 + d_2e_1 + d_1e_2 + d_3e_3 + d_0e_4)\beta + (d_0e_0 + d_4e_1 + d_3e_2 + d_2e_3 + d_1e_4)\beta^2 \\ &\quad + (d_3e_0 + d_1e_1 + d_4e_2 + d_0e_3 + d_2e_4)\beta^{2^2} + (d_1e_0 + d_0e_1 + d_2e_2 + d_4e_3 + d_3e_4)\beta^{2^3} \\ &\quad + (d_2e_0 + d_3e_1 + d_0e_2 + d_1e_3 + d_4e_4) && (16a) \\ &= (a_{1,2}b_{1,2} + a_{0,4}b_{0,4})\beta + (a_{2,3}b_{2,3} + a_{1,4}b_{1,4})\beta^2 + (a_{0,3}b_{0,3} + a_{2,4}b_{2,4})\beta^{2^2} \\ &\quad + (a_{0,1}b_{0,1} + a_{3,4}b_{3,4})\beta^{2^3} + (a_{0,2}b_{0,2} + a_{1,3}b_{1,3}), && (16b) \\ a_{j,l} &= d_j + d_l, \quad b_{j,l} = e_j + e_l \quad (0 \le j < l \le 4). && (16c) \end{aligned}$$

The critical path delay of the multiplication circuit given by Eq. (16b) is $1T_{\text{AND}} + 2T_{\text{XOR}}$. On the other hand, that given by Eq. (16a) is $1T_{\text{AND}} + 3T_{\text{XOR}}$. Thus, in principle, a multiplication in $\mathbb{F}_{2^4}$ should be calculated as Eq. (16b) (**Fig.** 2).

From here on, suppose that E is a non–zero constant element in $\mathbb{F}_{2^4}$, then this subsection considers a multiplication by the constant element E. When the Hamming weight of E is restricted to be equal to or less than 2, namely 1 or 2, E can be classified as **Table** 2. According to Eq. (16a), a multiplication $N = D \times E$ can be carried out with theoretically no delay when E belongs to the class (I) of **Table** 2, that is, the Hamming weight of E is 1. On the other hand, it can be calculated with $1T_{\text{XOR}}$ when E belongs to the class (II) of **Table** 2, that is, the Hamming weight of E is 2. For example, multiplications $N_0 = D \times (1, 0, 0, 0, 0)$ and $N_1 = D \times (1, 1, 0, 0, 0)$ are respectively given from Eq. (16a) as

$$N_0 = d_4\beta + d_0\beta^2 + d_3\beta^{2^2} + d_1\beta^{2^3} + d_2, \tag{17a}$$

$$N_1 = (d_2 + d_4)\beta + (d_0 + d_4)\beta^2 + (d_3 + d_1)\beta^{2^2} + (d_0 + d_1)\beta^{2^3} + (d_2 + d_3). \tag{17b}$$

A squaring $S = D^2$ can be carried out with theoretically no delay as

$$S = d_3\beta + d_0\beta^2 + d_1\beta^{2^2} + d_2\beta^{2^3} + d_4. \tag{18}$$

Table 2. Classification of non–zero elements in $\mathbb{F}_{2^4}$

Class	(I)	(II)
element in $\mathbb{F}^*_{2^4}$ †	(1,0,0,0,0) (0,1,0,0,0) (0,0,1,0,0) (0,0,0,1,0) (0,0,0,0,1)	(1,1,0,0,0), (1,0,1,0,0) (0,1,1,0,0), (0,1,0,1,0) (0,0,1,1,0), (0,0,1,0,1) (0,0,0,1,1), (1,0,0,1,0) (1,0,0,0,1), (0,1,0,0,1)
Hamming weight	1	2

† $(e_0, e_1, e_2, e_3, e_4)$ denotes an element E in Eq. (15).

Table 3. The critical path delay of each arithmetic operation circuit in $\mathbb{F}_{2^4}$

Implementation	Multiplication	Squaring	Inversion	Multiplication by the class (I) element	Multiplication by the class (II) element
Rudra al.'s [5]	$(1,3)^\dagger$	$(0,1)^\dagger$	$(2,2)^\dagger$	—	—
Jeon al.'s [11]					
With RRB	$(1,\mathbf{2})^\dagger$	$(0,\mathbf{0})^\dagger$		$(0,0)^\dagger$	$(0,1)^\dagger$

† (j,l) means $jT_{\mathrm{AND}} + lT_{\mathrm{XOR}}$.
‡ The delay when $T_{\mathrm{AND}} \geq T_{\mathrm{XOR}}$ is shown. That when $T_{\mathrm{AND}} \leq T_{\mathrm{XOR}}$ is given as $(1,3)$.

From here on, suppose that D is a non–zero element in $\mathbb{F}_{2^4}$, then an inversion $I = D^{-1}$ is given as follows (**Fig.** 3). See **Appendix** A about how to derive it.

$$\begin{aligned} I &= i_0\beta + i_1\beta^2 + i_2\beta^{2^2} + i_3\beta^{2^3} + i_4 \quad (i_j \in \mathbb{F}_2) \\ &= (a_{2,4} + a_{0,4}a_{1,4}a_{1,3})\beta + (a_{3,4} + a_{1,4}a_{2,4}a_{0,2})\beta^2 + (a_{0,4} + a_{2,4}a_{3,4}a_{1,3})\beta^{2^2} \\ &\quad + (a_{1,4} + a_{3,4}a_{0,4}a_{0,2})\beta^{2^3} + (a_{0,4}a_{2,4}\overline{a_{1,3}} + a_{1,4}a_{3,4}\overline{a_{0,2}}), \qquad (19a) \\ a_{j,l} &= (d_j + d_l) \quad (0 \leq j < l \leq 4), \qquad (19b) \end{aligned}$$

where $\overline{d}$ $(d \in \mathbb{F}_2)$ means "NOT d".

The critical path delay of each arithmetic operation circuit with RRB is given as **Table** 3. As shown in **Table** 3, compared to Rudra et al.'s [5] and Jeon et al.'s [11] implementations , RRB can reduce each critical path delay of a multiplication circuit and a squaring circuit in $\mathbb{F}_{2^4}$ by $1T_{\mathrm{XOR}}$.

3.2 2–nd Towering Field $\mathbb{F}_{(2^4)^2}$

Irreducible Polynomial and Normal Basis: In the same way as **Sec.** 3.1, this subsection first considers the setting of irreducible polynomial. Let a quadratic polynomial over $\mathbb{F}_{2^4}$ be described as

$$g(t) = t^2 + \mu t + \nu \quad (\mu, \nu \in (\mathbb{F}_{2^4} - \{0\})). \qquad (20)$$

In order that $g(t)$ is irreducible over $\mathbb{F}_{2^4}$, $g(t)$ needs to satisfy that $\mu^2/\nu \notin \mathbb{F}_{2^2}$. Suppose that γ is a zero of $g(t)$, then the sets $\{\gamma, \gamma^{16}\}$ and $\{1, \gamma\}$ respectively form a normal basis and a polynomial basis in $\mathbb{F}_{(2^4)^2}$. Among these bases, this subsection focuses on the normal basis only.

Arithmetic Operations: Let C denote an element in $\mathbb{F}_{(2^4)^2}$, $\mathbf{B}$ denote a basis conversion matrix from the $\mathbb{F}_{2^8}$ to its isomorphic towering field $\mathbb{F}_{(2^4)^2}$, and 'j' ($0 \le j < 256$) denote an element in $\mathbb{F}_{2^8}$ described by the integer style of **Table** 1. Then, C and 'j'$\mathbf{B}$ is represented with the normal basis $\{\gamma, \gamma^{16}\}$ as

$$C = D\gamma + E\gamma^{16}\ (D, E \in \mathbb{F}_{2^4}), \quad `j\text{'}\mathbf{B} = Q_j\gamma + R_j\gamma^{16}\ (Q_j, R_j \in \mathbb{F}_{2^4}), \quad (21)$$

where D, E, Q_j, and R_j are represented with RRB in $\mathbb{F}_{2^4}$. Then, a multiplication $W = C \times$ 'j'$\mathbf{B}$ is given as follows. See **Appendix** A about how to derive it.

$$W = Y\gamma + Z\gamma^{16}\ (Y, Z \in \mathbb{F}_{2^4}) = \{D\delta_j + E\epsilon_j\}\gamma + \{D\epsilon_j + E\eta_j\}\gamma^{16}, \quad (22a)$$

$$\delta_j = Q_j(\mu + \frac{\nu}{\mu}) + R_j\cdot\frac{\nu}{\mu}, \quad \epsilon_j = (Q_j + R_j)\cdot\frac{\nu}{\mu}, \quad \eta_j = Q_j\cdot\frac{\nu}{\mu} + R_j(\mu + \frac{\nu}{\mu}), \quad (22b)$$

where δ_j, ϵ_j, and η_j can be preliminarily calculated. According to **Tables** 2, 3, the critical path delay of the multiplication circuit given by Eq. (22a) is at most $\mathbf{2}T_{\text{XOR}}$ even if δ_j, ϵ_j, and η_j are assigned with arbitrary elements.

From here on, suppose that C is a non–zero element in $\mathbb{F}_{(2^4)^2}$, then with *Itoh–Tsujii inversion Algorithm* (ITA) [14], an inversion $X = C^{-1} = (C \cdot C^{16})^{-1}C^{16}$ is given as follows (**Fig**. 6(a)). Note that it is derived by generalizing the approach in [9], in detail, by appending a μ^2–multiplication in $\mathbb{F}_{2^4}$.

$$X = Y\gamma + Z\gamma^{16}\ (Y, Z \in \mathbb{F}_{2^4}) = \{E\gamma + D\gamma^{16}\}/\{DE\mu^2 + (D+E)^2\nu\}, \quad (23)$$

where each multiplication by μ^2 and ν can be carried out with theoretically no delay according to **Table** 3 when the following condition is satisfied.

Condition 1 *Both μ^2 and ν belong to the class (I) of* **Table** *1.*

Thus, this paper considers that both μ^2 and ν are assigned with the class (I) elements. Then, there exist **20** irreducible polynomials over $\mathbb{F}_{2^4}$ which satisfies **Cond**. 1, and the critical path delay of the inversion circuit in $\mathbb{F}_{(2^4)^2}$ is given as $4T_{\text{AND}} + 7T_{\text{XOR}}$ from **Table** 3 and **Fig**. 6(a). As shown in **Table** 4, the circuit of this work can carry out an inversion in the towering field isomorphic to the $\mathbb{F}_{2^8}$ faster than those of the others. On the other hand, the circuit size is given as **Table** 5 (before downsizing). As shown in **Table** 5, the inversion circuit in $\mathbb{F}_{(2^4)^2}$ of this work (before downsizing) uses more XOR gates than that of Jeon et al. Thus, the next subsection considers how to downsize the inversion circuit in $\mathbb{F}_{(2^4)^2}$.

3.3 Theoretically Downsizing the Inversion Circuit in $\mathbb{F}_{(2^4)^2}$

Focus on **Fig**. 6(a), then it is seeable that the wire (i) directly connects to the multiplication circuit (I) and (II), the wire (ii) connects through the μ^2–multiplication circuit to the multiplication circuit (I) and directly connects to the multiplication circuit (III), and the wire (iii) directly connects to the multiplication circuit (II) and (III). Thus, for the inversion circuit in $\mathbb{F}_{(2^4)^2}$, a part, namely 1–st part shown in **Fig.** 2(a), of each multiplication circuit in $\mathbb{F}_{2^4}$ can be shared with each other as **Fig.** 6(b). Then, the circuit size can be reduced by **30**XOR gates according to **Table** 5. As a result, the inversion circuit in $\mathbb{F}_{(2^4)^2}$ of this work (after downsizing) uses less logic gates than that of Jeon et al.

Table 4. The critical path delay of an inversion circuit in towering field

Towering field	Implementation	Critical path delay
$\mathbb{F}_{((2^2)^2)^2}$	Satoh and Morioka et al.'s [6,7]	$4T_{\text{AND}} + 17T_{\text{XOR}}$
	Mentens's et al. [8]	
	Canright's [9]	$4T_{\text{AND}} + 15T_{\text{XOR}}$
	Nogami et al.'s [10]	$4T_{\text{AND}} + 14T_{\text{XOR}}$
$\mathbb{F}_{(2^4)^2}$	Rudra et al.'s [5]	$4T_{\text{AND}} + 10T_{\text{XOR}}$
	Jeon et al.'s [11]	
	This work	$4T_{\text{AND}} + \mathbf{7}T_{\text{XOR}}$

Table 5. The number of logic gates for an inversion circuit in $\mathbb{F}_{(2^4)^2}$

Implementation	Before downsizing	After downsizing
Rudra et al.'s [5]	60AND + 72XOR	
Jeon et al.'s [11]	58AND + 67XOR + 2OR	
This work	42AND + 98XOR + 2XNOR	**42**AND + **68**XOR + **2**XNOR

4 Basis Conversion between $\mathbb{F}_{2^8}$ and $\mathbb{F}_{(2^4)^2}$

This section evaluates the calculation efficiencies given by basis conversion matrices for Eq. (3) (namely, SubBytes), Eq. (4) (namely, ShiftRows, MixColumns, and AddRoundKey), Eq. (7) (namely, InvShiftRows and InvSubBytes), and Eq. (8) (namely, InvMixColumns and AddRoundKey).

4.1 Calculation Efficiency of Eqs. (3) and (7)

This subsection considers each multiplication by $\bar{\mathbf{B}}\mathbf{A}\mathbf{B}$ and $\bar{\mathbf{B}}\bar{\mathbf{A}}\mathbf{B}$ in Eqs. (3) and (7), where $\mathbf{B}$, $\bar{\mathbf{B}}$, $\mathbf{A}$, and $\bar{\mathbf{A}}$ respectively denote a basis conversion matrix from the $\mathbb{F}_{2^8}$ to its isomorphic towering field $\mathbb{F}_{(2^4)^2}$, its inverse matrix, Affine transformation matrix, and the inverse Affine transformation matrix. In the case of adopting RRB described in **Sec.** 3.1, both conversion matrices $\bar{\mathbf{B}}\mathbf{A}\mathbf{B}$ and $\bar{\mathbf{B}}\bar{\mathbf{A}}\mathbf{B}$ from $\mathbb{F}_{(2^4)^2}$ over the $\mathbb{F}_{2^4}$ constructed by RRB to the same $\mathbb{F}_{(2^4)^2}$ are required. Actually, these conversion matrices are given by a basis conversion matrix $\mathbf{B}$ from the $\mathbb{F}_{2^8}$ to $\mathbb{F}_{(2^4)^2}$ over the $\mathbb{F}_{2^4}$ constructed by type–I ONB of Eq. (12) or the polynomial bases of Eq. (13) according to Eq. (14).

In order to show an example, suppose an extension field $\mathbb{F}_{2^4}$ constructed by type–I ONB $\{\beta, \beta^2, \beta^{2^2}, \beta^{2^3}\}$, a field $\mathbb{F}_{(2^4)^2}$ which 2–nd towers over the $\mathbb{F}_{2^4}$ with the normal basis $\{\gamma, \gamma^{16}\}$, and a basis conversion matrix $\mathbf{B}$ from the $\mathbb{F}_{2^8}$ to the $\mathbb{F}_{(2^4)^2}$. Then, $\bar{\mathbf{B}}\mathbf{A}\mathbf{B}$ in Eq. (3) is represented as the left–hand equation in Eq. (24), and an example of the $\bar{\mathbf{B}}\mathbf{A}\mathbf{B}$ is given as the right–hand equation in Eq. (24).

$$\bar{\mathbf{B}}\mathbf{A}\mathbf{B} = \left[\begin{array}{ccccc} u_{0,0} & u_{0,1} & u_{0,2} & \cdots & u_{0,6} \; u_{0,7} \\ \vdots & & & \ddots & \vdots \\ u_{3,0} & u_{3,1} & u_{3,2} & \cdots & u_{3,6} \; u_{3,7} \\ \hline v_{0,0} & v_{0,1} & v_{0,2} & \cdots & v_{0,6} \; v_{0,7} \\ \vdots & & & \ddots & \vdots \\ v_{3,0} & v_{3,1} & v_{3,2} & \cdots & v_{3,6} \; v_{3,7} \end{array}\right], \quad \bar{\mathbf{B}}\mathbf{A}\mathbf{B} = \left[\begin{array}{c|c|c|c|c|c|c|c} 1 & 1 & 1 & 1 & 1 & 1 & 1 & 1 \\ 0 & 1 & 0 & 1 & 1 & 1 & 1 & 1 \\ 0 & 0 & 0 & 1 & 0 & 1 & 1 & 0 \\ 0 & 0 & 0 & 0 & 0 & 1 & 0 & 0 \\ \hline 0 & 0 & 1 & 0 & 1 & 0 & 1 & 1 \\ 0 & 0 & 0 & 0 & 0 & 0 & 0 & 1 \\ 0 & 0 & 0 & 0 & 0 & 0 & 0 & 0 \\ 0 & 0 & 0 & 0 & 0 & 0 & 0 & 0 \end{array}\right]. \tag{24}$$

Let $C_{r-1,j,l}$ in Eq. (3) be corresponding to a non–zero element $C = D\gamma + E\gamma^{16}$ $(D, E \in \mathbb{F}_{2^4})$ which is the input of the inversion circuit of **Fig.** 6(b), and let $(C_{r-1,j,l})^{-1}$ in Eq. (3) be corresponding to $X = C^{-1} = Y\gamma + Z\gamma^{16}$ $(Y, Z \in \mathbb{F}_{2^4})$ which is the output of the inversion circuit of **Fig.** 6(b). In the case that the elements Y and Z in $\mathbb{F}_{2^4}$ are represented with RRB as shown in **Fig.** 6(b), converting the representations from RRB to type–I ONB is easy from Eq. (11a) as

$$\begin{aligned} Y &= y_0\beta + y_1\beta^2 + y_2\beta^{2^2} + y_3\beta^{2^3} + y_4 \\ &= (y_0 + y_4)\beta + (y_1 + y_4)\beta^2 + (y_2 + y_4)\beta^{2^2} + (y_3 + y_4)\beta^{2^3}, \end{aligned} \tag{25a}$$

$$\begin{aligned} Z &= z_0\beta + z_1\beta^2 + z_2\beta^{2^2} + z_3\beta^{2^3} + z_4 \\ &= (z_0 + z_4)\beta + (z_1 + z_4)\beta^2 + (z_2 + z_4)\beta^{2^2} + (z_3 + z_4)\beta^{2^3}. \end{aligned} \tag{25b}$$

Then, a multiplication by $\bar{\mathbf{B}}\mathbf{A}\mathbf{B}$ is given as Eq. (26), and the circuits of Eq. (26) is drawn as **Fig.** 1.

$$X\bar{\mathbf{B}}\mathbf{A}\mathbf{B} = \begin{bmatrix} y_0 + y_4 \\ y_0 + y_1 \\ (y_0 + y_4) + (z_0 + z_4) \\ (y_0 + y_1) + (y_2 + y_4) \\ (y_0 + y_1) + (z_0 + z_4) \\ (y_0 + y_1) + (y_2 + y_3) \\ \big((y_0 + y_1) + (y_2 + y_4)\big) + (z_0 + z_4) \\ (y_0 + y_1) + (z_0 + z_1) \end{bmatrix}^{T}. \tag{26}$$

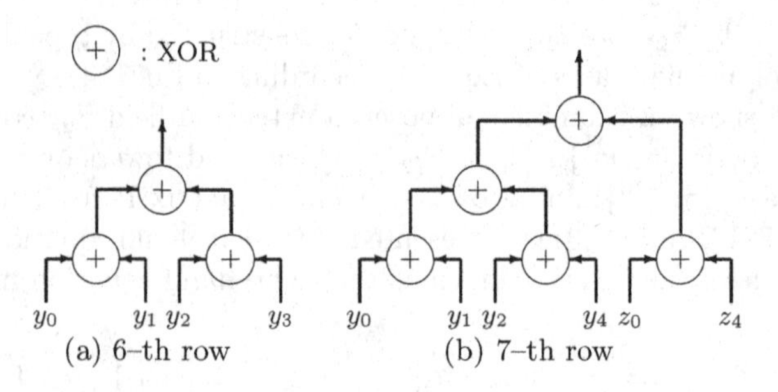

Fig. 1. Example images of circuits for Eq. (26)

According to the above consideration, a conversion matrix $\mathbf{\bar{B}AB}$ from the $\mathbb{F}_{(2^4)^2}$ over the $\mathbb{F}_{2^4}$ constructed by RRB to the same $\mathbb{F}_{(2^4)^2}$ over the $\mathbb{F}_{2^4}$ constructed by RRB (actually, type–I ONB) is obtained.

A row of $\mathbf{\bar{B}AB}$ can be represented with the following 2 vectors from Eq. (24).

$$U_j = [u_{j,0}\ u_{j,1}\ u_{j,2}\ u_{j,3}]^{\mathcal{T}}, \qquad V_j = [v_{j,0}\ v_{j,1}\ v_{j,2}\ v_{j,3}]^{\mathcal{T}}. \tag{27}$$

Let $\mathrm{Hw}(U)$ denote the number of "1" in the vector U, namely the Hamming weight of U. According to Eq. (26) and **Fig.** 1, the critical path delay of the circuit multiplying $\mathbf{\bar{B}AB}$ is equal to or less than $2T_{\mathrm{XOR}}$ when all vectors U_j and V_j $(0 \leq j < 8)$ satisfy that $\mathrm{Hw}(U_j){:}\mathrm{Hw}(V_j) \neq 3{:}1, 1{:}3$, and $\mathrm{Hw}(U_j)+\mathrm{Hw}(V_j) \leq 4$; otherwise, it is $3T_{\mathrm{XOR}}$. The probability when all column vectors of $\mathbf{\bar{B}AB}$ satisfy that $\mathrm{Hw}(U_j){:}\mathrm{Hw}(V_j) \neq 3{:}1, 1{:}3$, and $\mathrm{Hw}(U_j) + \mathrm{Hw}(V_j) \leq 4$ is given as

$$({}_8\mathrm{C}_0 + {}_8\mathrm{C}_1 + {}_8\mathrm{C}_2 + {}_8\mathrm{C}_3 + {}_4\mathrm{C}_4 \cdot {}_4\mathrm{C}_0 + {}_4\mathrm{C}_2 \cdot {}_4\mathrm{C}_2 + {}_4\mathrm{C}_0 \cdot {}_4\mathrm{C}_4)^8 / 2^{8\times 8} \approx 0.47\%. \tag{28}$$

Note that the above probability is not strictly accurate because a basis conversion matrix must be a regular matrix.

On the other hand , the above consideration of a multiplication by $\mathbf{\bar{B}AB}$ in Eq. (3) is also available for a multiplication by $\mathbf{\bar{B}\bar{A}B}$ in Eq. (7).

4.2 Calculation Efficiency of Eqs. (4) and (8)

The calculation circuit of Eq. (4a) is shown in **Fig.** 4. Naturally, the calculation circuit of Eq. (4b) can be drawn in the same way as **Fig.** 4. According to **Fig.** 4, the calculation efficiency of Eq. (4) depends on the element '2'**B** or '3'**B** in $\mathbb{F}_{(2^4)^2}$. In more detail, when a multiplication by either '2'**B** or '3'**B** can be carried out in $1T_{\mathrm{XOR}}$, the critical path delay of the calculation circuit of Eq. (4) is $3T_{\mathrm{XOR}}$; otherwise, it is $4T_{\mathrm{XOR}}$ since each multiplication by '2'**B** and '3'**B** needs at most $2T_{\mathrm{XOR}}$ according to **Sec.** 3.2.

On the other hand, the calculation efficiency of Eq. (8) depends on the elements '14'**B**, '11'**B**, '13'**B**, and '9'**B** in $\mathbb{F}_{(2^4)^2}$. This paper proposes how to find the **B** such that the critical path delay of the calculation circuit of Eq. (8) is $4T_{\mathrm{XOR}}$. In order to achieve the above proposal, according to Eq. (22a), both an element among δ_{14}, δ_{11}, δ_{13}, δ_9, ϵ_{14}, ϵ_{11}, ϵ_{13} and ϵ_9 of Eq. (22b), and an element among ϵ_{14}, ϵ_{11}, ϵ_{13}, ϵ_9, η_{14}, η_{11}, η_{13} and η_9 of Eq. (22b) must be a zero element or the class (I) element of **Table** 2. For example, when ϵ_9 is a zero element or the class (I) element, the calculation of Eq. (8) can be carried out as **Fig.** 5, where $D_{j,l}$ and $E_{j,l}$ denote elements in $\mathbb{F}_{2^4}$ which satisfy that $G_{r,j,l} = D_{j,l}\gamma + E_{j,l}\gamma^{16}$, $Y_{j,l}$ and $Z_{j,l}$ denote elements in $\mathbb{F}_{2^4}$ which satisfy that $C_{r-1,j,l} = Y_{j,l}\gamma + Z_{j,l}\gamma^{16}$, and $U_{j,l}$ and $V_{j,l}$ denote elements in $\mathbb{F}_{2^4}$ which satisfy that $J_{r,j,l} = U_{j,l}\gamma + V_{j,l}\gamma^{16}$.

4.3 More Miscellaneously Mixed Basis (MMMB)

This paper tries for the following goals.

Goal 1: Each multiplication by $\bar{\mathbf{B}}\mathbf{A}\mathbf{B}$ and $\bar{\mathbf{B}}\bar{\mathbf{A}}\mathbf{B}$ in Eqs. (3) and (7) is carried out in $2T_{\mathrm{XOR}}$.
Goal 2: The calculation of either Eq. (4a) or Eq. (4b) is carried out in $3T_{\mathrm{XOR}}$.
Goal 3: The calculation of Eq. (8) is carried out in $4T_{\mathrm{XOR}}$.

In order to achieve the above goals, it is important that an efficient basis conversion matrix $\mathbf{B}$ among a lot of prepared basis conversion matrices $\mathbf{B}$s is selectable. As an efficient technique to prepare more $\mathbf{B}$s, Nogami et al. have proposed *Mixed Bases* (MB) [10], which is applied to an implementation with $\mathbb{F}_{((2^2)^2)^2}$ in [10]. This subsection first considers to apply MB to an implementation with $\mathbb{F}_{(2^4)^2}$.

For a multiplication in $\mathbb{F}_{(2^4)^2}$ in Eq. (8), consider the following multiplication instead of Eq. (22a). See **Appendix** A about how to derive it.

$$W = Y + Z\gamma \ (Y, Z \in \mathbb{F}_{2^4}) \ = \{D\delta_j + E\epsilon_j\}\gamma + \{D\zeta_j + E\eta_j\}\gamma^{16}, \tag{29a}$$

$$\delta_j = (Q_j + R_j)\nu, \quad \epsilon_j = Q_j\nu + R_j(\mu^2 + \nu), \quad \zeta_j = Q_j\mu, \quad \eta_j = R_j\mu, \tag{29b}$$

where δ_j, ϵ_j, ζ_j, and η_j can be preliminarily calculated. In Eq. (29a), the normal basis $\{\gamma, \gamma^{16}\}$ is adopted for the input in the same way of Eq. (22a). On the other hand, the polynomial basis $\{1, \gamma\}$ is adopted for the output instead of the normal basis $\{\gamma, \gamma^{16}\}$. The critical path delay of this multiplication circuit in $\mathbb{F}_{(2^4)^2}$ is considered in the same way of that of Eq. (22a) (See **Sec.** 4.2). This multiplication circuit in $\mathbb{F}_{(2^4)^2}$ can provide conversion matrices $\bar{\mathbf{B}}\bar{\mathbf{A}}\mathbf{B}$s from $\mathbb{F}_{(2^4)^2}$ 2–nd towering with not only the normal basis $\{\gamma, \gamma^{16}\}$ but also the polynomial basis $\{1, \gamma\}$. However, the number of $\bar{\mathbf{B}}\bar{\mathbf{A}}\mathbf{B}$s prepared by this technique is not enough to perfectly achieve the above goals. Thus, this paper improves MB.

As described in **Sec.** 4.1, in the case that $\mathbb{F}_{2^4}$ is constructed by RRB, the basis conversion matrices $\mathbf{B}$s when $\mathbb{F}_{2^4}$ are constructed by type–I ONB of Eq. (12) and the polynomial bases of Eq. (13) are available. In more detail, a combination of two bases among the bases of Eqs. (12), (13) can be used to represent an element in $\mathbb{F}_{(2^4)^2}$. Let C denote an element in $\mathbb{F}_{(2^4)^2}$. For example, consider the combination of the normal basis $\{\beta, \beta^2, \beta^{2^2}, \beta^{2^3}\}$ and the polynomial basis $\{1, \beta, \beta^2, \beta^3\}$, then C is represented with the combination as

$$C = (d_0\beta + d_1\beta^2 + d_2\beta^{2^2} + d_3\beta^{2^3})\gamma + (e_0 + e_1\beta + e_2\beta^2 + e_3\beta^3)\gamma^{16} \ (d_j, e_j \in \mathbb{F}_2). \tag{30}$$

By only adopting the combinations as above, $20 \times 5 \times 5 \times 5 \times 5 = \mathbf{12,500}$ kinds of $\bar{\mathbf{B}}\mathbf{AB}$s and $\bar{\mathbf{B}}\bar{\mathbf{A}}\mathbf{B}$s can be respectively prepared. In this paper, the technique to adopt different bases for the input and output of arithmetic operation in $\mathbb{F}_{(2^4)^2}$ and to use a combination of different bases in $\mathbb{F}_{2^4}$ is especially called More Miscellaneously Mixed Bases (MMMB).

Actually, by using MMMB, some $\bar{\mathbf{B}}\mathbf{AB}$s and $\bar{\mathbf{B}}\bar{\mathbf{A}}\mathbf{B}$s to achieve **Goal 1**, and some **B**s to achieve **Goal 3** can be found; however, no '2'**B**s and '3'**B**s to achieve **Goal 2** can be found. Thus, in this case, the calculation delay of Eq. (4) becomes $4T_{\mathrm{XOR}}$, not $3T_{\mathrm{XOR}}$. This issue will be kept as a future work.

By adopting RRB and MMMB as described in this paper, the critical path delays of the encryption and decryption procedures of AES algorithm are shown as **Tables** 6, 7. Then, each round of the encryption procedure can be carried out in $\mathbf{4}T_{\mathrm{AND}} + \mathbf{13}T_{\mathrm{XOR}}$. On the other hand, each round of the decryption procedure also can be carried out in $\mathbf{4}T_{\mathrm{AND}} + \mathbf{13}T_{\mathrm{XOR}}$.

Table 6. The critical path delay of the encryption procedure of AES

Implementaion	SubBytes		MixColumns	AddRoundKey
	Inversion	Others		
Rudra et al.'s [5]	$(4, 10)^\dagger$	no data	$(0, 7)^\dagger$	$(0, 1)^\dagger$
Satoh and Morioka et al.'s [6,7]	$(4, 17)^\dagger$			
Jeon et al.'s [11]	$(4, 10)^\dagger$	$(0, 11)^\dagger$		
This work	$(4, \mathbf{7})^\dagger$	$(0, \mathbf{2})^\dagger$	$(0, \mathbf{4})^\dagger$	

† (j, l) means $jT_{\mathrm{AND}} + lT_{\mathrm{XOR}}$.

Table 7. The critical path delay of the decryption procedure of AES

Implementaion	SubBytes		MixColumns	AddRoundKey
	Inversion	Others		
Jeon et al.'s [11]	$(4, 10)^\dagger$	$(0, 10)^\dagger$	$(0, 7)^\dagger$	$(0, 1)^\dagger$
This work	$(4, \mathbf{7})^\dagger$	$(0, \mathbf{2})^\dagger$	$(0, \mathbf{4})^\dagger$	

† (j, l) means $jT_{\mathrm{AND}} + lT_{\mathrm{XOR}}$.

5 Conclusion and Future Works

This paper proposed RRB to make arithmetic operations in towering field $\mathbb{F}_{(2^4)^2}$ isomorphic to the AES original $\mathbb{F}_{2^8}$ more efficient, and MMMB to provide efficient basis conversion matrix from the $\mathbb{F}_{2^8}$ to $\mathbb{F}_{(2^4)^2}$. As a result, this paper theoretically showed that both of the encryption and decryption procedures of AES algorithm can be carried out in the critical path delay $\mathbf{4}T_{\mathrm{AND}} + \mathbf{13}T_{\mathrm{XOR}}$.

On the other hand, the authors hold a lot of agendas, for example, as

FW1: An acceleration of a multiplication by either '2'**B** or '3'**B**
FW2: An implementaion of *key expansion* as described in [6]
FW3: An actual hardware implementation of this paper's approach
FW4: To report the evaluations of the above implementation such as the hardware size, the memory requirement, the power consumption and the security vulnerabilities
FW5: Countermeasures for *side–channel attacks* such as applying *masking* [15]

Acknowledgment. This research is supported by "Strategic information and COmmunications R&D Promotion programmE (SCOPE)" from Ministry of internal Affairs and Communications (MIC), Japan.

References

1. National Institute of Standards and Technology (NIST), Advanced Encryption Standard (AES), FIPS publication 197 (2001), `http://csrc.nist.gov/publications/fips/fips197/fips-197.pdf`
2. Daemen, J., Rijmen, V.: AES Proposal: Rijndael. AES Algorithm (Rijndael) Information (1999), `http://csrc.nist.gov/archive/aes/rijndael/Rijndael-ammended.pdf`
3. Matsui, M.: Linear Cryptanalysis Method for DES Cipher. In: Helleseth, T. (ed.) EUROCRYPT 1993. LNCS, vol. 765, pp. 386–397. Springer, Heidelberg (1994)
4. Paar, C.: Efficient VLSI Architectures for Bit–Parallel Computation in Galois Fields. PhD thesis, Institute for Experimental Mathematics, University of Essen, Germany (1994)
5. Rudra, A., Dubey, P.K., Jutla, C.S., Kumar, V., Rao, J.R., Rohatgi, P.: Efficient Rijndael Encryption Implementation with Composite Field Arithmetic. In: Koç, Ç.K., Naccache, D., Paar, C. (eds.) CHES 2001. LNCS, vol. 2162, pp. 171–184. Springer, Heidelberg (2001)
6. Satoh, A., Morioka, S., Takano, K., Munetoh, S.: A Compact Rijndael Hardware Architecture with S-Box Optimization. In: Boyd, C. (ed.) ASIACRYPT 2001. LNCS, vol. 2248, pp. 239–254. Springer, Heidelberg (2001)
7. Morioka, S., Satoh, A.: An Optimized S-Box Circuit Architecture for Low Power AES Design. In: Kaliski Jr., B.S., Koç, Ç.K., Paar, C. (eds.) CHES 2002. LNCS, vol. 2523, pp. 172–186. Springer, Heidelberg (2003)
8. Mentens, N., Batina, L., Preneel, B., Verbauwhede, I.: A Systematic Evaluation of Compact Hardware Implementations for the Rijndael S-Box. In: Menezes, A. (ed.) CT-RSA 2005. LNCS, vol. 3376, pp. 323–333. Springer, Heidelberg (2005)
9. Canright, D.: A Very Compact S-Box for AES. In: Rao, J.R., Sunar, B. (eds.) CHES 2005. LNCS, vol. 3659, pp. 441–455. Springer, Heidelberg (2005)
10. Nogami, Y., Nekado, K., Toyota, T., Hongo, N., Morikawa, Y.: Mixed Bases for Efficient Inversion in $\mathbb{F}$_$((2^2)^2)$_2 and Conversion Matrices of SubBytes of AES. In: Mangard, S., Standaert, F.-X. (eds.) CHES 2010. LNCS, vol. 6225, pp. 234–247. Springer, Heidelberg (2010)
11. Jeon, Y., Kim, Y., Lee, D.: A Compact Memory-free Architecture for the AES Algorithm Using Resource Sharing Methods. Journal of Circuits, Systems, and Computers 19(5), 1109–1130 (2010)

12. Mullin, R., Onyszchuk, I., Vanstone, S., Wilson, R.: Optimal Normal Bases in GF(p^n). Discrete Applied Mathematics 22(2), 149–161 (1988)
13. Nogami, Y., Saito, A., Morikawa, Y.: Finite Extension Field with Modulus of All–One Polynomial and Representation of Its Elements for Fast Arithmetic Operations. IEICE Transactions E86-A(9), 2376–2387 (2003)
14. Itoh, T., Tsujii, S.: A Fast Algorithm for Computing Multiplicative Inverse in GF(2^m) Using Normal Basis. Information and Computation 78(3), 171–177 (1988)
15. Canright, D., Batina, L.: A Very Compact "Perfectly Masked" S-Box for AES. In: Bellovin, S.M., Gennaro, R., Keromytis, A.D., Yung, M. (eds.) ACNS 2008. LNCS, vol. 5037, pp. 446–459. Springer, Heidelberg (2008)

A Derivation of Eqs. (19), (22), and (29)

Eq. (19) is derived with ITA [14] as

$$\begin{aligned}
I &= D^{-1} = (D\cdot D^4)^{-1}D^4 = (D\cdot D^{2^2})^{-1}D^{2^2}\\
&= \{(d_0\beta + d_1\beta^2 + d_2\beta^{2^2} + d_3\beta^{2^3} + d_4)(d_2\beta + d_3\beta^2 + d_0\beta^{2^2} + d_1\beta^{2^3} + d_4)\}^2\\
&\quad\times(d_2\beta + d_3\beta^2 + d_0\beta^{2^2} + d_1\beta^{2^3} + d_4) \qquad (\because \text{Eq. (18)},\ \ D\cdot D^{2^2}\in\mathbb{F}_{2^2})\\
&= \{(d_1d_2+d_2d_0+d_0d_3+d_3d_4+d_4d_1)(\beta+\beta^{2^2})\\
&\qquad + (d_0d_1+d_1d_3+d_3d_2+d_2d_4+d_4d_0)(\beta^2+\beta^{2^3}) + (d_0+d_1+d_2+d_3+d_4)\}\\
&\quad\times(d_2\beta + d_3\beta^2 + d_0\beta^{2^2} + d_1\beta^{2^3} + d_4) \qquad (\because \text{Eqs. (16a), (18)})\\
&= (a_{2,4}+a_{0,4}a_{1,4}a_{1,3})\beta + (a_{3,4}+a_{1,4}a_{2,4}a_{0,2})\beta^2 + (a_{0,4}+a_{2,4}a_{3,4}a_{1,3})\beta^{2^2}\\
&\quad + (a_{1,4}+a_{3,4}a_{0,4}a_{0,2})\beta^{2^3} + (a_{0,4}a_{2,4}\overline{a_{1,3}}+a_{1,4}a_{3,4}\overline{a_{0,2}}) \ (\because \text{Eqs. (16a)}), \quad (31a)\\
a_{j,l} &= (d_j + d_l) \ \ (0\le j < l \le 4), \qquad (31b)
\end{aligned}$$

On the other hand, because γ and γ^{16} in Eq. (22a) are zeros of $g(t)$ in Eq. (20), the following relations are obtained with the *Vieta's formulas.*

$$\gamma + \gamma^{16} = \mu, \qquad \gamma\cdot\gamma^{16} = \nu = \frac{\nu}{\mu}\cdot(\gamma+\gamma^{16}). \qquad (32)$$

Thus, Eq. (22) is derived as

$$\begin{aligned}
W &= C\times \text{`}j\text{'}\mathbf{B} = (D\gamma + E\gamma^{16})(Q_j\gamma + R_j\gamma^{16})\\
&= (D+E)(Q_j+R_j)(\gamma\cdot\gamma^{16}) + DQ_j(\gamma+\gamma^{16})\gamma + ER_j(\gamma+\gamma^{16})\gamma^{16}\\
&= (D+E)(Q_j+R_j)\cdot\frac{\nu}{\mu}\cdot(\gamma+\gamma^{16}) + DQ_j\cdot\mu\cdot\gamma + ER_j\cdot\mu\cdot\gamma^{16} \qquad (\because \text{Eq. (32)})\\
&= \{D\delta_j + E\epsilon_j\}\gamma + \{D\epsilon_j + E\eta_j\}\gamma^{16}, \qquad (33a)\\
\delta_j &= Q_j(\mu+\frac{\nu}{\mu}) + R_j\cdot\frac{\nu}{\mu},\ \ \epsilon_j = (Q_j+R_j)\cdot\frac{\nu}{\mu},\ \ \eta_j = Q_j\cdot\frac{\nu}{\mu} + R_j(\mu+\frac{\nu}{\mu}). \quad (33b)
\end{aligned}$$

On the other hand, Eq. (29) is derived in the same way.

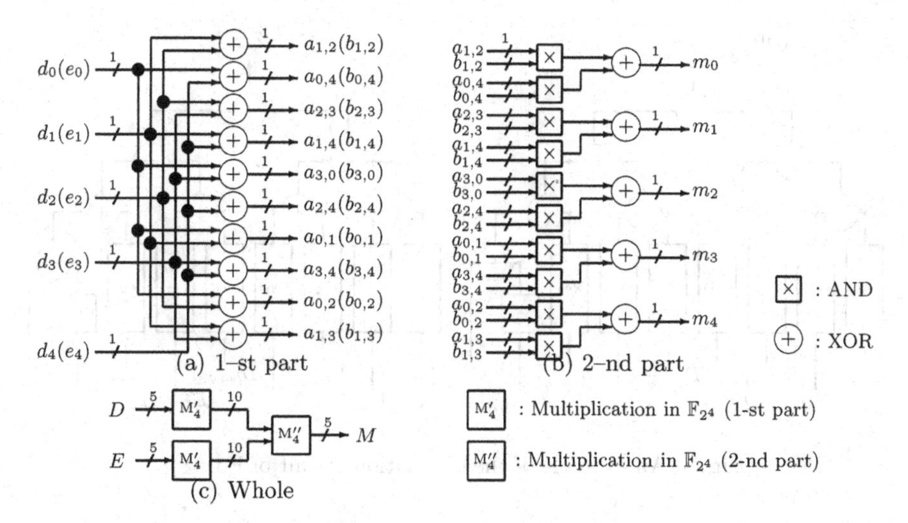

Fig. 2. The multiplication circuit adopting RRB in $\mathbb{F}_{2^4}$

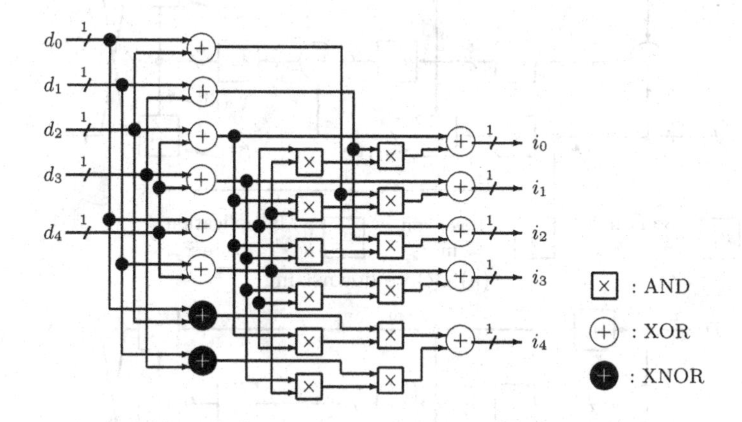

Fig. 3. The inversion circuit adopting RRB in $\mathbb{F}_{2^4}$

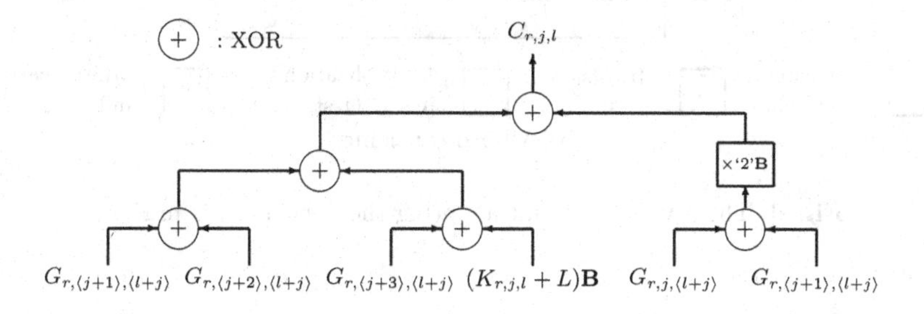

Fig. 4. The calculation circuit of Eq. (4a)

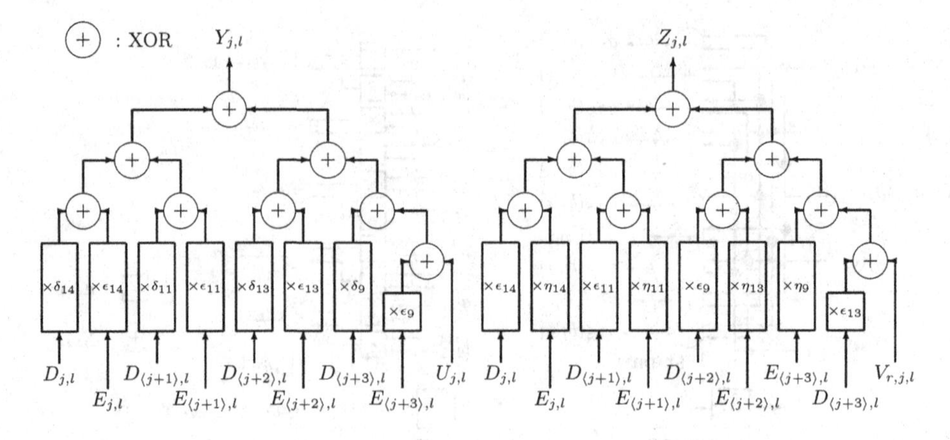

Fig. 5. An example of the calculation circuit of Eq. (8)

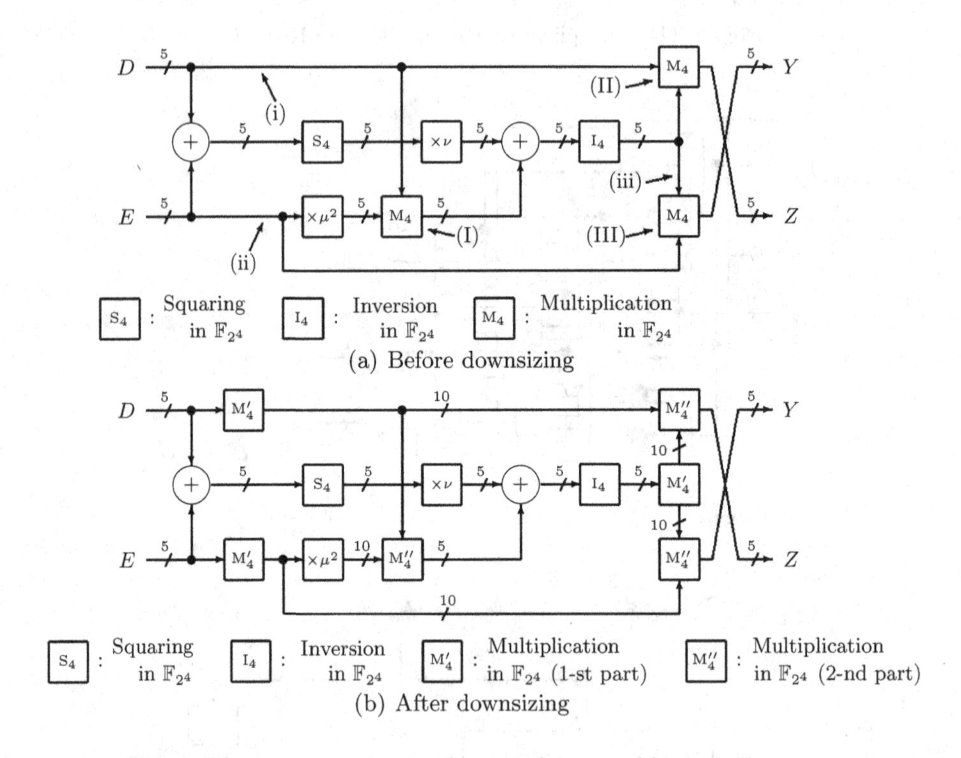

Fig. 6. The inversion circuit adopting the normal basis in $\mathbb{F}_{(2^4)^2}$

One-Round Authenticated Key Exchange with Strong Forward Secrecy in the Standard Model against Constrained Adversary

Kazuki Yoneyama

NTT Secure Platform Laboratories
3-9-11 Midori-cho Musashino-shi Tokyo 180-8585, Japan
yoneyama.kazuki@lab.ntt.co.jp

Abstract. Forward secrecy (FS) is a central security requirement of authenticated key exchange (AKE). Especially, strong FS (sFS) is desirable because it can guarantee security against a very realistic attack scenario that an adversary is allowed to be active in the target session. However, most of AKE schemes cannot achieve sFS, and currently known schemes with sFS are only proved in the random oracle model. In this paper, we propose a generic construction of AKE protocol with sFS in the standard model against a constrained adversary. The constraint is that session-specific intermediate computation results (i.e., session state) cannot be revealed to the adversary for achieving sFS, that is shown to be inevitable by Boyd and González Nieto. However, our scheme maintains weak FS (wFS) if session state is available to the adversary. Thus, our scheme satisfies one of strongest security definitions, the CK$^+$ model, which includes wFS and session state reveal. The main idea to achieve sFS is to use signcryption KEM while the previous CK$^+$ secure construction uses ordinary KEM. We show a possible instantiation of our construction from Diffie-Hellman problems.

Keywords: authenticated key exchange, strong forward secrecy, signcryption.

1 Introduction

1.1 Background

Authenticated key exchange (AKE) is one of most important cryptographic protocols in the real world applications. The goal of standard two-party AKE is to provide a common secret *session key* between two-parties with mutual authentication. Each party publishes long-term public information (called a *static public key*), and keeps corresponding secret information (called a *static secret key*). The static public key is expected to be certified with a party's identity through an infrastructure such as PKI. When a party wants to establish a session key with a peer, the party initiates a key exchange *session*, sends some message (called *ephemeral public key*) generated from corresponding temporary secret information (called *ephemeral secret key*) to the peer, and computes an intermediate information (called *session state*) from static public keys, static secret keys, ephemeral public keys and ephemeral secret keys. Note that the session state contains the ephemeral secret key. Both parties then derive a *session key* from these keys

G. Hanaoka and T. Yamauchi (Eds.): IWSEC 2012, LNCS 7631, pp. 69–86, 2012.

and session states with a function called the *key derivation function*. If a party does not have the correct static secret key corresponding to the certified static public key, any information of the session key is not leaked. AKE is practically used to establish secure channels (e.g., the handshake protocol in SSL/TLS).

Various security properties are required for AKE schemes such as impersonation resilience, known key security, secret key exposure resilience, etc. Forward secrecy (FS) [1] is one of such basic security properties. FS implies that an adversary cannot recover a session key of a completed session (i.e., a session in which the session key was already established) even if static secret keys are compromised. There are two strength criteria of FS: perfect or non-perfect, and weak or strong.

- *perfect vs. non-perfect.* We say that an AKE scheme is with perfect FS (PFS) if FS is satisfied even when *both* static secret keys of the initiator and the responder are compromised. Conversely, we say that an AKE scheme satisfies non-perfect FS if FS is satisfied even only when the static secret key of either the initiator or the responder is compromised. PFS guarantees very strong secrecy in future because ephemeral secret keys are removed after completion of the session and there is no problem against leakage of all static secret keys.
- *weak vs. strong.* We say that an AKE scheme is with strong FS (sFS) if FS is satisfied even when the adversary is *active* in the target session. 'Active' means that the adversary is allowed to modify messages to the owner of the target session. Conversely, we say that an AKE scheme is with weak FS (wFS) if FS is satisfied when the adversary must be *passive* in the target session. sFS is exactly desirable in real world applications because the adversary should be allowed to be active in any session.

Thus, the strongest and most desirable level of FS is strong perfect FS (sPFS).

1.2 Motivating Problem

Provable security [2–7] for AKE has been actively discussed for two decades. Many *two-pass* MQV-type AKE schemes [8, 9, 4, 5, 10–14, 6, 15, 16, 7] achieve provable security. However, most of such AKE schemes do not satisfy sFS (some of such schemes only satisfy wPFS). Krawczyk [4] gives an intuitive reason of difficulty to achieve sFS with two-pass protocols by showing the following generic attack; An adversary generates an ephemeral public and secret key pair, and sends the ephemeral public key to the owner (U_A) of the target session. After completion of the session, the adversary obtains the static secret key of the peer (U_B) of the session. Then, the adversary can derive the session key because all secret information to generate the session key is obtained. Recently, Boyd and González Nieto [17] show rigorous impossibility to achieve sFS with *one-round* protocols when leakage of an ephemeral secret key of U_B in any session occurs. One-round protocols mean that the initiator and the responder can send their messages independently and simultaneously in two-pass protocols. On the other hand, they also show that secure one-round scheme with sFS is possible in a constrained model in which the reveal of ephemeral secret keys of U_B in any session is not allowed. For example, some schemes [9, 18, 17, 19, 16, 20] satisfy sFS with one-round protocols against the constrained adversary. However, two [9, 18] of such schemes do not satisfy even wFS against an ephemeral secret key exposure attack. The scheme in [17] is not proved to be secure against maximal exposure attacks (MEX) (i.e., an adversary can reveal any pair of secret static keys and ephemeral secret keys of the initiator and the responder in any session except for both the static and ephemeral secret keys of the initiator or the responder) due to the security model. The other schemes [19, 16, 20] satisfies both sPFS and resistance to MEX; but, they are proved in the random oracle (RO) model and a special signature scheme is necessary to construct them. Thus, for the construction of a one-round AKE scheme with sFS without ROs (i.e., in the standard model (StdM)) is an unresolved open problem.

1.3 Our Contribution

We achieve the first one-round AKE scheme with sPFS in the StdM against the constrained adversary. Specifically, we give a generic construction of AKE, and it can be instantiated with Diffie-Hellman (DH) type assumptions. Our construction also satisfies one of the 'strongest' models, CK^+ model [4, 7], for AKE as well as sPFS. The CK^+ model contains all known security properties of AKE except sFS as follows: Even though an adversary can reveal any non-trivial[1] combination of ephemeral secret keys and static secret keys, any information of the session key is not leaked. For example, if static secret keys of parties in the target session are revealed, the situation corresponds to wPFS. Our scheme is based on the two-pass generic construction [7] (FSXY construction) that is a CK^+ secure AKE scheme. An intuitive protocol of the FSXY construction is as follows: An initiator and a responder exchange ciphertexts of a chosen ciphertext secure (IND-CCA) KEM. Also, the initiator sends a session-specific public key of a semantically secure (IND-CPA) KEM, and the responder computes and sends a ciphertext with the public key. Then, they share three KEM keys, and derive a session key with a strong randomness extractor, and a pseudo-random function (PRF).

The main idea of our construction is using a *signcryption KEM*. Signcryption [21] provides the combined functionality of signatures and encryption with higher efficiency than simply combining signature and encryption. While the FSXY construction uses a IND-CCA KEM, a IND-CPA KEM, a strong randomness extractor, and a PRF as building blocks, we use an insider chosen ciphertext secure in the dynamic multi-user model (dM-IND-iCCA) and strong unforgeability against insider chosen message attacks in the dynamic multi-user model (dM-sUF-iCMA) signcryption KEM instead of the IND-CCA KEM. Intuitively, signcryption KEM can prevent an adversary from modifying ciphertexts in the target session; thus, we can achieve FS even if the adversary is active in the target session. A subtle point is that signcryption KEM must be secure against *insider attacks* (i.e., an adversary can use sender's secret for attacking confidentiality and receiver's secret for attacking unforgeability) because the security model of AKE allows an adversary to obtain static secret keys of each party. Also, security in the dynamic multi-user model is necessary because each party may execute multiple sessions with different parties. The existence of an efficient dM-IND-iCCA and dM-sUF-iCMA signcryption KEM has been proposed from the DH assumptions [22].

[1] If both the static key and the ephemeral key of a party in the target session are revealed, the adversary trivially obtains the session key for any protocol. Thus, the CK^+ model prohibits such a reveal.

Moreover, we introduce a new notion named *KEM with public-key-independent-ciphertext* (PKIC-KEM). We say a KEM scheme is PKIC-KEM if the ciphertext is independent of the public key. A typical example is the ElGamal KEM; ciphertext g^r can be generated independently with public key g^a (though the generation of KEM key g^{ra} still needs the public key). In the FSXY construction, an initiator sends a session-specific public key of the IND-CPA KEM, and a responder computes a ciphertext and a KEM key with the public key and sends the ciphertext. Thus, unfortunately, it is not one-round protocol because the responder cannot generate the ciphertext of the IND-CPA KEM until receiving the public key sent by the initiator. We can resolve this problem with PKIC-KEM. We use an IND-CPA secure PKIC-KEM instead of the IND-CPA KEM.

To the best of our knowledge, our generic construction provides a first CK^+ secure one-round AKE protocol with sFS in the StdM even against the constrained adversary.

We also extend the CK^+ model to a model which guarantees the CK^+ security with sPFS, for proving security. We call the extended model CK^+-$\mathrm{sFS}^{\mathrm{NSR}}$ model. The modification is minor; that is, the case of sPFS is added to adversary's behavior. We must constrain the adversary to obtain ephemeral secret keys of the peer of the target session in any session due to impossibility of [17]. The sPFS part of the CK^+-$\mathrm{sFS}^{\mathrm{NSR}}$ model is same as the model in [17]. Therefore, this model satisfies all security requirements of the CK^+ model without the constraint. Also, sPFS is guaranteed if session states are protected. This is very reasonable in reality; that is, when the system requires a high level security (including sFS), session states will be stored in some tamper-proof area of storages.

2 Security Model

In this section, we define the CK^+-$\mathrm{sFS}^{\mathrm{NSR}}$ model that adds sFS against the constrained adversary to the CK^+ model [7]. The difference between these models is in the case that an adversary is active in the test session and obtains the static secret key of the peer of the test session after the completion of the test session. In the CK^+ model, security is not guaranteed in this situation (i.e., no guarantee of sFS). Conversely, in the CK^+-$\mathrm{sFS}^{\mathrm{NSR}}$ model, security is guaranteed in this situation. Note that we show a model specified to one-round protocols for simplicity. It can be trivially extended to any round protocol.

We denote a party by U_i, and party U_i and other parties are modeled as probabilistic polynomial-time (PPT) Turing machines w.r.t. security parameter κ. For party U_i, we denote static secret (public) key by s_i (S_i) and ephemeral secret (public) key by x_i (X_i). Party U_i generates its own keys, s_i and S_i, and the static public key S_i is linked with U_i's identity in some systems like PKI.

Session. An invocation of a protocol is called a *session*. Session activation is done by an incoming message of the forms $(\Pi, \mathcal{I}, U_A, U_B)$ or $(\Pi, \mathcal{R}, U_B, U_A)$, where we equate Π with a protocol identifier, $\mathcal{I}$ and $\mathcal{R}$ with role identifiers, and U_A and U_B with user identifiers. If U_A is activated with $(\Pi, \mathcal{I}, U_A, U_B)$, then U_A is called the session *initiator*. If U_B is activated with $(\Pi, \mathcal{R}, U_B, U_A)$, then U_B is called the session *responder*. The initiator U_A outputs X_A, receives an incoming message of the form $(\Pi, \mathcal{I}, U_A, U_B, X_B)$

from the responder U_B, and computes the session key SK. On the contrary, the responder U_B outputs X_B, receives an incoming message of the form $(\Pi, \mathcal{R}, U_B, U_A, X_A)$ from the initiator U_A, and computes the session key SK.

If U_A is the initiator of a session, the session is identified by $\mathsf{sid} = (\Pi, \mathcal{I}, U_A, U_B, X_A)$ or $\mathsf{sid} = (\Pi, \mathcal{I}, U_A, U_B, X_A, X_B)$. If U_B is the responder of a session, the session is identified by $\mathsf{sid} = (\Pi, \mathcal{R}, U_B, U_A, X_B)$ or $\mathsf{sid} = (\Pi, \mathcal{R}, U_B, U_A, X_A, X_B)$. We say that U_A is the *owner* of session sid, if the third coordinate of session sid is U_A. We say that U_A is the *peer* of session sid, if the fourth coordinate of session sid is U_A. We say that a session is *completed* if its owner computes the session key. The *matching session* of $(\Pi, \mathcal{I}, U_A, U_B, X_A, X_B)$ is session $(\Pi, \mathcal{R}, U_B, U_A, X_A, X_B)$ and vice versa.

Adversary. The adversary $\mathcal{A}$, which is modeled as a probabilistic polynomial-time (PPT) Turing machine, controls all communications between parties including session activation by performing the following adversary query.

- Send(message): The message has one of the following forms: $(\Pi, \mathcal{I}, U_A, U_B)$, $(\Pi, \mathcal{R}, U_B, U_A)$, $(\Pi, \mathcal{I}, U_A, U_B, X_B)$, or $(\Pi, \mathcal{R}, U_B, U_A, X_A)$. The adversary $\mathcal{A}$ obtains the response from the party.

To capture leakage of secret information, the adversary $\mathcal{A}$ is allowed to issue the following queries.

- KeyReveal(sid): The adversary $\mathcal{A}$ obtains the session key SK for the session sid if the session is completed.
- StateReveal(sid): The adversary $\mathcal{A}$ obtains the session state of the owner of session sid if the session is not completed (the session key is not established yet). The session state includes all ephemeral secret keys and intermediate computation results except for immediately erased information but does not include the static secret key.
- Corrupt(U_i): This query allows the adversary $\mathcal{A}$ to obtain all static secret keys of the party U_i. If a party is corrupted by a Corrupt(U_i) query issued by the adversary $\mathcal{A}$, then we call the party U_i *dishonest*. If not, we call the party *honest*.

Freshness. For the security definition, we need the notion of freshness.

Definition 1 (Freshness). *Let* $\mathsf{sid}^* = (\Pi, \mathcal{I}, U_A, U_B, X_A, X_B)$ *or* $(\Pi, \mathcal{R}, U_A, U_B, X_B, X_A)$ *be a completed session between honest users* U_A *and* U_B. *If the matching session exists, then let* $\overline{\mathsf{sid}^*}$ *be the matching session of* sid^*. *We say session* sid^* *is* fresh *if none of the following conditions hold:*

1. *The adversary* $\mathcal{A}$ *issues* KeyReveal(sid^*), *or* KeyReveal($\overline{\mathsf{sid}^*}$) *if* $\overline{\mathsf{sid}^*}$ *exists,*
2. $\overline{\mathsf{sid}^*}$ *exists and the adversary* $\mathcal{A}$ *makes either of the following queries*
   - StateReveal(sid^*) *or* StateReveal($\overline{\mathsf{sid}^*}$),
3. $\overline{\mathsf{sid}^*}$ *does not exist and the adversary* $\mathcal{A}$ *makes the following query*
   - StateReveal(sid^*).

Security Experiment. For the security definition, we consider the following security experiment. Initially, the adversary $\mathcal{A}$ is given a set of honest users and makes any sequence of the queries described above. During the experiment, the adversary $\mathcal{A}$ makes the following query.

- $\mathsf{Test}(\mathsf{sid}^*)$: Here, sid^* must be a fresh session. Select random bit $b \in_U \{0, 1\}$, and return the session key held by sid^* if $b = 0$, and return a random key if $b = 1$.

The experiment continues until the adversary $\mathcal{A}$ makes a guess b'. The adversary $\mathcal{A}$ *wins* the game if the test session sid^* is still fresh and if the guess of the adversary $\mathcal{A}$ is correct, i.e., $b' = b$. The advantage of the adversary $\mathcal{A}$ in the AKE experiment with the PKI-based AKE protocol Π is defined as $\mathrm{Adv}_{\Pi}^{\mathrm{AKE}}(\mathcal{A}) = \Pr[\mathcal{A}\ wins] - \frac{1}{2}$. We define the security as follows.

Definition 2 (Security). *We say that a PKI-based AKE protocol Π is* secure in the CK^+-$\mathrm{sFS}^{\mathrm{NSR}}$ model *if the following conditions hold:*

1. *If two honest parties complete matching sessions, then, except with negligible probability, they both compute the same session key.*
2. *For any PPT bounded adversary $\mathcal{A}$, $\mathrm{Adv}_{\Pi}^{\mathrm{AKE}}(\mathcal{A})$ is negligible in security parameter κ for the test session sid^*,*
   - *(a) if $\overline{\mathsf{sid}^*}$ does not exist, and the static secret key of the owner of sid^* is given to $\mathcal{A}$. Also, the static secret key of the peer of sid^* is given to $\mathcal{A}$ after completion of sid^*. The adversary is not allowed* $\mathsf{StateReveal}$ *query to any session between the owner and the peer of sid^*.*[2]
   - *(b) if $\overline{\mathsf{sid}^*}$ does not exist, and the static secret key of the owner of sid^* is given to $\mathcal{A}$.*
   - *(c) if $\overline{\mathsf{sid}^*}$ does not exist, and the ephemeral secret key of sid^* is given to $\mathcal{A}$.*
   - *(d) if $\overline{\mathsf{sid}^*}$ exists, and the static secret key of the owner of sid^* and the ephemeral secret key of $\overline{\mathsf{sid}^*}$ are given to $\mathcal{A}$.*
   - *(e) if $\overline{\mathsf{sid}^*}$ exists, and the ephemeral secret key of sid^* and the ephemeral secret key of $\overline{\mathsf{sid}^*}$ are given to $\mathcal{A}$.*
   - *(f) if $\overline{\mathsf{sid}^*}$ exists, and the static secret key of the owner of sid^* and the static secret key of the peer of sid^* are given to $\mathcal{A}$.*
   - *(g) if $\overline{\mathsf{sid}^*}$ exists, and the ephemeral secret key of sid^* and the static secret key of the peer of sid^* are given to $\mathcal{A}$.*

The definition is identical to the CK^+ model except item 2.a. Thus, security properties included in the CK^+ model are also included in the CK^+-$\mathrm{sFS}^{\mathrm{NSR}}$ model. Specifically, items 2.d and 2.g correspond to resistance to KCI (i.e., given a static secret key an adversary cannot impersonate some honest party in order to fool the owner of the leaked secret key), item 2.f corresponds to wPFS, and items 2.b, 2.c and 2.e correspond to resistance to MEX. Item 2.a is newly considered, and corresponds to sPFS against the constrained adversary.

[2] This constraint is due to impossibility in [17] and is the same as the model in [17].

3 Generic AKE Construction with sPFS from Signcryption KEM

In this section, we propose a generic construction GC-sFS of CK^+-sFS^{NSR}-secure one-round AKE.

3.1 Preliminaries

Security Notions of KEM with Public-Key-Independent-Ciphertext. Here, we introduce syntax of PKIC-KEM schemes. Then, we show the definition of IND-CPA security for PKIC-KEM, and min-entropy of KEM keys as follows.

Definition 3 (Syntax of PKIC-KEM). *A PKIC-KEM scheme consists of the following 4-tuple* (KeyGen, EnCapC, EnCapK, DeCap)*:*

$(ek, dk) \leftarrow \mathsf{KeyGen}(1^\kappa; r_g)$: *a key generation algorithm which on inputs* 1^κ, *where* κ *is the security parameter and* r_g *is randomness in space* $\mathcal{RS}_G$, *outputs a pair of keys* (ek, dk).

$CT \leftarrow \mathsf{EnCapC}(r_e)$: *a ciphertext generation algorithm which outputs ciphertext* $CT \in \mathcal{CS}$ *on inputs public parameters, where* r_e *is randomness in space* $\mathcal{RS}_E$, *and* $\mathcal{CS}$ *is a ciphertext space.*

$K \leftarrow \mathsf{EnCapK}_{ek}(CT, r_e)$: *an encryption algorithm which takes as inputs encapsulation key* ek, *ciphertext* CT, *and randomness* r_e, *outputs KEM key* $K \in \mathcal{KS}$, *where* r_e *is randomness used in* EnCapC, *and* $\mathcal{KS}$ *is a KEM key space.*

$K \leftarrow \mathsf{DeCap}_{dk}(CT)$: *a decryption algorithm which takes as inputs decapsulation key* dk *and ciphertext* $CT \in \mathcal{CS}$, *and outputs KEM key* $K \in \mathcal{KS}$.

Definition 4 (IND-CPA Security). *A PKIC-KEM scheme is IND-CPA-secure if the following property holds for security parameter* κ*; For any PPT adversary* $\mathcal{A} = (\mathcal{A}_1, \mathcal{A}_2)$, $\mathbf{Adv}^{\text{ind-cpa}} = |\Pr[r_g \leftarrow \mathcal{RS}_G; (ek, dk) \leftarrow \mathsf{KeyGen}(1^\kappa; r_g); state \leftarrow \mathcal{A}_1(ek); b \leftarrow \{0, 1\}; r_e \leftarrow \mathcal{RS}_E; CT_0^* \leftarrow \mathsf{EnCapC}(r_e); K_0^* \leftarrow \mathsf{EnCapK}_{ek}(CT_0^*, r_e); K_1^* \leftarrow \mathcal{KS}; b' \leftarrow \mathcal{A}_2(ek, (K_b^*, CT_0^*), state); b' = b] - 1/2| \leq negl$, *where state is state information that* $\mathcal{A}$ *wants to preserve from* $\mathcal{A}_1$ *to* $\mathcal{A}_2$.

Definition 5 (Min-Entropy of KEM Key). *We say a PKIC-KEM scheme is k-min-entropy PKIC-KEM if for any ek, for distribution* $D_{\mathcal{KS}}$ *of variable K defined by* $CT \leftarrow \mathsf{EnCapC}(r_e)$, $K \leftarrow \mathsf{EnCapK}_{ek}(CT, r_e)$, *and random* $r_e \in \mathcal{RS}_E$, $H_\infty(D_{\mathcal{KS}}) \geq k$ *holds, where* H_∞ *denotes min-entropy.*

Security Notions of Signcryption KEM. Here, we recall the definition of dM-IND-iCCA and dM-sUF-iCMA security for signcryption KEM, and min-entropy of KEM keys as follows.

Definition 6 (Syntax of Signcryption KEM). *A signcryption KEM scheme consists of the following 4-tuple* (SKeyGen, RKeyGen, SC, USC)*:*

$(pk_S, sk_S) \leftarrow \mathsf{SKeyGen}(1^\kappa; r_S)$: *a key generation algorithm for sender* U_S *which on inputs* 1^κ, *where* κ *is the security parameter and* r_S *is randomness in space* $\mathcal{RS}_{SG}$, *outputs a pair of keys* (pk_S, sk_S).

$(pk_R, sk_R) \leftarrow \mathsf{RKeyGen}(1^\kappa; r_R)$: *a key generation algorithm for receiver* U_R *which on inputs* 1^κ*, where* κ *is the security parameter and* r_R *is randomness in space* $\mathcal{RS}_{RG}$*, outputs a pair of keys* (pk_R, sk_R)*.*

$(K, CT) \leftarrow \mathsf{SC}_{sk_S, pk_R}(m; r_e)$: *a signcryption algorithm which takes as inputs sender's secret key* sk_S*, receiver's public key* pk_R*, and message* m*, outputs KEM key* $K \in \mathcal{KS}$ *and ciphertext* $CT \in \mathcal{CS}$*, where* r_e *is randomness in space* $\mathcal{RS}_{SE}$*,* $\mathcal{KS}$ *is a KEM key space, and* $\mathcal{CS}$ *is a ciphertext space.*

$K/\perp \leftarrow \mathsf{USC}_{sk_R, pk_S}(m, CT)$: *an unsigncryption algorithm which takes as inputs receiver's secret key* sk_R*, sender's public key* pk_S*, message* m*, and ciphertext* $CT \in \mathcal{CS}$*, and outputs KEM key* $K \in \mathcal{KS}$ *or reject symbol* $\perp$*.*

Definition 7 (dM-IND-iCCA Security for Signcryption KEM). *A signcryption KEM scheme is dM-IND-iCCA secure if the following property holds for security parameter* κ*; For any PPT adversary* $\mathcal{S} = (\mathcal{S}_1, \mathcal{S}_2)$*,* $\mathbf{Adv}^{\text{dm-ind-icca}} = |\Pr[r_R \leftarrow \mathcal{RS}_{RG}; (pk_R, sk_R) \leftarrow \mathsf{RKeyGen}(1^\kappa, r_R); (m^*, pk_S^*, sk_S^*, state) \leftarrow \mathcal{S}_1^{\mathcal{UO}(sk_R,\cdot,\cdot,\cdot)}(pk_R); b \leftarrow \{0,1\}; r_e \leftarrow \mathcal{RS}_{SE}; (K_0^*, CT_0^*) \leftarrow \mathsf{SC}_{sk_S^*, pk_R}(m^*; r_e); K_1^* \leftarrow \mathcal{KS}; b' \leftarrow \mathcal{S}_2^{\mathcal{UO}(sk_R,\cdot,\cdot,\cdot)}(pk_S^*, sk_S^*, pk_R, K_b^*, CT_0^*, m^*, state); b' = b] - 1/2| \leq negl$*, where* $\mathcal{UO}$ *is the unsigncryption oracle who outputs K on input* (pk_S, m, CT) *with respect to* sk_R*,* $\mathcal{KS}$ *is the KEM key space and state is state information that* $\mathcal{S}$ *wants to preserve from* $\mathcal{S}_1$ *to* $\mathcal{S}_2$*.* $\mathcal{A}$ *cannot submit the ciphertext* $CT = CT_0^*$ *to* $\mathcal{UO}$*.*

Definition 8 (dM-UF-iCMA Security for Signcryption KEM). *A signcryption KEM scheme is dM-sUF-iCMA secure if the following property holds for security parameter* κ*; For any PPT adversary* $\mathcal{F}$*,* $\mathbf{Adv}^{\text{dm-suf-icma}} = \Pr[r_S \leftarrow \mathcal{RS}_{SG}; (pk_S, sk_S) \leftarrow \mathsf{SKeyGen}(1^\kappa, r_S); (pk_R^*, sk_R^*, m^*, CT^*) \leftarrow \mathcal{F}^{\mathcal{SO}(sk_S,\cdot,\cdot)}(pk_S); K^* \leftarrow \mathsf{USC}_{sk_R^*, pk_S}(m^*, CT^*) \wedge K^*(\neq \perp) \in \mathcal{KS}] \leq negl$*, where* $\mathcal{SO}$ *is the signcryption oracle who outputs* (K, CT) *on input* (pk_R, m) *with respect to* sk_S*,* $\mathcal{KS}$ *is the KEM key space.* $\mathcal{F}$ *cannot output* (pk_R^*, m^*, CT^*) *such that* CT^* *is the output of* $\mathcal{SO}$ *on input* (pk_R^*, m^*)*.*

Definition 9 (Min-Entropy of Signcryption KEM Key). *A signcryption KEM scheme is k-min-entropy signcryption KEM if for any* (sk_S, pk_R)*, for distribution* $D_{\mathcal{KS}}$ *of variable K defined by* $(K, CT) \leftarrow \mathsf{SC}_{sk_S, pk_R}(m, r_e)$ *and random* $r_e \in \mathcal{RS}_{SE}$*,* $H_\infty(D_{\mathcal{KS}}) \geq k$ *holds, where* H_∞ *denotes min-entropy.*

Security Notions of Randomness Extractor and Pseudo-Random Function. Let $Ext : S \times X \rightarrow Y$ be a function with finite seed space S, finite domain X, and finite range Y.

Definition 10 (Strong Randomness Extractor [7]). *We say that function Ext is a strong randomness extractor, if for any distribution* D_X *over X with* $H_\infty(D_X) \geq k$*,* $\Delta((U_S, Ext(U_S, D_X)), (U_S, U_Y)) \leq negl$ *holds, where both* U_S *in* $(U_S, Ext(U_S, D_X))$ *denotes the same random variable,* Δ *denotes statistical distance,* U_S, U_X, U_Y *denotes uniform distribution over* S, X, Y *respectively,* $|X| = n \geq k$*,* $|Y| = k$*, and* $|S| = d$*.*

Let κ be a security parameter and $\mathsf{F} = \{F_\kappa : Dom_\kappa \times \mathcal{FS}_\kappa \rightarrow Rng_\kappa\}_\kappa$ be a function family with a family of domains $\{Dom_\kappa\}_\kappa$, a family of key spaces $\{\mathcal{FS}_\kappa\}_\kappa$ and a family of ranges $\{Rng_\kappa\}_\kappa$.

Definition 11 (Pseudo-Random Function [7]). *We say that function family* $\mathsf{F} = \{F_\kappa\}_\kappa$ *is the PRF family, if for any PPT distinguisher* $\mathcal{D}$, $\mathbf{Adv}^{\mathrm{prf}} = |\Pr[\mathcal{D}^{F_\kappa(\cdot)} \rightarrow 1] - \Pr[\mathcal{D}^{RF_\kappa(\cdot)} \rightarrow 1]| \leq negl$, *where* $RF_\kappa : Dom_\kappa \rightarrow Rng_\kappa$ *is a truly random function.*

3.2 Construction

Here, we propose a new generic construction of PKI-based AKE, which is secure in the CK$^+$-sFS^{NSR} model in the standard model.

Design Principle. Our construction is an extension of the FSXY construction which is based on an IND-CCA secure KEM, an IND-CPA secure KEM, PRFs, and strong randomness extractors. Their construction achieves the CK$^+$ security with two techniques: *twisted PRF* trick and *session-specific* key generation.

The twisted PRF trick is effective for achieving resistance to MEX. Two PRFs (F, F') with reversing keys are used; that is, we choose two ephemeral keys (r, r') and compute $F_\sigma(r) \oplus F'_{r'}(\sigma)$, where σ is the static secret key. It is especially effective in the following two scenarios: leakage of both ephemeral secret keys of the initiator and the responder, and leakage of the static secret key of the initiator and the ephemeral secret key of the responder (i.e., corresponding to KCI). If (r, r') is leaked, $F_\sigma(r)$ cannot be computed without knowing σ. Similarly, if σ is leaked, $F'_{r'}(\sigma)$ cannot be computed without knowing r'. In their construction, the output of the twisted PRF is used as randomness for the encapsulation algorithm.

Also, generation of session-specific decapsulation and encapsulation keys are effective for achieving wPFS. The initiator sends the temporary encapsulation key to the responder, the responder encapsulates a KEM key with the temporary encapsulation key, and the initiator decapsulates the ciphertext. Since this procedure does not depend on the static secret keys, the KEM key is hidden even if both static secret keys of the initiator and the responder are leaked.

A problem on the FSXY construction is that it is not one-round protocol (i.e, the responder cannot send a message until receiving the message from the initiator). If we use an IND-CPA secure PKIC-KEM instead of the IND-CPA secure KEM for session-specific key generation, the responder can generate the ephemeral public key without knowing the public key in the ephemeral public key of the initiator. Thus, our construction achieves one-round protocol.

The other problem is that, if an adversary is active in the test session (i.e., a situation according to sFS), the FSXY construction is insecure as follows; First, the adversary encapsulates a KEM key with the encapsulation keys of the owner of the test session and sends ciphertexts as impersonating the peer. Next, the adversary obtains the decapsulation key of the peer after completion of the test session and decapsulates the ciphertext sent from the owner. Then, the adversary obtains all KEM keys and easily derives the session key. Thus, the FSXY construction does not satisfy the CK$^+$-sFS^{NSR} security.

The main idea to achieve CK$^+$-sFS^{NSR} security is to use a dM-IND-iCCA and dM-sUF-iCMA secure signcryption KEM instead of an IND-CCA secure KEM. Security against insider attacks is necessary because we must consider cases that an adversary obtains static secret keys of parties in the test session (i.e., 2.a, 2.b, 2.d, and 2.g in

Definition 2). Also, the multi-user setting is necessary to prove security because each party may send ciphertexts with different public keys under a secret key, and we must simulate such a situation in the security proof. In our construction, parties signcrypt the public key or the ciphertext of PKIC-KEM, and exchange ciphertexts of signcryption KEM. If an adversary tries to modify ciphertexts as impersonating the peer of the test session like the above attack to the FSXY construction, ciphertexts is rejected by the owner of the test session because of dM-sUF-iCMA security. Also, the adversary cannot obtain the secret key of the peer *before* completion of the test session. Thus, there is no way to modify ciphertexts even if the adversary is active in the test session.

Generic Construction GC-sFS. The protocol of GC-sFS from signcryption KEM (SKeyGen, RKeyGen, SC, USC) and PKIC-KEM (KeyGen, EnCapC, EnCapK, DeCap) is as follows.

Public Parameters. Let κ be the security parameter, $F : \{0,1\}^* \times \mathcal{FS} \rightarrow \mathcal{RS}_E$, $F' : \{0,1\}^* \times \mathcal{FS} \rightarrow \mathcal{RS}_E$, and $G : \{0,1\}^* \times \mathcal{FS} \rightarrow \{0,1\}^\kappa$ be pseudo-random functions, where $\mathcal{FS}$ is the key space of PRFs ($|\mathcal{FS}| = \kappa$), $\mathcal{RS}_E$ is the randomness space of SC, and $\mathcal{RS}_G$ is the randomness space of SKeyGen and RKeyGen, and let $Ext : \mathcal{SS} \times \mathcal{KS} \rightarrow \mathcal{FS}$ be a strong randomness extractor with randomly chosen seed $s \in \mathcal{SS}$, where $\mathcal{SS}$ is the seed space and $\mathcal{KS}$ is the KEM key space. These are provided as some of the public parameters.

Secret and Public Keys. Party U_I randomly selects $\sigma_I \in \mathcal{FS}$, $r_{IS} \in \mathcal{RS}_{SG}$ and $r_{IR} \in \mathcal{RS}_{RG}$, and runs the key generation algorithms $(pk_{IS}, sk_{IS}) \leftarrow \mathsf{SKeyGen}(1^\kappa, r_{IS})$ and $(pk_{IR}, sk_{IR}) \leftarrow \mathsf{RKeyGen}(1^\kappa, r_{IR})$, where $\mathcal{RS}_{SG}$ is the randomness space of SKeyGen and $\mathcal{RS}_{RG}$ is the randomness space of RKeyGen. Party U_I's static secret and public keys are $((sk_{IS}, sk_{IR}, \sigma_I), (pk_{IS}, pk_{IR}))$.

Key Exchange. Party U_A with secret and public keys $((sk_{AS}, sk_{AR}, \sigma_A), (pk_{AS}, pk_{AR}))$, and who is the initiator, and party U_B with secret and public keys $((sk_{BS}, sk_{BR}, \sigma_B), (pk_{BS}, pk_{BR}))$, and who is the responder, perform the following two-pass key exchange protocol.

1. Party U_A randomly chooses ephemeral secret keys $r_{A,1}, r'_{A,1} \in \mathcal{FS}$ and $r_{A,2} \in \mathcal{RS}_G$. Party U_A computes $(ek_A, dk_A) \leftarrow \mathsf{KeyGen}(1^\kappa, r_{A,2})$ and $(CT_A, K_A) \leftarrow \mathsf{SC}_{sk_{AS}, pk_{BR}}(ek_A; F_{\sigma_A}(r_{A,1}) \oplus F'_{r'_{A,1}}(\sigma_A))$, and sends (U_A, U_B, CT_A, ek_A) to party U_B.
2. party U_B randomly chooses ephemeral secret keys $r_{B,1}, r'_{B,1} \in \mathcal{FS}$ and $r_{B,2} \in \mathcal{RS}_E$. Party U_B computes $CT_{B,2} \leftarrow \mathsf{EnCapC}(r_{B,2})$ and $(CT_{B,1}, K_{B,1}) \leftarrow \mathsf{SC}_{sk_{BS}, pk_{AR}}(CT_{B,2}; F_{\sigma_B}(r_{B,1}) \oplus F'_{r'_{B,1}}(\sigma_B))$ and, sends $(U_A, U_B, CT_{B,1}, CT_{B,2})$ to party U_A.
3. Upon receiving $(U_A, U_B, CT_{B,1}, CT_{B,2})$, party U_A computes $K_{B,1} \leftarrow \mathsf{USC}_{sk_{AR}, pk_{BS}}(CT_{B,2}, CT_{B,1})$, $K_{B,2} \leftarrow \mathsf{DeCap}_{dk_A}(CT_{B,2})$, $K'_1 \leftarrow Ext(s, K_A)$, $K'_2 \leftarrow Ext(s, K_{B,1})$ and $K'_3 \leftarrow Ext(s, K_{B,2})$, sets the session transcript $\mathsf{ST} = (U_A, U_B, pk_{AS}, pk_{AR}, pk_{BS}, pk_{BR}, ek_A, CT_A, CT_{B,1}, CT_{B,2})$ and the session key $SK = G_{K'_1}(\mathsf{ST}) \oplus G_{K'_2}(\mathsf{ST}) \oplus G_{K'_3}(\mathsf{ST})$, completes the session, and erases all session states.

4. Upon receiving (U_A, U_B, CT_A, ek_A), party U_B computes $K_A \leftarrow \mathsf{USC}_{sk_{BR}, pk_{AS}}(ek_A, CT_A)$, $K_{B,2} \leftarrow \mathsf{EnCapK}_{ek_A}(CT_{B,2}, r_{B,2})$, $K'_1 \leftarrow Ext(s, K_A)$, $K'_2 \leftarrow Ext(s, K_{B,1})$ and $K'_3 \leftarrow Ext(s, K_{B,2})$, sets the session transcript $\mathsf{ST} = (U_A, U_B, pk_{AS}, pk_{AR}, pk_{BS}, pk_{BR}, ek_A, CT_A, CT_{B,1}, CT_{B,2})$ and the session key $SK = G_{K'_1}(\mathsf{ST}) \oplus G_{K'_2}(\mathsf{ST}) \oplus G_{K'_3}(\mathsf{ST})$, completes the session, and erases all session states.

The session state of a session owned by U_A contains ephemeral secret keys $(r_{A,1}, r'_{A,1}, r_{A,2})$, KEM keys $(K_A, K_{B,1}, K_{B,2})$, outputs of the extractor (K'_1, K'_2, K'_3) and outputs of PRFs (i.e., $F_{\sigma_A}(r_{A,1}), F'_{r'_{A,1}}(\sigma_A), G_{K'_1}(\mathsf{ST}), G_{K'_2}(\mathsf{ST})$, and $G_{K'_3}(\mathsf{ST})$). Similarly, the session state of a session owned by U_B contains ephemeral secret keys $(r_{B,1}, r'_{B,1}, r_{B,2})$, decapsulated KEM keys $(K_A, K_{B,1}, K_{B,2})$, outputs of the extractor (K'_1, K'_2, K'_3) and outputs of PRFs (i.e., $F_{\sigma_B}(r_{B,1}), F'_{r'_{B,1}}(\sigma_B), G_{K'_1}(\mathsf{ST}), G_{K'_2}(\mathsf{ST})$, and $G_{K'_3}(\mathsf{ST})$).

Security. We show the following theorem.

Theorem 1. *If* (SKeyGen, RKeyGen, SC, USC) *is dM-IND-iCCA and dM-sUF-iCMA secure signcryption KEM and is κ-min-entropy signcryption KEM,* (KeyGen, EnCapC, EnCapK, DeCap) *is IND-CPA secure PKIC-KEM and is κ-min-entropy PKIC-KEM, F, F′ and G are PRFs, and Ext is a strong randomness extractor, then AKE scheme* GC-sFS *is* CK$^+$-sFS^{NSR}*-secure.*

First, we give an overview of the security proof for the case that the test session has a non-matching session.

We have to consider the following six leakage patterns in the CK$^+$-sFS^{NSR} security model:

1. The owner of sid* is the initiator, and the static secret key of the initiator is leaked. Also, the static secret key of the peer is leaked after completion of sid*.
2. The owner of sid* is the responder, and the static secret key of the initiator is leaked. Also, the static secret key of the peer is leaked after completion of sid*.
3. The owner of sid* is the initiator, and the static secret keys of the initiator is leaked.
4. The owner of sid* is the responder, and the static secret keys of the responder is leaked.
5. The owner of sid* is the initiator, and the ephemeral secret keys of sid* is leaked.
6. The owner of sid* is the responder, and the ephemeral secret keys of sid* is leaked.

The proof outline is similar to that in [7] except events 1 and 2. Thus, we show the proof sketch of event 1. (Event 2 is almost same as event 1.) We suppose that party U_A is the owner of sid* and U_A believes that the peer of sid* is U_B. Note that the adversary obtains $(sk_{AS}, sk_{AR}, \sigma_A)$, but $(sk_{BS}, sk_{BR}, \sigma_B)$ is not leaked before starting sid*. Also, the adversary is not allowed StateReveal query to any session between U_A and U_B.

We transform the CK$^+$-sFS^{NSR} security game into the game that the session key in the test session is randomly distributed. First, we change the game as the adversary wins if a forgery event with respect to $CT_{B,1}$ occurs. This event occurs only with negligible probability from the dM-sUF-iCMA security of (SKeyGen, RKeyGen, SC, USC). Though the adversary may forward $(CT_{B,1}, CT_{B,2})$ in a session between U_A and U_B other than sid*, $K_{B,2}$ is not leaked from IND-CPA security of (KeyGen, EnCapC, EnCapK,

DeCap) because StateReveal query to such sessions is not allowed. Thus, the adversary cannot obtain $K_{B,2}$ after this transformation. Second, we change the output of $\mathsf{EnCapK}_{ek_A}(CT_{B,2}, r_{B,2})$ into a random key; therefore, the input of Ext is randomly distributed and has sufficient min-entropy. Third, we change the output of Ext into randomly chosen values; therefore, the key of one of the PRFs (corresponding to the output of $\mathsf{EnCapK}_{ek_A}(CT_{B,2}, r_{B,2})$) is randomly distributed. Finally, we change this PRF into a random function. Therefore, the session key in the test session is randomly distributed; thus, there is no advantage to the adversary.

Proof. In the experiment of CK$^+$-sFS^{NSR} security, we suppose that sid^* is the session identity for the test session, and that there are N users and at most ℓ sessions are activated. Let κ be the security parameter, and let $\mathcal{A}$ be a PPT (in κ) bounded adversary. Suc denotes the event that $\mathcal{A}$ wins. We consider the following events that cover all cases of the behavior of $\mathcal{A}$.

- Let E_1 be the event that the test session sid^* has no matching session $\overline{\mathsf{sid}^*}$, the owner of sid^* is the initiator, and the static secret key of the initiator is given to $\mathcal{A}$. Also, the static secret key of the peer of sid^* is given to $\mathcal{A}$ after completion of sid^*. The adversary is not allowed StateReveal query to any session between the owner and the peer of sid^*.
- Let E_2 be the event that the test session sid^* has no matching session $\overline{\mathsf{sid}^*}$, the owner of sid^* is the responder, and the static secret key of the responder is given to $\mathcal{A}$. Also, the static secret key of the peer of sid^* is given to $\mathcal{A}$ after completion of sid^*. The adversary is not allowed StateReveal query to any session between the owner and the peer of sid^*.
- Let E_3 be the event that the test session sid^* has no matching session $\overline{\mathsf{sid}^*}$, the owner of sid^* is the initiator and the static secret key of the initiator is given to $\mathcal{A}$.
- Let E_4 be the event that the test session sid^* has no matching session $\overline{\mathsf{sid}^*}$, the owner of sid^* is the initiator and the ephemeral secret key of sid^* is given to $\mathcal{A}$.
- Let E_5 be the event that the test session sid^* has no matching session $\overline{\mathsf{sid}^*}$, the owner of sid^* is the responder and the static secret key of the responder is given to $\mathcal{A}$.
- Let E_6 be the event that the test session sid^* has no matching session $\overline{\mathsf{sid}^*}$, the owner of sid^* is the responder and the ephemeral secret key of sid^* is given to $\mathcal{A}$.
- Let E_7 be the event that the test session sid^* has matching session $\overline{\mathsf{sid}^*}$, and both static secret keys of the initiator and the responder are given to $\mathcal{A}$.
- Let E_8 be the event that the test session sid^* has matching session $\overline{\mathsf{sid}^*}$, and both ephemeral secret keys of sid^* and $\overline{\mathsf{sid}^*}$ are given to $\mathcal{A}$.
- Let E_9 be the event that the test session sid^* has matching session $\overline{\mathsf{sid}^*}$, and the static secret key of the owner of sid^* and the ephemeral secret key of $\overline{\mathsf{sid}^*}$ are given to $\mathcal{A}$.
- Let E_{10} be the event that the test session sid^* has matching session $\overline{\mathsf{sid}^*}$, and the ephemeral secret key of sid^* and the static secret key of the owner of $\overline{\mathsf{sid}^*}$ are given to $\mathcal{A}$.

To finish the proof, we investigate events $E_i \wedge Suc$ $(i = 1, \ldots, 10)$ that cover all cases of event Suc.

Due to space limitations we only show the proof of the event $E_1 \wedge Suc$ because this event and event $E_2 \wedge Suc$ contain significant difference with the proof of the FSXY construction [7]. Proofs of $E_1 \wedge Suc$ and $E_2 \wedge Suc$ are essentially same.

We change the interface of oracle queries and the computation of the session key. These instances are gradually changed over hybrid experiments, depending on specific sub-cases. In the last hybrid experiment, the session key in the test session does not contain information of the bit b. Thus, the adversary clearly only output a random guess. We denote these hybrid experiments by $\mathbf{H}_0, \ldots, \mathbf{H}_6$ and the advantage of the adversary $\mathcal{A}$ when participating in experiment $\mathbf{H}_i$ by $\mathbf{Adv}(\mathcal{A}, \mathbf{H}_i)$.

Hybrid Experiment $\mathbf{H}_0$: This experiment denotes the real experiment for CK^+-$\mathrm{sFS}^{\mathrm{NSR}}$ security and in this experiment the environment for $\mathcal{A}$ is as defined in the protocol. Thus, $\mathbf{Adv}(\mathcal{A}, \mathbf{H}_0)$ is the same as the advantage of the real experiment.

Hybrid Experiment $\mathbf{H}_1$: In this experiment, if session identities in two sessions are identical, the experiment halts.

When two ciphertexts from different randomness are identical and two public keys from different randomness are identical, session identities in two sessions are also identical. In the dM-IND-iCCA secure signcryption KEM, such an event occurs with negligible probability. Thus, $|\mathbf{Adv}(\mathcal{A}, \mathbf{H}_1) - \mathbf{Adv}(\mathcal{A}, \mathbf{H}_0)| \leq negl$.

Hybrid Experiment $\mathbf{H}_2$: In this experiment, the experiment selects party U_A and integer $i \in [1, \ell]$ randomly in advance. If $\mathcal{A}$ poses Test query to a session except i-th session of U_A, the experiment halts.

Since guess of the test session matches with $\mathcal{A}$'s choice with probability $1/N^2\ell$, $\mathbf{Adv}(\mathcal{A}, \mathbf{H}_2) \geq 1/N^2\ell \cdot \mathbf{Adv}(\mathcal{A}, \mathbf{H}_1)$.

Hybrid Experiment $\mathbf{H}_3$: In this experiment, we consider a forgery event E_F, and if E_F occurs, we regard the adversary successful and the experiment aborts. E_F occurs if $\mathcal{A}$ sends CT'_1 and CT'_2 as part of an ephemeral public key of U_B in the test session such that

- $K' \leftarrow \mathsf{USC}_{sk_{AR},pk_{BS}}(CT'_2, CT'_1)$ and $K' \neq \perp$,
- (CT'_1, CT'_2) was not contained in any output by previous $\mathsf{Send}(\Pi, \mathcal{R}, U_B, U_A)$ queries, and
- U_A completes the test session.

Since $\mathcal{A}$ cannot obtain sk_{AR} before completion of the test session, the only way E_F occurs is to forge (CT'_1, CT'_2). Thus, from the Difference Lemma $|\Pr[E_3 \wedge Suc] - \Pr[E_2 \wedge Suc]| \leq \Pr[E_F]$ and $|\mathbf{Adv}(\mathcal{A}, \mathbf{H}_3) - \mathbf{Adv}(\mathcal{A}, \mathbf{H}_2)| \leq \Pr[E_F]$.

We construct a dM-sUF-iCMA forger $\mathcal{F}$ from $\mathcal{A}$ such that E_F occurs with non-negligible probability. $\mathcal{F}$ performs the following steps.

Init. $\mathcal{F}$ receives pk^*_S as a challenge.

Setup. $\mathcal{F}$ chooses pseudo-random functions $F : \{0,1\}^* \times \mathcal{FS} \rightarrow \mathcal{RS}_E$, $F' : \{0,1\}^* \times \mathcal{FS} \rightarrow \mathcal{RS}_E$ and $G : \{0,1\}^* \times \mathcal{FS} \rightarrow \{0,1\}^k$, where $\mathcal{FS}$ is the key space of PRFs, and a strong randomness extractor $Ext : \mathcal{SS} \times \mathcal{KS} \to \mathcal{FS}$ with a randomly chosen seed $s \in \mathcal{SS}$. These are provided as a part of the public parameters. Also, $\mathcal{F}$ sets all N users' static secret and public keys except U_B. $\mathcal{F}$ selects $\sigma_I \in \mathcal{FS}$, $r_{IS} \in \mathcal{RS}_{SG}$, and $r_{IR} \in \mathcal{RS}_{RG}$ randomly, and runs the key generation algorithms $(pk_{IS}, sk_{IS}) \leftarrow \mathsf{SKeyGen}(1^\kappa, r_{IS})$ and $(pk_{IR}, sk_{IR}) \leftarrow \mathsf{RKeyGen}(1^\kappa, r_{IR})$ and U_I's static secret and public keys are $((sk_{IS}, sk_{IR}\sigma_I), (pk_{IS}, pk_{IR})$. For U_B, $\mathcal{F}$ sets $pk_{BS} := pk_S^*$. sk_{BR} and pk_{BR} are legitimately generated.

Simulation. $\mathcal{F}$ maintains the list $\mathcal{L}_{SK}$ that contains queries and answers of KeyReveal. $\mathcal{F}$ simulates oracle queries by $\mathcal{A}$ as follows.

1. $\mathsf{Send}(\Pi, \mathcal{I}, U_P, U_{\bar{P}})$: $\mathcal{F}$ computes the ephemeral public key (CT_P, ek_P) obeying the protocol, returns it and records $(\Pi, \mathcal{I}, U_P, U_{\bar{P}}, (CT_P, ek_P))$.
2. $\mathsf{Send}(\Pi, \mathcal{R}, U_{\bar{P}}, U_P)$: If $\bar{P} = B$, $\mathcal{F}$ computes $CT_{B,2}$, poses (pk_{PR}, CT_B) to signcryption oracle $\mathcal{SO}$, obtains $(K_{B,1}, CT_{B,1})$. Then, $\mathcal{F}$ sets the ephemeral public key $(CT_{B,1}, CT_{B,2})$, returns the ephemeral public key, and records $(\Pi, \mathcal{R}, U_B, U_P, (CT_{B,1}, CT_{B,2}))$. Otherwise, $\mathcal{F}$ computes the ephemeral public key $(CT_{\bar{P},1}, CT_{\bar{P},2})$, returns the ephemeral public key, and records $(\Pi, \mathcal{R}, U_{\bar{P}}, U_P, (CT_{\bar{P},1}, CT_{\bar{P},2}))$.
3. $\mathsf{Send}(\Pi, \mathcal{I}, U_P, U_{\bar{P}}, (CT_{\bar{P},1}, CT_{\bar{P},2}))$: If $P = A$, $\bar{P} = B$, the session is i-th session of A, $K_{B,1} \leftarrow \mathsf{USC}_{sk_{AR}, pk_{BS}}(CT_{B,2}, CT_{B,1})$ and $K_{B,1} \neq \perp$, and $(CT_{B,1}, CT_{B,2})$ was not contained in any output by previous $\mathsf{Send}(\Pi, \mathcal{R}, U_B, U_A)$ queries, then $\mathcal{F}$ outputs $(pk_{AR}, CT_{B,2}, CT_{B,1})$ as a forgery. Else if $(\Pi, \mathcal{I}, U_P, U_{\bar{P}}, (CT_P, ek_P))$ is not recorded, $\mathcal{F}$ records the session $(\Pi, \mathcal{I}, U_P, U_{\bar{P}}, (CT_P, ek_P), (CT_{\bar{P},1}, CT_{\bar{P},2}))$ is not completed. Otherwise, $\mathcal{F}$ computes the session key SK obeying the protocol, and records $(\Pi, \mathcal{I}, U_P, U_{\bar{P}}, (CT_P, ek_P), (CT_{\bar{P},1}, CT_{\bar{P},2}))$ as the completed session and SK in the list $\mathcal{L}_{SK}$.
4. $\mathsf{Send}(\Pi, \mathcal{R}, U_{\bar{P}}, U_P, (CT_P, ek_P))$: If $(\Pi, \mathcal{R}, U_{\bar{P}}, U_P, (CT_{\bar{P},1}, CT_{\bar{P},2}))$ is not recorded, $\mathcal{F}$ records the session $(\Pi, \mathcal{R}, U_{\bar{P}}, U_P, (CT_P, ek_P), (CT_{\bar{P},1}, CT_{\bar{P},2}))$ is not completed. Otherwise, $\mathcal{F}$ computes the session key SK obeying the protocol, and records $(\Pi, \mathcal{R}, U_{\bar{P}}, U_P, (CT_P, ek_P), (CT_{\bar{P},1}, CT_{\bar{P},2}))$ as the completed session and SK in the list $\mathcal{L}_{SK}$.
5. KeyReveal(sid):
   (a) If the session sid is not completed, $\mathcal{F}$ returns an error message.
   (b) Otherwise, $\mathcal{F}$ returns the recorded value SK.
6. StateReveal(sid): $\mathcal{F}$ responds the ephemeral secret key and intermediate computation results of sid as the definition.
7. Corrupt(U_P): $\mathcal{F}$ responds the static secret key of U_P as the definition.
8. Test(sid): $\mathcal{F}$ responds to the query as the definition.
9. If $\mathcal{A}$ outputs a guess b', $\mathcal{F}$ aborts.

Analysis. If E_F occurs with non-negligible probability, a successful forgery $(CT_{B,1}, CT_{B,2})$ is contained in a $\mathsf{Send}(\Pi, \mathcal{I}, U_P, U_{\bar{P}}, (CT_{\bar{P},1}, CT_{\bar{P},2})$ query with non-negligible probability. Thus, $\mathcal{F}$ can also output a successful forgery $(pk_{AR}, CT_{B,2}, CT_{B,1})$ with non-negligible probability. If the advantage of $\mathcal{F}$ is negligible, then E_F occurs with negligible probability, and $|\mathbf{Adv}(\mathcal{A}, \mathbf{H}_3) - \mathbf{Adv}(\mathcal{A}, \mathbf{H}_2)| \leq \Pr[E_F] = negl$.

Hybrid Experiment $\mathbf{H}_4$: In this experiment, the computation of $K^*_{B,2}$ in the test session is changed. Instead of computing $K^*_{B,2} \leftarrow \mathsf{EnCapK}_{ek_A}(CT^*_{B,2}, r^*_{B,2})$), it is changed as choosing $K^*_{B,2} \leftarrow \mathcal{KS}$ randomly, where we suppose that U_B is the intended partner of U_A in the test session.

We construct an IND-CPA adversary $\mathcal{S}$ for (KeyGen, EnCapC, EnCapK, DeCap) from $\mathcal{A}$ in $\mathbf{H}_3$ or $\mathbf{H}_4$. $\mathcal{S}$ performs the following steps.

Init. $\mathcal{S}$ receives the public key ek^* as a challenge. Also, $\mathcal{S}$ receives the challenge (K^*, CT^*) for the IND-CPA game.

Setup. $\mathcal{S}$ chooses pseudo-random functions $F : \{0,1\}^* \times \mathcal{FS} \rightarrow \mathcal{RS}_E$, $F' : \{0,1\}^* \times \mathcal{FS} \rightarrow \mathcal{RS}_E$ and $G : \{0,1\}^* \times \mathcal{FS} \rightarrow \{0,1\}^k$, where $\mathcal{FS}$ is the key space of PRFs, and a strong randomness extractor $Ext : \mathcal{SS} \times \mathcal{KS} \rightarrow \mathcal{FS}$ with a randomly chosen seed $s \in \mathcal{SS}$. These are provided as a part of the public parameters. Also, $\mathcal{S}$ sets all N users' static secret and public keys. $\mathcal{F}$ selects $\sigma_I \in \mathcal{FS}$, $r_{IS} \in \mathcal{RS}_{SG}$, and $r_{IR} \in \mathcal{RS}_{RG}$ randomly, and runs the key generation algorithms $(pk_{IS}, sk_{IS}) \leftarrow \mathsf{SKeyGen}(1^\kappa, r_{IS})$ and $(pk_{IR}, sk_{IR}) \leftarrow \mathsf{RKeyGen}(1^\kappa, r_{IR})$ and U_I's static secret and public keys are $((sk_{IS}, sk_{IR}\sigma_I), (pk_{IS}, pk_{IR})$.

Simulation. $\mathcal{S}$ maintains the list $\mathcal{L}_{SK}$ that contains queries and answers of KeyReveal. $\mathcal{S}$ simulates oracle queries by $\mathcal{A}$ as follows.

1. Send($\Pi, \mathcal{I}, U_P, U_{\bar{P}}$): If $P = A$ and $\bar{P} = B$, the session is i-th session of A, then $\mathcal{S}$ sets $ek_A := ek^*$, computes CT_A, and returns (U_A, U_B, CT_A, ek_A) and records $(\Pi, \mathcal{I}, U_A, U_B, (CT_A, ek_A))$. Otherwise, $\mathcal{S}$ computes the ephemeral public key (CT_P, ek_P) obeying the protocol, returns it and records $(\Pi, \mathcal{I}, U_P, U_{\bar{P}}, (CT_P, ek_P))$.
2. Send($\Pi, \mathcal{R}, U_{\bar{P}}, U_P$): If $P = A$ and $\bar{P} = B$, the session is i-th session of A, then $\mathcal{S}$ sets $CT_{B,2} := CT^*$, computes $CT_{B,1}$, and returns $(U_B, U_A, CT_{B,1}, CT_{B,2})$ and records $(\Pi, \mathcal{R}, U_B, U_A, (CT_{B,1}, CT_{B,2}))$. Otherwise, $\mathcal{S}$ computes the ephemeral public key $(CT_{\bar{P},1}, CT_{\bar{P},2})$, returns the ephemeral public key, and records $(\Pi, \mathcal{R}, U_{\bar{P}}, U_P, (CT_{\bar{P},1}, CT_{\bar{P},2}))$.
3. Send($\Pi, \mathcal{I}, U_P, U_{\bar{P}}, (CT_{\bar{P},1}, CT_{\bar{P},2})$): If $P = A$ and $\bar{P} = B$, the session is i-th session of A, then $\mathcal{S}$ sets $K_{B,2} := K^*$, computes the session key SK^* obeying the protocol, and records $(\Pi, \mathcal{I}, U_A, U_B, (CT_A, ek_A), (CT_{B,1}, CT_{B,2}))$ as the completed session and SK^* in the list $\mathcal{L}_{SK}$. Else if $(\Pi, \mathcal{I}, U_P, U_{\bar{P}}, (CT_P, ek_P))$ is not recorded, $\mathcal{S}$ records the session $(\Pi, \mathcal{I}, U_P, U_{\bar{P}}, (CT_P, ek_P), (CT_{\bar{P},1}, CT_{\bar{P},2}))$ is not completed. Otherwise, $\mathcal{S}$ computes the session key SK obeying the protocol, and records $(\Pi, \mathcal{I}, U_P, U_{\bar{P}}, (CT_P, ek_P), (CT_{\bar{P},1}, CT_{\bar{P},2}))$ as the completed session and SK in the list $\mathcal{L}_{SK}$.
4. Send($\Pi, \mathcal{R}, U_{\bar{P}}, U_P, (CT_P, ek_P)$): If $(\Pi, \mathcal{R}, U_{\bar{P}}, U_P, (CT_{\bar{P},1}, CT_{\bar{P},2}))$ is not recorded, $\mathcal{F}$ records the session $(\Pi, \mathcal{R}, U_{\bar{P}}, U_P, (CT_P, ek_P), (CT_{\bar{P},1}, CT_{\bar{P},2}))$ is not completed. Otherwise, $\mathcal{F}$ computes the session key SK obeying the protocol, and records $(\Pi, \mathcal{R}, U_{\bar{P}}, U_P, (CT_P, ek_P), (CT_{\bar{P},1}, CT_{\bar{P},2}))$ as the completed session and SK in the list $\mathcal{L}_{SK}$.
5. KeyReveal(sid):
   (a) If the session sid is not completed, $\mathcal{S}$ returns an error message.
   (b) Otherwise, $\mathcal{S}$ returns the recorded value SK.

6. StateReveal(sid): $\mathcal{S}$ responds the ephemeral secret key and intermediate computation results of sid as the definition. Note that the StateReveal query is not posed to the test session from the freshness definition.
7. Corrupt(U_P): $\mathcal{S}$ responds the static secret key of U_P as the definition.
8. Test(sid): $\mathcal{S}$ responds to the query as the definition.
9. If $\mathcal{A}$ outputs a guess b', $\mathcal{S}$ outputs b'.

Analysis. For $\mathcal{A}$, the simulation by $\mathcal{S}$ is same as the experiment $\mathbf{H}_3$ if the challenge is (K_0^*, CT_0^*). Otherwise, the simulation by $\mathcal{S}$ is same as the experiment $\mathbf{H}_4$. Also, both $K_{B,2}^*$ in two experiments have κ-min-entropy because (KeyGen, EnCapC, EnCapK, DeCap) is κ-min-entropy PKIC-KEM. Thus, if the advantage of $\mathcal{S}$ is negligible, then $|\mathbf{Adv}(\mathcal{A}, \mathbf{H}_4) - \mathbf{Adv}(\mathcal{A}, \mathbf{H}_3)| \leq negl.$

Hybrid Experiment $\mathbf{H}_5$: In this experiment, the computation of $K_3'^*$ in the test session is changed. Instead of computing $K_3'^* \leftarrow Ext(s, K_{B,2}^*)$, it is changed as choosing $K_3'^* \in \mathcal{FS}$ randomly.

Since $K_{B,2}^*$ is randomly chosen in $\mathbf{H}_4$, it has sufficient min-entropy. Thus, by the definition of the strong randomness extractor, $|\mathbf{Adv}(\mathcal{A}, \mathbf{H}_5) - \mathbf{Adv}(\mathcal{A}, \mathbf{H}_4)| \leq negl.$

Hybrid Experiment $\mathbf{H}_6$: In this experiment, the computation of SK in the test session is changed. Instead of computing $SK = G_{K_1'}(\mathsf{ST}) \oplus G_{K_2'}(\mathsf{ST}) \oplus G_{K_3'}(\mathsf{ST})$, it is changed as $SK = G_{K_1'}(\mathsf{ST}) \oplus G_{K_2'}(\mathsf{ST}) \oplus x$ where $x \in \{0,1\}^\kappa$ is chosen randomly and we suppose that U_B is the intended partner of U_A in the test session.

We construct a distinguisher $\mathcal{D}'$ between PRF $F^* : \{0,1\}^* \times \mathcal{FS} \to \{0,1\}^k$ and a random function RF from $\mathcal{A}$ in $\mathbf{H}_5$ or $\mathbf{H}_6$. The construction and analysis of $\mathcal{D}'$ is similar to that in the proof in [7]. Thus, we omit it due to space limitations, and if the advantage of $\mathcal{D}'$ is negligible, then $|\mathbf{Adv}(\mathcal{A}, \mathbf{H}_6) - \mathbf{Adv}(\mathcal{A}, \mathbf{H}_5)| \leq negl.$

In $\mathbf{H}_6$, the session key in the test session is perfectly randomized. Thus, $\mathcal{A}$ cannot obtain any advantage from Test query.

Therefore, $\mathbf{Adv}(\mathcal{A}, \mathbf{H}_6) = 0$ and $\Pr[E_1 \wedge Suc]$ is negligible.

□

3.3 Instantiation

We can achieve the first DH-based AKE schemes from the generic construction GC-sFS in Section 3. For example, we can apply an efficient dM-IND-iCCA and dM-sUF-iCMA secure signcryption KEM [22] from the decisional bilinear DH assumption and the q-strong DH assumption. The ciphertext overhead of the best scheme in [22] is only $4|p|$, where $|p|$ is the length of a group element. The computational cost is 4 regular exponentiations for signcryption, and 1 regular exponentiation, 1 multi-exponentiation and 2 paring computations for unsigncryption. Also, IND-CPA secure PKIC-KEM is instantiated with the ElGamal KEM under the decisional DH assumption. Communication complexity (for two parties) of this instantiation is $10|p|$, where $|p|$ is the length of a group element. Computational complexity (for two parties) of this instantiation is 4

Table 1. Comparison of previous schemes and an instantiation of our scheme

	Model	Resource	Assumption	Computation (#parings+#[multi,regular]-exp,)	Communication complexity	
HMQV [4]	CK^+	ROM	GDH & KEA1	0 + [2, 2]	$2\|p\|$	512
FSXY [7]	CK^+	StdM	DDH	0 + [4, 12]	$8\|p\|$	2048
MAC(NAXOS) [17]	CK & sFS†	ROM	GDH	0 + [0, 8]	$3\|p\|$	768
SIG(NAXOS) [20]	eCK & sFS‡	ROM	GDH & CDH	4 + [0, 10]	$4\|p\|$	1024
Ours	CK^+-sFS^{NSR}	StdM	DDH & DBDH & q-SDH	4 + [2, 14]	$10\|p\|$	2560

† against the constrained adversary
‡ against a constrained but more powerful adversary than MAC(NAXOS)

CDH means the Computational Diffie-Hellman assumption. DDH means the Decisional Diffie-Hellman assumption. GDH means the Gap Diffie-Hellman assumption. DBDH means the Decisional Bilinear Diffie-Hellman assumption. q-SDH means the q-strong Diffie-Hellman assumption. KEA1 means the Knowledge-of-Exponent assumption. For concreteness the expected ciphertext overhead for a 128-bit implementation is also given. Note that computational costs are estimated without any pre-computation technique.

parings, 2 multi-exponentiations and 14 regular exponentiations (all symmetric operations such as hash function/KDF/PRF/MAC and multiplications are ignored). We show a comparison between this instantiation and previous schemes in Table 1. Note that we use the GDH signature [23] as a deterministic and strongly unforgeable signature scheme in the instantiation of SIG(NAXOS) [20].

We can easily show that these schemes are κ-min-entropy signcryption KEM. The signcryption scheme in [22] uses tag-based KEM version of the Boyen-Mei-Waters PKE [24]. Thus, The KEM key consists of $e(g_1, g_2)^{\alpha r} \in G_T$, where G_T is a finite cyclic group of order prime p with bilinear pairing, $e(g_1, g_2)^{\alpha}$ is part of public keys, and r is uniformly chosen randomness, and $|r|$ is 2κ. Thus, $e(g_1, g_2)^{\alpha r}$ has min-entropy larger than κ.

Acknowledgement. We thank Dario Fiore for helpful comments about this work.

References

1. Diffie, W., van Oorschot, P.C., Wiener, M.J.: Authentication and Authenticated Key Exchanges. Des. Codes Cryptography 2(2), 107–125 (1992)
2. Bellare, M., Rogaway, P.: Entity Authentication and Key Distribution. In: Stinson, D.R. (ed.) CRYPTO 1993. LNCS, vol. 773, pp. 232–249. Springer, Heidelberg (1994)
3. Canetti, R., Krawczyk, H.: Analysis of Key-Exchange Protocols and Their Use for Building Secure Channels. In: Pfitzmann, B. (ed.) EUROCRYPT 2001. LNCS, vol. 2045, pp. 453–474. Springer, Heidelberg (2001)
4. Krawczyk, H.: HMQV: A High-Performance Secure Diffie-Hellman Protocol. In: Shoup, V. (ed.) CRYPTO 2005. LNCS, vol. 3621, pp. 546–566. Springer, Heidelberg (2005)
5. LaMacchia, B.A., Lauter, K., Mityagin, A.: Stronger Security of Authenticated Key Exchange. In: Susilo, W., Liu, J.K., Mu, Y. (eds.) ProvSec 2007. LNCS, vol. 4784, pp. 1–16. Springer, Heidelberg (2007)
6. Sarr, A.P., Elbaz-Vincent, P., Bajard, J.-C.: A New Security Model for Authenticated Key Agreement. In: Garay, J.A., De Prisco, R. (eds.) SCN 2010. LNCS, vol. 6280, pp. 219–234. Springer, Heidelberg (2010)

7. Fujioka, A., Suzuki, K., Xagawa, K., Yoneyama, K.: Strongly Secure Authenticated Key Exchange from Factoring, Codes, and Lattices. In: Fischlin, M., Buchmann, J., Manulis, M. (eds.) PKC 2012. LNCS, vol. 7293, pp. 467–484. Springer, Heidelberg (2012); See also Cryptology ePrint Archive-2012/211
8. Law, L., Menezes, A., Qu, M., Solinas, J.A., Vanstone, S.A.: An Efficient Protocol for Authenticated Key Agreement. Des. Codes Cryptography 28(2), 119–134 (2003)
9. Jeong, I.R., Katz, J., Lee, D.-H.: One-Round Protocols for Two-Party Authenticated Key Exchange. In: Jakobsson, M., Yung, M., Zhou, J. (eds.) ACNS 2004. LNCS, vol. 3089, pp. 220–232. Springer, Heidelberg (2004)
10. Ustaoglu, B.: Obtaining a secure and efficient key agreement protocol from (H)MQV and NAXOS. Des. Codes Cryptography 46(3), 329–342 (2008)
11. Boyd, C., Cliff, Y., González Nieto, J.M., Paterson, K.G.: Efficient One-Round Key Exchange in the Standard Model. In: Mu, Y., Susilo, W., Seberry, J. (eds.) ACISP 2008. LNCS, vol. 5107, pp. 69–83. Springer, Heidelberg (2008)
12. Boyd, C., Cliff, Y., González Nieto, J.M., Paterson, K.G.: One-round key exchange in the standard model. IJACT 1(3), 181–199 (2009)
13. Kim, M., Fujioka, A., Ustaoğlu, B.: Strongly Secure Authenticated Key Exchange without NAXOS' Approach. In: Takagi, T., Mambo, M. (eds.) IWSEC 2009. LNCS, vol. 5824, pp. 174–191. Springer, Heidelberg (2009)
14. Sarr, A.P., Elbaz-Vincent, P., Bajard, J.-C.: A Secure and Efficient Authenticated Diffie–Hellman Protocol. In: Martinelli, F., Preneel, B. (eds.) EuroPKI 2009. LNCS, vol. 6391, pp. 83–98. Springer, Heidelberg (2010)
15. Fujioka, A., Suzuki, K.: Designing Efficient Authenticated Key Exchange Resilient to Leakage of Ephemeral Secret Keys. In: Kiayias, A. (ed.) CT-RSA 2011. LNCS, vol. 6558, pp. 121–141. Springer, Heidelberg (2011)
16. Cremers, C.J.F., Feltz, M.: One-round Strongly Secure Key Exchange with Perfect Forward Secrecy and Deniability. Cryptology ePrint Archive: 2011/300 (2011)
17. Boyd, C., González Nieto, J.M.: On Forward Secrecy in One-Round Key Exchange. In: IMA Int. Conf. 2011, pp. 451–468 (2011)
18. Gennaro, R., Krawczyk, H., Rabin, T.: Okamoto-Tanaka Revisited: Fully Authenticated Diffie-Hellman with Minimal Overhead. In: Zhou, J., Yung, M. (eds.) ACNS 2010. LNCS, vol. 6123, pp. 309–328. Springer, Heidelberg (2010)
19. Huang, H.: Strongly Secure One Round Authenticated Key Exchange Protocol with Perfect Forward Security. In: Boyen, X., Chen, X. (eds.) ProvSec 2011. LNCS, vol. 6980, pp. 389–397. Springer, Heidelberg (2011)
20. Cremers, C., Feltz, M.: Beyond eCK: Perfect Forward Secrecy under Actor Compromise and Ephemeral-Key Reveal. In: Foresti, S., Yung, M., Martinelli, F. (eds.) ESORICS 2012. LNCS, vol. 7459, pp. 734–751. Springer, Heidelberg (2012)
21. Zheng, Y.: Digital Signcryption or How to Achieve Cost (Signature & Encryption) << Cost(Signature) + Cost(Encryption). In: Kaliski Jr., B.S. (ed.) CRYPTO 1997. LNCS, vol. 1294, pp. 165–179. Springer, Heidelberg (1997)
22. Chiba, D., Matsuda, T., Schuldt, J.C.N., Matsuura, K.: Efficient Generic Constructions of Signcryption with Insider Security in the Multi-user Setting. In: Lopez, J., Tsudik, G. (eds.) ACNS 2011. LNCS, vol. 6715, pp. 220–237. Springer, Heidelberg (2011)
23. Boneh, D., Lynn, B., Shacham, H.: Short Signatures from the Weil Pairing. In: Boyd, C. (ed.) ASIACRYPT 2001. LNCS, vol. 2248, pp. 514–532. Springer, Heidelberg (2001)
24. Boyen, X., Mei, Q., Waters, B.: Direct chosen ciphertext security from identity-based techniques. In: ACM Conference on Computer and Communications Security 2005, pp. 320–329 (2005)

Compact Stateful Encryption Schemes with Ciphertext Verifiability

S. Sree Vivek, S. Sharmila Deva Selvi, and C. Pandu Rangan

Theoretical Computer Science Lab.,
Department of Computer Science and Engineering,
Indian Institute of Technology Madras, India
{sharmila,svivek,prangan}@cse.iitm.ac.in

Abstract. Increasingly wider deployment of encryption schemes call for schemes possessing additional properties such as randomness re-use, compactness and ciphertext verifiability. While novel approaches such as stateful encryption schemes contributes for randomness re-use (to save computational efforts), the requirements such as ciphertext verifiability leads to increase in the size of ciphertext. Thus, it is interesting and challenging to design stateful encryption schemes that offer ciphertext verifiability and result in compact ciphertexts. We propose two new stateful public key encryption schemes with ciphertext verifiability. Our schemes offer more compact ciphertexts when compared to all existing stateful public key encryption schemes with ciphertext verifiability. Our first scheme is based on the SDH assumption and the second scheme is based on the CDH assumption. We have proved both the schemes in the random oracle model.

Keywords: Stateful Public Key Encryption, Adaptive Chosen Ciphertext Security (CCA), Compact Ciphertext with Ciphertext Verification, Random Oracle model.

1 Introduction

For any public key encryption scheme, the difference between the size of the ciphertext and the size of the message is referred to as its *Ciphertext Overhead.* An encryption scheme is said to generate compact ciphertext if the overhead is utmost the size of one element in the underlying group. Needless to say, compact ciphertexts are very useful in bandwidth-critical environments [3,4]. In general, when we design encryption schemes with stronger security properties, we tend to loose compactness and often arrive at ciphertexts that have large overheads. However, in the recent past, several researchers have successfully designed CCA secure encryption schemes (stronger notion of security for encryption schemes) that result in compact ciphertexts [13,6,7,3,4]. While these schemes yield compact ciphertexts, they lack an important property which we refer as *Ciphertext Verifiability.* We briefly describe about this property and its importance below.

G. Hanaoka and T. Yamauchi (Eds.): IWSEC 2012, LNCS 7631, pp. 87–104, 2012.

For the public key encryption schemes that are used in important applications such as key transport, electronic auction etc, the encryption scheme must provide a guarantee that the ciphertext (and thus the message contained in the ciphertext) was not altered during transit. If such a guarantee is not available, it may lead to unacceptable situations. For example, suppose a user A wishes to safely send a key value key to user B and use key as ephemeral/session key for some further interaction with B. A may use the public key of B and encrypt key and send the ciphertext c to B. If no verification mechanism is available and if c is altered to c' (by the adversary or by transmission error) and if c' is decrypted to key', B would simply assume that key' is the key that A wished to send to him. This would cause further interactions between A and B impossible and this is clearly undesirable. A similar scenario can be imagined in a KEM/DEM scheme if modified ciphertexts are used to recover keys. It is not hard to imagine the possibility of change of bid values in e-auctions/e-tendering, where the altered ciphertext getting decrypted to a value different from the value actually meant by the sender.

Hence, it is important that the encryption schemes provide 'ciphertext verifiability' in addition to all the other desirable properties such as compactness and CCA security. By ciphertext verifiability we mean a testing process that is integrated in the decryption algorithm which identifies if the received ciphertext is a tweaked one or not. If the test fails, the receiver infers that the ciphertext is corrupted during transmission and rejects it. If the test passes, the receiver considers the message constructed by the decryption algorithm as a valid message. The ability to distinguish a tweaked ciphertext from a genuine ciphertext is an important property for decryption algorithm and see [15] by Pass et al. for a formal and rigorous treatment of the same.

One of the effective strategies to save computational effort needed for encryption is to re-use the randomness used for encryption between the same pair of (sender, receiver) across different messages. For this purpose, we consider the encryption process happening in a session where a session consists of sending some fixed number of messages (say one million). All messages in the same session will use the same random value and this saves efforts related to random number generation and computations involving only those random numbers in each encryption. For different sessions, we of course use different random value. A session is recognized by the state. Thus, the concept of stateful encryption, introduced by Bellare et al. [5] is very useful in the contexts where low power devises are involved. There are only two stateful PKI based encryption schemes available in the literature [5], [4]. Wile the scheme in [5] offers cipher text verification implicitly, it is not compact and the scheme in [4] is compact but not ciphertext verifiable. Thus, we have addressed the interesting question that asks to design a stateful encryption scheme that is compact and supporting ciphertext verifiability in the PKI model.

Related Work: There are several CCA secure encryption schemes available in the literature. Some of them are customized designs [1,6], some are based on transforming a CPA secure system to a CCA secure system [10,9,12], some are based on KEM/DEM (Key Encapsulation Mechanism/Data Encapsulation Mechanism) [8,11,13] and some are based on Tag-KEM/DEM framework [2]. However, none of them produced compact ciphertext and this prompted researchers to design afresh CCA secure encryption schemes outputting compact ciphertexts. Several new and interesting ideas emerged in the past, resulting in schemes reported in [13,6,7,3,4]. Though these schemes output compact ciphertext and CCA security, none of them offer ciphertext verifiability.

Our Contribution: There are two contributions in this paper. First, we design a new PKI based stateful public key encryption scheme ($\mathcal{N} - \mathcal{SPKE}_1$), whose security is based on the SDH problem. Our second contribution is a stateful public key encryption scheme ($\mathcal{N} - \mathcal{SPKE}_2$), whose security is based on CDH problem but with the same ciphertext overhead as ($\mathcal{N} - \mathcal{SPKE}_1$). The ciphertext overhead of these two schemes are slightly higher than that of the $\mathcal{SPKE}$ scheme proposed in [4]. The ciphertext overhead of the $\mathcal{SPKE}$ scheme in [4] is one group element and another element with λ bits, where λ is greater than 128-bits. In our schemes we include an integer value called as `encryption-count` which represents the encryption number. That is, we index each encryption performed during a session using an integer counter. At the start of each session, the value of `encryption-count` is initialized to 1 and incremented each time an encryption is performed during the session. If we consider that *one million* encryption operations are to be done in a session, the `encryption-count` ranges from 1-bit to 20-bits utmost. This also contributes to the ciphertext overhead of the scheme. Thus, the ciphertext overhead of our scheme is one group element, one element of size 128-bits and an `encryption-count`. With this overhead, it is possible to offer ciphertext verifiability and this is the highlighting difference of our scheme. The sender has to just increment the index after each encryption and store only the incremented value (utmost 20-bits) and does not need to remember the indices that are used previously in the session. Thus, this will not lead to big storage overhead. It is possible to use the folkloric construction of appending 80-bits of known value (usually 80-bits of 0's) to the plaintext while encrypting it and checking whether the decryption of the ciphertext produces a message with those 80-bits at the end to ensure ciphertext verifiability. However, the size of this value is lower bound by 80-bits, where as in our construction, the index is upper bound by 20-bits (for 2^{20} encryption) and hence can take a value starting from $1-$bit, which is a considerable reduction for resource constrained devices. This makes our construction more attractive.

2 Preliminaries, Frameworks and Security Models

We use Computational Diffie Hellman Problem (CDH) and Strong Diffie Hellman Problem (SDH) [3] to establish the security of the schemes.

Definition 1. *(Computational Diffie Hellman Problem (CDH)): Let κ be the security parameter and $\mathbb{G}$ be a multiplicative group of order q, where $|q| = \kappa$. Given $(g, g^a, g^b) \in_R \mathbb{G}^4$, the computational Diffie Hellman problem is to compute $g^{ab} \in \mathbb{G}$.*

The advantage of an adversary $\mathcal{A}$ in solving the computational Diffie Hellman problem is defined as the probability with which $\mathcal{A}$ solves the above computational Diffie Hellman problem.

$$Adv_{\mathcal{A}}^{CDH} = Pr[\mathcal{A}(g, g^a, g^b) = g^{ab}]$$

The computational Diffie Hellman assumption holds in $\mathbb{G}$ if for all polynomial time adversaries $\mathcal{A}$, the advantage $Adv_{\mathcal{A}}^{CDH}$ is negligible.

Definition 2. ***(Strong Diffie Hellman Problem (SDH) as given in [3]):*** *Let κ be the security parameter and $\mathbb{G}$ be a multiplicative group of order q, where $|q| = \kappa$. Given $(g, g^a, g^b) \in_R \mathbb{G}^3$ and access to a Decision Diffie Hellman (DDH) oracle $\mathcal{DDH}_{g,a}(.,.)$ which on input g^b and g^c outputs* `True` *if and only if $g^{ab} = g^c$, the strong Diffie Hellman problem is to compute $g^{ab} \in \mathbb{G}$.*

The advantage of an adversary $\mathcal{A}$ in solving the strong Diffie Hellman problem is defined as the probability with which $\mathcal{A}$ solves the above strong Diffie Hellman problem.

$$Adv_{\mathcal{A}}^{SDH} = Pr[\mathcal{A}(g, g^a, g^b) = g^{ab} | \mathcal{DDH}_{g,a}(.,.)]$$

The strong Diffie Hellman assumption holds in $\mathbb{G}$ if for all polynomial time adversaries $\mathcal{A}$, the advantage $Adv_{\mathcal{A}}^{SDH}$ is negligible.
Note: In pairing groups (also known as gap groups), the DDH oracle can be efficiently instantiated and hence the strong Diffie Hellman problem is equivalent to the Gap Diffie Hellman problem [14].

Definition 3. ***Stateful Public Key Encryption ($\mathcal{SPKE}$):***

A stateful public key encryption scheme $\mathcal{SPKE}$ is a tuple of five polynomial time algorithms `Setup`, `Key Generation`, `New State`, `Encryption` and `Decryption` (all are randomized algorithms except the last) such that:

`Setup`: This algorithm is run by an authority to generate the system parameters *params*.

`Key Generation`: This algorithm takes the system parameters *params* as input and outputs a pair of keys (sk, pk), namely the private key and the public key. This algorithm can be denoted as $(sk, pk) \leftarrow$ `Key Generation`$(params)$.

`New State`: This algorithm is run by any one who wants to encrypt the message, to generate a fresh state information st by taking *params* as input.

`Encryption`: As mentioned before, when a sender wants to send several messages to a receiver, he schedules the encryption in to sessions. In each session, a sender may wish to send some specific number of messages and this count is maintained by a variable called `encryption-count`. Each session has an associated state and each encryption in a session has an associated `encryption-count`. The `encryption-count` value is incremented by one for each encryption done in a session where the index is initiated to 1 at the beginning of each session. The index number is also sent as a component of the ciphertext. Thus, the extended form of encryption algorithm may be specified as $(c,$ `encryption-count`$) \leftarrow$ `Encryption`$(params, st, pk, m,$ `encryption-count`$)$.

`Decryption`: This algorithm takes the private key sk and a ciphertext c as input and executes two sub-algorithms $\overline{\texttt{Decryption}}$ and $\overline{\texttt{Verify}}$.

- Execute $m \leftarrow \overline{\texttt{Decryption}}(params, sk, c)$ and obtain the message m.
- Using m and c, perform $\{\texttt{True}, \texttt{False}\} \leftarrow \overline{\texttt{Verify}}(c, m)$.
- If the output of the $\overline{\texttt{Verify}}$ algorithm is `True`, output m as the message. If it outputs `False`, reject the ciphertext.

Note that in order to capture the notion of ciphertext verifiability, we have split the decryption algorithm into these two sub-algorithms.

Remark: We omit the `Public Key Check` algorithm in our paper and hence our framework has one less algorithm from the actual definition in [5]. This is because public key check is concerned with all Public Key Infrastructure (PKI) based encryption schemes. It is mandatory for a sender to perform this check in order to verify whether the components of public keys are elements of the underlying group and they comply with the system. Few checks like this are sometimes required for the security of standard schemes.

Definition 4. ***Game for CCA Security of Stateful PKE*** $(\mathcal{SPKE}_{\mathcal{A}}^{CCA}(\kappa))$***:*** *The game for CCA security of a stateful public key encryption scheme is between a challenger* $\mathcal{C}$ *and an adversary* $\mathcal{A}$*. Note that with out loss of generality we accept only the public keys that are valid, in the game. Public keys those are not well-formed will be rejected by public key check algorithm which we do not make explicit in our proofs. The game follows:*

Setup: $\mathcal{C}$ generates the system parameters $params$, generates a key pair $(sk, pk) \leftarrow$ `Key Generation`(κ) and $prams, pk$ are given to $\mathcal{A}$. (It should be noted that since $\mathcal{A}$ knows $params$, $\mathcal{A}$ could generate any number of private key / public key pairs but $\mathcal{A}$ does not know sk which is the private key corresponding to pk).

Phase I: $\mathcal{A}$ is given oracle access to the following oracles:

- `Encryption`$(params, st_i, m_j)$: $\mathcal{A}$ can make encryption queries for a message m_j in the state st_i, where ($j = 1$ to $\hat{m}$), ($i = 1$ to $\hat{n}$) and $\hat{m}$, $\hat{n}$ are the upper bounds for the number of messages that can be encrypted in a state and total number of states respectively. Note that encryption with respect to the public keys those are valid and passes the public key validity check alone are allowed.
- `Decryption`$(params, sk, c)$: Decryption for any ciphertext c can be queried by $\mathcal{A}$, irrespective of the state information, $\mathcal{C}$ should be able to provide the decryption.

Challenge: $\mathcal{A}$ gives $\mathcal{C}$ two messages m_0 and m_1 of the same length. $\mathcal{C}$ chooses a random bit $\beta \leftarrow \{0, 1\}$ and generates the challenge ciphertext $c^* \leftarrow$ `Encryption` $(params, st^*, pk, m_\beta)$ and gives it to $\mathcal{A}$.

Phase II: $\mathcal{A}$ continues to get oracle access to all ciphertexts for any message including m_0 and m_1 for the state information st^* through the encryption oracle `Encryption`$(params,\ st^*, pk, m_j)$, where $j \leq \hat{m}$. $\mathcal{A}$ also gets access to the

`Decryption` oracle, where it is allowed to query the decryption of any ciphertext $c \neq c^*$.

Guess: $\mathcal{A}$ outputs a bit β' finally.

$\mathcal{C}$ outputs 1, if $\beta = \beta'$ and 0 otherwise. A stateful public key encryption scheme $\mathcal{SPKE}$ has indistinguishable encryption under adaptive chosen ciphertext attack (CCA) if for all probabilistic polynomial time adversaries $\mathcal{A}$, there exists a negligible function $negl(.)$ such that:

$$Pr[\mathcal{SPKE}_{\mathcal{A}}^{CCA}(\kappa) \rightarrow 1] \leq \frac{1}{2} + negl(\kappa)$$

3 Stateful Public Key Encryption Scheme ($\mathcal{N} - \mathcal{SPKE}_1$)

In this section, we propose a compact CCA secure public key encryption scheme which provides shorter ciphertext and is stateful, in the sense that the same randomness can be used across a session that typically comprises encrypting different messages to the same receiver during the session. The ciphertext overhead of our scheme is slightly higher than the recent stateful public key encryption scheme reported in [4] with the added advantage that the ciphertext is verifiable after the decryption process. The main thing to be noticed is that this ciphertext verifiability property comes with almost the same computational complexity as the scheme in [4] and one more exponentiation for decryption which is strictly due to the additional verifiability property of our scheme. The description of the new stateful public key encryption scheme with verifiable ciphertext follows:

`Setup(κ)`: Let κ be the security parameter and $\mathbb{G}$ be a group of prime order q. Choose a generator $g \in_R \mathbb{G}$. Let $F : \mathbb{G} \rightarrow \{0,1\}^{\lambda}$, $G : \mathbb{G} \times \mathbb{G} \times \{0,1\}^{l_m} \times \{0,1\}^{\mu} \rightarrow \{0,1\}^{\lambda}$ and $H : \mathbb{G} \times \mathbb{G} \times \{0,1\}^{\lambda} \times \{0,1\}^{\mu} \rightarrow \{0,1\}^{l_m}$ be three cryptographic hash functions, where l_m represents the size of the message and μ is the size of the `encryption-count` used in the scheme. Here λ is a parameter such that any computation involving 2^{λ} or more steps is considered in-feasible in practice and the hash functions G and F offers collision resistance, first and second pre-image resistance with a range of λ-bits. Typically `encryption-count` may be a number from 1 to 2^{20} (this supports one million encryption per session) and hence the size of `encryption-count` will be utmost 20-bits. Set the system parameters as $params = \langle \kappa, q, g, \mathbb{G}, F, G, H, \rangle$.

`Key Generation(`*params*`)`: Choose $x \in_R \mathbb{Z}_q$ and compute $h = g^x$. The private key of the user is $sk = x$ and the public keys are $pk = \langle g, h \rangle$.

`New State(`*params*`)`: Let i represent the index of the current state and hence the current state will be referred as st_i. The sender generates the state information as follows:

- Choose $r_i \in_R \mathbb{Z}_q$
- Compute $u_i = F(g^{r_i})$
- Compute $s_i = r_i u_i$
- Compute $v_i = g^{s_i}$

The state information $st_i = \langle u_i, v_i, s_i \rangle$.

Encryption($params, st_i, pk, m$): Let encryption-count be a number which represents the invocation number of the encryption algorithm in the ith session. So during the start of each session, the value of encryption-count is initialized to 1 and incremented each time an encryption is performed during the session. The sender generates the ciphertext with params, state information, public key and the message as follows:

- Set $c_1 = v_i$
- Compute $w = h^{s_i}$
- Compute $c_2 = G(c_1, w, m, \texttt{encryption-count}) \oplus u_i$
- Compute $c_3 = H(c_1, w, c_2, \texttt{encryption-count}) \oplus m$

The ciphertext $c = \langle c_1, c_2, c_3, \texttt{encryption-count} \rangle$. We emphasize that the maximum number of encryption to be performed in a session will be determined by the sender. Thus, encryption-count is a user determined integer value and to perform one million encryption operations in a session, the value of index may be utmost 2^{20}. Hence, encryption-count may typically be a value from $1 \leq \texttt{encryption-count} \leq 2^{20}$ and thus of size less than 20-bits.

Decryption($params, sk, c$): The receiver decrypts the ciphertext with the private key by performing the following:

Decryption($params, sk, c$) :

- Compute $w' = c_1^{sk}$
- Compute $m' = c_3 \oplus H(c_1, w', c_2, \texttt{encryption-count})$

Verify(c, m') :

- Compute $u' = c_2 \oplus G(c_1, w', m', \texttt{encryption-count})$
- Check whether $u' \stackrel{?}{=} F(c_1^{(u')^{-1}})$.

If the Verify algorithm outputs True, return m', else return $\perp$.

Correctness: We have to show that the u' computed by the decryption algorithm passes the verification test $u' \stackrel{?}{=} F(c_1^{(u')^{-1}})$, if $u' = u_i = F(g^{r_i})$.

$$\begin{aligned} RHS = F(c_1^{(u')^{-1}}) &= F(v_i^{(u')^{-1}}) = F(g^{s_i (u')^{-1}}) = F(g^{r_i u_i (u')^{-1}}) \\ &= F(g^{r_i}) \text{ (If } u' = u_i = F(g^{r_i})) \\ &= u' = LHS \end{aligned}$$

Thus, the decryption will hold if $u' = u_i = F(g^{r_i})$.

Theorem 1. *The compact stateful public key encryption scheme $\mathcal{N} - \mathcal{SPKE}_1$ is IND-CCA secure in the random oracle model if the SDH problem is hard in $\mathbb{G}$. More specifically, if $\mathbb{G}$ is a $(t, \epsilon) - SDH$ group of order q then the $\mathcal{N} - \mathcal{SPKE}_1$ scheme is $(t', q_D, q_H, q_G, \epsilon')$-secure against IND-CCA adversary where $\epsilon' \geq \epsilon$ and*

$$t' \leq t - C_{\mathbb{G}}(q_H + q_G + 3q_E + 3q_D)$$

Proof: Let κ be the security parameter and $\mathbb{G}$ be a multiplicative group of order q, where $|q| = \kappa$. The challenger $\mathcal{C}$ is challenged with an instance of the SDH problem, say $(g, g^a, g^b) \in_R \mathbb{G}^3$ and access to a DDH oracle $\mathcal{DDH}_{g,a}(.,.)$ which on input g^b and g^c outputs True if and only if $g^{ab} = g^c$. Consider an adversary $\mathcal{A}$, who is capable of breaking the IND-CCA security of the scheme $\mathcal{N} - \mathcal{SPKE}_1$. $\mathcal{C}$ can make use of $\mathcal{A}$ to compute g^{ab}, by playing the following interactive game with $\mathcal{A}$.

Setup: $\mathcal{C}$ begins the game by setting up the system parameters as in the $\mathcal{N} - \mathcal{SPKE}_1$ scheme by performing the following:

- Sets the public key $h = g^a$ (where g^a is taken from the SDH instance).
- Hence, the private key is a implicitly.

$\mathcal{C}$ gives $\mathcal{A}$ the public keys $pk = \langle g, h \rangle$ and $\mathcal{C}$ also designs the three cryptographic hash functions F, G and H as random oracles $\mathcal{O}_F$, $\mathcal{O}_G$ and $\mathcal{O}_H$. $\mathcal{C}$ maintains three lists L_F, L_G and L_H in order to consistently respond to the queries to the random oracles $\mathcal{O}_F$, $\mathcal{O}_G$ and $\mathcal{O}_H$ respectively. A typical entry in list $L_{\hat{h}}$ will have the input parameters of hash functions $\hat{h}$ (for $\hat{h} = F, G$ and H) followed by the corresponding hash value returned as the response to the hash oracle query. In order to generate stateful encryption, $\mathcal{C}$ generates $\hat{n}$ tuple of state information and stores them in a state list L_{st}. Each tuple in the list corresponds to a state information. This is done as follows.

- For $i = 1$ to $\hat{n}$, $\mathcal{C}$ performs the following:
  - Choose $r_i \in_R \mathbb{Z}_q$, compute $k_i = g^{r_i}$, choose $u_i \in_R \mathbb{Z}_q$ and adds the tuple $\langle k_i, u_i \rangle$ in the list L_F, compute $s_i = r_i u_i$ and compute $v_i = g^{s_i}$.
  - The state information $st_i = \langle u_i, v_i, s_i, \texttt{encryption-count}_i = 1 \rangle$.
  - Store the tuple st_i in list L_{st}.

The game proceeds as per the $\mathcal{SPKE}_{\mathcal{A}}^{CCA}(\kappa)$ game.

Phase I: $\mathcal{A}$ performs a series of queries to the oracles provided by $\mathcal{C}$. The descriptions of the oracles and the responses given by $\mathcal{C}$ to the corresponding oracle queries by $\mathcal{A}$ are described below:

$\mathcal{O}_F(k \in \mathbb{G})$: To respond to this query, $\mathcal{C}$ checks whether a tuple of the form $\langle k, u \rangle$ exists in the list L_F. If a tuple of this form exists, $\mathcal{C}$ returns the corresponding u, else chooses $u \in_R \mathbb{Z}_q$, adds the tuple $\langle k, u \rangle$ to the list L_F and returns u to $\mathcal{A}$.

$\mathcal{O}_G(c_1 \in \mathbb{G}, w \in \mathbb{G}, m \in \{0,1\}^{l_m}, \texttt{encryption-count} \in \{0,1\}^{\mu})$: To respond to this query, $\mathcal{C}$ checks whether a tuple of the form $\langle c_1, w, m, \texttt{encryption-count}, h_1 \rangle$ exists in the list L_G. If a tuple of this form exists, $\mathcal{C}$ returns the corresponding h_1, else chooses $h_1 \in_R \{0,1\}^{\lambda}$, adds the tuple $\langle c_1, w, m, \texttt{encryption-count}, h_1 \rangle$ to the list L_G and returns h_1 to $\mathcal{A}$.

$\mathcal{O}_H(c_1 \in \mathbb{G}, w \in \mathbb{G}, c_2 \in \{0,1\}^{\lambda}, \texttt{encryption-count} \in \{0,1\}^{\mu})$: To respond to this query, $\mathcal{C}$ checks whether a tuple of the form $\langle c_1, w, c_2, \texttt{encryption-count}, h_2 \rangle$ exists in the list L_H. If a tuple of this form exists, $\mathcal{C}$ returns the corresponding h_2, else chooses $h_2 \in_R \{0,1\}^{l_m}$, adds the tuple $\langle c_1, w, c_2, \texttt{encryption-count}, h_2 \rangle$ to the list L_H and returns h_2 to $\mathcal{A}$.

$\mathcal{O}_{Encryption}(st_i, m_j)$: $\mathcal{A}$ may perform encryption with respect to any state information st_i, chosen by $\mathcal{C}$. $\mathcal{C}$ performs the following to encrypt the message m_j with respect to the state information st_i, where $i = 1$ to $\hat{n}$, where $\hat{n}$ is bound by the total number of states and $j = 1$ to $\hat{m}$ is bound by the number of messages that can be encrypted in one session:

- $\mathcal{C}$ retrieves the tuple st_i of the form $\langle u_i, v_i, s_i, \texttt{encryption-count}_i \rangle$ from L_{st}, sets $c_1 = v_i$, computes $w = h^{s_i}$.
- Chooses $h_1 \in_R \{0,1\}^\lambda$, adds the tuple $\langle c_1, w, m_j, \texttt{encryption-count}_i, h_1 \rangle$ to the list L_G and computes $c_2 = h_1 \oplus u_i$.
- Chooses $h_2 \in_R \{0,1\}^{l_m}$, adds the tuple $\langle c_1, w, c_2, \texttt{encryption-count}_i, h_2 \rangle$ to the list L_H and computes $c_3 = h_2 \oplus m_j$.
- Returns $c = \langle c_1, c_2, c_3 \rangle$ as the ciphertext, increments $\texttt{encryption-count}_i$ and updates the state information st_i.

$\mathcal{O}_{Decryption}(c)$: $\mathcal{C}$ does the following to decrypt $c = \langle c_1, c_2, c_3, \texttt{encryption-count} \rangle$:

- Retrieve the tuple $\langle c_1, w, c_2, \texttt{encryption-count}, h_2 \rangle$ from list L_H such that the output of the DDH oracle query $\mathcal{DDH}_{g,a}(w, c_1)$ is True and compute $m' = c_3 \oplus h_2$.
- Check whether a tuple of the form $\langle c_1, w, m, \texttt{encryption-count}, h_1 \rangle$, where w is the same as the w value retrieved from the tuple in the list L_H and m is equal to m' computed in the above step appears in the list L_G. If such a tuple appears, retrieve h_1 and compute $u' = c_2 \oplus h_1$.
- Check whether a tuple of the form $\langle k, u \rangle$, where $k = c_1^{u'^{-1}}$ and $u = u'$ appears in list L_F,
- If any of the required tuples did not appear in the lists L_F, L_G or L_H return $\perp$.

Challenge: At the end of ***Phase I***, $\mathcal{A}$ produces two messages m_0 and m_1 of equal length. $\mathcal{C}$ randomly chooses a bit $\beta \in_R \{0,1\}$ and computes a ciphertext c^* by performing the following steps:

- Choose $u \in_R \{0,1\}^\lambda$ and add the tuple $\langle g^b, u \rangle$ to the list L_F.
- Set $\texttt{encryption-count}^* = 1$ and compute $c_1^* = g^{bu}$.
- Choose $h_1 \in_R \{0,1\}^\lambda$, add the tuple $\langle c_1^*, -, m_\beta, \texttt{encryption-count}^*, h_1 \rangle$ in the list L_G and compute $c_2^* = h_1 \oplus u$.
- Choose $h_2 \in_R \{0,1\}^{l_m}$, add the tuple $\langle c_1^*, -, c_2, \texttt{encryption-count}^*, h_2 \rangle$ in the list L_H. and compute $c_3^* = h_2 \oplus m_\beta$.
- The state information $st^* = \langle u^* = u, v^* = g^{bu}, s^* = -, \texttt{encryption-count}^* \rangle$

Now, $c^* = \langle c_1^*, c_2^*, c_3^*, \texttt{encryption-count}^* \rangle$ is sent to $\mathcal{A}$ as the challenge ciphertext.

Phase II: $\mathcal{A}$ performs the second phase of interaction, where it makes polynomial number of queries to the oracles provided by $\mathcal{C}$ with the following condition:

- $\mathcal{A}$ should not query the $\mathcal{O}_{Decryption}$ oracle with c^* as input.

– $\mathcal{A}$ continues to get oracle access to all the oracles. It can also get the encryption for any message including m_0 and m_1 for the state information st^* through the encryption oracle $\texttt{Encryption}(params, st^*, pk, m_j)$.

The simulation of the $\mathcal{O}_G$, $\mathcal{O}_H$, $\mathcal{O}_{Encryption}$ and $\mathcal{O}_{Decryption}$ oracles are not same as in Phase I and hence we provide the details below:

$\mathcal{O}_G(c_1 \in \mathbb{G}, w \in \mathbb{G}, m \in \{0,1\}^{l_m}, \texttt{encryption-count} \in \{0,1\}^{\mu})$: To respond to this query, $\mathcal{C}$ performs the following:

– Check whether a tuple of the form $\langle c_1, w, m, \texttt{encryption-count}, h_1\rangle$ exists in the list L_G. If a tuple of this form exists, return the corresponding h_1, else,
  - If $c_1 = c_1^*$ then check with the DDH oracle whether $\mathcal{DDH}_{g,a}(w, c_1)$ is $\texttt{True}$. If the output is $\texttt{True}$, return $w^{u^{*-1}}$ as the solution to the SDH problem instance.
  - Else, choose $h_1 \in_R \{0,1\}^{\lambda}$, add the tuple $\langle c_1, w, m, \texttt{encryption-count}, h_1\rangle$ to the list L_G and return h_1 to $\mathcal{A}$.

$\mathcal{O}_H(c_1 \in \mathbb{G}, w \in \mathbb{G}, c_2 \in \{0,1\}^{\lambda}, \texttt{encryption-count} \in \{0,1\}^{\mu})$: To respond to this query, $\mathcal{C}$ performs the following:

– Check whether a tuple of the form $\langle c_1, w, c_2, \texttt{encryption-count}, h_2\rangle$ exists in the list L_H. If a tuple of this form exists, $\mathcal{C}$ returns the corresponding h_2, else,
  - If $c_1 = c_1^*$ then check with the DDH oracle whether $\mathcal{DDH}_{g,a}(w, c_1)$ is $\texttt{True}$. If the output is $\texttt{True}$, return $w^{u^{*-1}}$ as the solution to the SDH problem instance.
  - Else, choose $h_2 \in_R \{0,1\}^{l_m}$, add the tuple $\langle w, c_2, \texttt{encryption-count}, h_2\rangle$ to the list L_H and return h_2 to $\mathcal{A}$.

$\mathcal{O}_{Encryption}(st_i, m_j)$: $\mathcal{A}$ may perform encryption with respect to any state information st_i including st^*, chosen by $\mathcal{C}$. $\mathcal{C}$ performs the following to encrypt the message m_j with respect to the state information st_i:

– If $st_i \neq st^*$ then encryption is done as in Phase I
– If $st_i = st^*$ then perform the following:
  - Retrieve the tuple st^* of the form $st^* = \langle u^* = u, v^* = g^{bu}, s^* = -, \texttt{encryption-count}^*\rangle$ from L_{st} and set $c_1 = v^*$.
  - Choose $h_1 \in_R \{0,1\}^{\lambda}$, add the tuple $\langle c_1, -, m_j, \texttt{encryption-count}^*, h_1\rangle$ to the list L_G and compute $c_2 = h_1 \oplus u^*$.
  - Choose $h_2 \in_R \{0,1\}^{l_m}$, add the tuple $\langle c_1, -, c_2, \texttt{encryption-count}^*, h_2\rangle$ to the list L_H and compute $c_3 = h_2 \oplus m_j$.
  - Return $c = \langle c_1, c_2, c_3\rangle$ as the ciphertext, increment $\texttt{encryption-count}^*$ and update the state information st^*.

$\mathcal{O}_{Decryption}(c)$: $\mathcal{C}$ does the following to decrypt $c=\langle c_1, c_2, c_3, \texttt{encryption-count}\rangle$:

If $c_1 \neq c_1^*$ then decryption is done as in Phase - I.
If $c_1 = c_1^*$ then perform the following:

- Retrieve the tuple of the form $\langle c_1, w, c_2, \texttt{encryption-count}, h_2 \rangle$ from list L_H, such that the output of the DDH oracle query, $\mathcal{DDH}_{g,a}(w, c_1)$ is $\texttt{True}$. If the retrieved tuple is of the form $\langle c_1, -, c_2, \texttt{encryption-count}, h_2 \rangle$ then it was the tuple generated by $\mathcal{C}$ during an encryption oracle query in the phase II. Note that $\mathcal{C}$ can even work consistently with the tuple of this form. In this case, $\mathcal{C}$ chooses the value h_2 without consulting the DDH oracle. Compute $m' = c_3 \oplus h_2$.
- Check whether a tuple of the form $\langle c_1, w, m, \texttt{encryption-count}, h_1 \rangle$, where w is the same as the w value retrieved from the tuple in the list L_H and m is equal to m' computed in the above step appears in the list L_G. If such a tuple appears, retrieve h_1 and compute $u' = c_2 \oplus h_1$. (Note that even in this case $\mathcal{C}$ works consistently with the tuple of the form $\langle c_1, -, m, \texttt{encryption-count}, h_1 \rangle$)
- Check whether a tuple of the form $\langle k, u \rangle$, where $k = c_1^{u'^{-1}}$ and $u = u'$ appears in list L_F,
- If any of the required tuples did not appear in the lists L_F, L_G or L_H return $\perp$.
- If in the process a tuple of the form $\langle c_1, w, c_2, \texttt{encryption-count}, h_2 \rangle$ appeared in the list L_G and a tuple of the form $\langle c_1, w, m, \texttt{encryption-count}, h_1 \rangle$ appeared in the list L_H with $\mathcal{DDH}_{g,a}(w, c_1)$ is $\texttt{True}$, then output w as the output to the SDH problem.

Lemma 1. *The decryption oracle responds correctly to well-formed ciphertexts and rejects invalid ciphertexts.*

Proof: Let us consider $c = \langle c_1, c_2, c_3, \texttt{encryption-count} \rangle$ is a well-formed ciphertext. In order to construct c, $\mathcal{A}$ should have done the following:

- Chosen $r \in_R \mathbb{Z}_q$ and queried the $\mathcal{O}_F$ oracle with $k = g^r$. Thus a tuple of the form $\langle k, u \rangle$ should appear in L_F.
- $\mathcal{A}$ should have computed $c_1 = g^{ru}$, $w = h^{ru}$ and queried the $\mathcal{O}_G$ oracle with $\langle c_1, w, m, \texttt{encryption-count} \rangle$ as input and received h_1 corresponding to this input.
- $\mathcal{A}$ should have computed $c_2 = h_1 \oplus u$ and queried the $\mathcal{O}_H$ oracle with $\langle c_1, w, c_2, \texttt{encryption-count} \rangle$ as input and received h_2 corresponding to this input.

During the decryption, $\mathcal{C}$ retrieves the corresponding tuples, one from the lists L_G and L_H for which both the w values are same and checks whether the output of the DDH oracle query, $\mathcal{DDH}_{g,a}(w, c_1)$ is $\texttt{True}$. For a well formed ciphertext, this check holds because,

$$c_1 = g^{ru} \tag{1}$$

$$w = h^{ru} = g^{aru} \tag{2}$$

From equations (1) and (2) it is clear that for a well formed ciphertext, this check holds and working with the corresponding h_1 and h_2 will properly yield the message during decryption. Else, the ciphertext will be rejected. □

Guess: At the end of ***Phase II***, $\mathcal{A}$ produces a bit β' to $\mathcal{C}$, but $\mathcal{C}$ ignores the response and performs the following to output the solution for the SDH problem instance.

- When a query to the $\mathcal{O}_G$ oracle, with $(c_1, w, m, \texttt{encryption-count})$ as input is made, $\mathcal{C}$ computes $g' = w^{u^{*-1}} = g^{ab}$ and checks whether $\mathcal{DDH}_{g,a}(g', g^b) \stackrel{?}{=}$ `True`. Alternatively, $\mathcal{C}$ can also perform the same with $\mathcal{O}_H$ oracle queries.
- Outputs the corresponding g' value for which the above check holds as the solution for the SDH problem instance.

Since there is no *Abort* during the simulation, $\mathcal{C}$ obtains the solution to the SDH problem with almost the same advantage of $\mathcal{A}$ in the IND-CCA game. Let q_G be the number of $\mathcal{O}_G$ oracle queries, q_H be the number of $\mathcal{O}_H$ oracle queries, q_E be the number of $\mathcal{O}_{Encryption}$ oracle queries and q_D the number of $\mathcal{O}_{Decryption}$ oracle queries. The maximum number of queries that are made to $\mathcal{O}_G$ oracle and $\mathcal{O}_H$ oracle is $q_G+q_E+q_D$ and $q_H+q_E+q_D$ respectively. The total number of queries made to the $\mathcal{O}_G$, $\mathcal{O}_H$, $\mathcal{O}_{Encryption}$ and $\mathcal{O}_{Decryption}$ oracle is $[q_G + q_E + q_D] + [q_H + q_E + q_D] + [q_E] + [q_D] = q_G + q_H + 3q_E + 3q_D$. Thus, if there exists an algorithm $\mathcal{A}$ that $(t', q_D, q_H, q_G, \epsilon')$-breaks the IND-CCA security of $\mathcal{N} - \mathcal{SPKE}_1$ scheme, then there exists an algorithm $\mathcal{C}$ that (t, ϵ)-breaks the SDH problem in $\mathbb{G}$, where $\epsilon' \geq \epsilon$ and

$$t' \leq t - C_{\mathbb{G}}(q_G + q_H + 3q_E + 3q_D)$$

$C_{\mathbb{G}}$ is a constant that depends on the group $\mathbb{G}$. ■

4 Stateful Public Key Encryption Scheme ($\mathcal{N} - \mathcal{SPKE}_2$)

In this section, we propose a compact CCA secure public key encryption scheme whose security is based on the CDH problem.

`Setup(`κ`)`: Same as the `Setup(.)` algorithm of $\mathcal{N} - \mathcal{SPKE}_1$.

`Key Generation(`$params$`)`: Choose $x, y \in_R \mathbb{Z}_q$, compute $g_1 = g^x$ and $g_2 = g^y$. The private key of the user is $sk = \langle x, y\rangle$ and the public keys are $pk = \langle g, g_1, g_2\rangle$.

`New State(`$params$`)`: Same as the `New State(.)` algorithm of $\mathcal{N} - \mathcal{SPKE}_1$.

`Encryption(`$params, st_i, pk, m$`)`: Let `encryption-count` be a number as defined in $\mathcal{N} - \mathcal{SPKE}_1$. The sender generates the ciphertext as follows:

- Set $c_1 = v_i$, compute $w_1 = g_1^{s_i}$ and $w_2 = g_2^{s_i}$
- Compute $c_2 = G(c_1, w_1, m, \texttt{encryption-count}) \oplus u_i$
- Compute $c_3 = H(c_1, w_2, c_2, \texttt{encryption-count}) \oplus m$

The ciphertext $c = \langle c_1, c_2, c_3, \texttt{encryption-count}\rangle$.

$\underline{\texttt{Decryption}(params, sk, c)}$: The receiver decrypts the ciphertext with the private key by performing the following:

$\underline{\texttt{Decryption}(params, sk, c)}$: Compute $w'_1 = c_1^x$, $w'_2 = c_1^y$ and $m' = c_3 \oplus H(c_1, w'_2, c_2, \texttt{encryption-count})$

$\underline{\texttt{Verify}(c, m')}$: Compute $u' = c_2 \oplus G(c_1, w'_1, m', \texttt{encryption-count})$ and check whether $u' \stackrel{?}{=} F(c_1^{(u')^{-1}})$.

If the $\underline{\texttt{Verify}}$ algorithm outputs True, return m', else return $\perp$.

Theorem 2. *The compact stateful public key encryption scheme $\mathcal{N}-\mathcal{SPKE}_2$ is IND-CCA secure in the random oracle model if the CDH problem is hard in $\mathbb{G}$. More specifically, if $\mathbb{G}$ is a $(t,\epsilon)-CDH$ group of order q then the $\mathcal{N}-\mathcal{SPKE}_2$ scheme is $(t', q_D, q_H, q_G, \epsilon')$-secure against IND-CCA adversary where $\epsilon' \geq \epsilon$ and*

$$t' \leq t - C_{\mathbb{G}}(q_H + q_G + 3q_E + 3q_D)$$

Let κ be the security parameter and $\mathbb{G}$ be a multiplicative group of order q, where $|q| = \kappa$. The challenger $\mathcal{C}$ is challenged with an instance of the CDH problem, say $(g, g^a, g^b) \in_R \mathbb{G}^3$. Consider an adversary $\mathcal{A}$, who is capable of breaking the IND-CCA security of the scheme $\mathcal{N}-\mathcal{SPKE}_2$. $\mathcal{C}$ can make use of $\mathcal{A}$ to compute g^{ab}, by playing the following interactive game with $\mathcal{A}$. The proof revolves around the technique of [7].

Setup: $\mathcal{C}$ chooses $z_1, z_2 \in_R \mathbb{Z}_q$, sets the public key $g_1 = g^a$ (where g^a is taken from the CDH instance) and computes $g_2 = g^{z_1}/g^{az_2}$. Therefore, the private keys are a and $(z_1 - az_2)$ implicitly. $\mathcal{C}$ gives $\mathcal{A}$ the public keys $pk = \langle g, g_1, g_2 \rangle$ and designs the three cryptographic hash functions F, G and H as random oracles $\mathcal{O}_F$, $\mathcal{O}_G$ and $\mathcal{O}_H$ as in Theorem 1. In order to generate stateful encryption, $\mathcal{C}$ generates $\hat{n}$ tuples of state information and stores them in a state list L_{st} as in Theorem 1. The game proceeds as per the $\mathcal{SPKE}_{\mathcal{A}}^{CCA}(\kappa)$ game.

Phase I: $\mathcal{A}$ performs a series of queries to the oracles provided by $\mathcal{C}$. The descriptions of the hash oracles and the responses given by $\mathcal{C}$ to the corresponding queries by $\mathcal{A}$ are similar to the simulation in Theorem 1.

$\mathcal{O}_{Encryption}(st_i, m_j)$: Similar to the simulation in Theorem 1.

$\mathcal{O}_{Decryption}(c)$: $\mathcal{C}$ does the following to decrypt $c = \langle c_1, c_2, c_3, \texttt{encryption-count} \rangle$:

- Retrieve the tuples of the form $\langle c_1, w_1, m, \texttt{encryption-count}, h_1 \rangle$ from the list L_G. Consider that there are $\hat{n}_G$ such tuples. Choose the corresponding (w_{1i}, h_{1i}) values, for $i = 1$ to $\hat{n}_G$.
- Retrieve the tuples of the form $\langle c_1, w_2, c_2, \texttt{encryption-count}, h_2 \rangle$ from the list L_H. Consider that there are $\hat{n}_H$ such tuples. Choose the corresponding (w_{2j}, h_{2j}) values, for $j = 1$ to $\hat{n}_H$.
- For $i = 1$ to $\hat{n}_G$
  - For $j = 1$ to $\hat{n}_H$
    - Check whether $w_{2j} \stackrel{?}{=} c_1^{z_1}/w_{1i}^{z_2}$.
    - If the check holds for some index $\hat{i}$ and $\hat{j}$, choose the corresponding $h_{1\hat{i}}$ and $h_{2\hat{j}}$. If the check does not hold for any tuple then reject the ciphertext c.

- Compute $m' = c_3 \oplus h_{2\hat{j}}$.
- Retrieve the value m from the tuple $\langle c_1, w_{1\hat{i}}, m, \texttt{encryption-count}, h_{1\hat{i}} \rangle$ in the list L_G.
- If $(m = m')$, then compute $u' = c_2 \oplus h_{1\hat{i}}$, else reject the ciphertext c..
- Check whether a tuple of the form $\langle k, u \rangle$, where $k = c_1^{u'^{-1}}$ and $u = u'$ appears in list L_F. If it appears accept m' and return it as the message.
- If any of the required tuples did not appear in L_F, L_G or L_H, return $\perp$.

The proof for consistency of the decryption oracle is given in the full version.

Challenge: At the end of ***Phase I***, $\mathcal{A}$ produces two messages m_0 and m_1 of equal length. $\mathcal{C}$ randomly chooses a bit $\beta \in_R \{0,1\}$ and computes a ciphertext c^* by performing the following steps:

- Choose $u \in_R \{0,1\}^\lambda$ and add the tuple $\langle g^b, u \rangle$ to the list L_F.
- Set $\texttt{encryption-count}^* = 1$ and compute $c_1^* = g^{bu}$
- Choose $h_1 \in_R \{0,1\}^\lambda$, add the tuple $\langle c_1^*, -, m_\beta, \texttt{encryption-count}^*, h_1 \rangle$ in the list L_G and compute $c_2^* = h_1 \oplus u$.
- Choose $h_2 \in_R \{0,1\}^{l_m}$, add the tuple $\langle c_1^*, -, c_2, \texttt{encryption-count}^*, h_2 \rangle$ in the list L_H and compute $c_3^* = h_2 \oplus m_\beta$.
- The state information $st^* = \langle u^* = u, v^* = g^{bu}, s^* = -, \texttt{encryption-count}^* \rangle$

Now, $c^* = \langle c_1^*, c_2^*, c_3^*, \texttt{encryption-count}^* \rangle$ is sent to $\mathcal{A}$ as the challenge ciphertext.

Phase II: $\mathcal{A}$ performs the second phase of interaction, where it makes polynomial number of queries to the oracles provided by $\mathcal{C}$ with the following condition:

- $\mathcal{A}$ should not query the $\mathcal{O}_{Decryption}$ oracle with c^* as input.
- $\mathcal{A}$ continues to get oracle access to all the oracles. It can also get the encryption for any message including m_0 and m_1 for the state information st^* through the encryption oracle $\texttt{Encryption}(params, st^*, pk, m_j)$.

The simulation of the $\mathcal{O}_G$, $\mathcal{O}_H$, $\mathcal{O}_{Encryption}$ and $\mathcal{O}_{Decryption}$ oracles are not same as in Phase I and hence we provide the details below:

$\mathcal{O}_G(c_1 \in \mathbb{G}, w_1 \in \mathbb{G}, m \in \{0,1\}^{l_m}, \texttt{encryption-count} \in \{0,1\}^\mu)$: To respond to this query, $\mathcal{C}$ performs the following:

- If $c_1 \neq c_1^*$ then
  - If a tuple of the form $\langle c_1, w_1, m, \texttt{encryption-count}, h_1 \rangle$ exists in the list L_G, return the corresponding h_1.
  - Else, choose $h_1 \in_R \{0,1\}^\lambda$, add the tuple $\langle c_1, w_1, m, \texttt{encryption-count}, h_1 \rangle$ to the list L_G and return h_1 to $\mathcal{A}$.
- If $c_1 = c_1^*$ then
  - If a tuple of the form $\langle c_1, w_2, c_2, \texttt{encryption-count}, h_2 \rangle$ exists in the list L_H, check whether $w_2 \stackrel{?}{=} c_1^{z_1}/w_1^{z_2}$. If the check holds then return $w_1^{u^{*-1}}$ as the solution to the CDH problem instance.
  - If a tuple of the form $\langle c_1, w_2, c_2, \texttt{encryption-count}, h_2 \rangle$ does not exist in the list L_H perform the following:

- * Choose $h_1 \in_R \{0,1\}^\lambda$.
- * Add the tuple $\langle c_1, w_1, m, \texttt{encryption-count}, h_1\rangle$ to the list L_G.
- * Return h_1 to $\mathcal{A}$.

$\mathcal{O}_H(c_1 \in \mathbb{G}, w_2 \in \mathbb{G}, c_2 \in \{0,1\}^\lambda, \texttt{encryption-count} \in \{0,1\}^\mu)$: To respond to this query, $\mathcal{C}$ performs the following:

- If $c_1 \neq c_1^*$ then
  - If a tuple of the form $\langle c_1, w_2, c_2, \texttt{encryption-count}, h_2\rangle$ exists in the list L_H, return the corresponding h_2.
  - Else, choose $h_2 \in_R \{0,1\}^{l_m}$, add the tuple $\langle c_1, w_2, c_2, \texttt{encryption-count}, h_2\rangle$ to the list L_H and return h_2 to $\mathcal{A}$.
- If $c_1 = c_1^*$ then
  - If a tuple of the form $\langle c_1, w_1, m, \texttt{encryption-count}, h_1\rangle$ exists in the list L_G, check whether $w_2 \stackrel{?}{=} c_1^{z_1}/w_1^{z_2}$. If the check holds then return $w_1^{u^{*-1}}$ as the solution to the CDH problem instance.
  - If a tuple of the form $\langle c_1, w_1, m, \texttt{encryption-count}, h_2\rangle$ does not exist in the list L_G perform the following:
    - * Choose $h_2 \in_R \{0,1\}^{l_m}$.
    - * Add the tuple $\langle c_1, w_2, c_2, \texttt{encryption-count}, h_2\rangle$ to the list L_H.
    - * Return h_2 to $\mathcal{A}$.

$\mathcal{O}_{Encryption}(st_i, m_j)$: Similar to the simulation in Theorem 1. $\mathcal{O}_{Decryption}(c)$: In the case where $(c_1 \neq c_1^*)$, $\mathcal{C}$ responds as in phase I. If $(c_1 = c_1^*)$, $\mathcal{C}$ performs the following to decrypt the ciphertext $c = \langle c_1, c_2, c_3, \texttt{encryption-count}\rangle$:

- Retrieve the tuples of the form $\langle c_1, w_1, m, \texttt{encryption-count}, h_1\rangle$ from the list L_G. Consider that there are $\hat{n}_G$ such tuples. Choose the corresponding (w_{1i}, h_{1i}) values, for $i = 1$ to $\hat{n}_G$. (If the retrieved tuple is of the form $\langle c_1, -, m, \texttt{encryption-count}, h_1\rangle$ then it was the tuple generated by $\mathcal{C}$ during an encryption oracle query in phase II. Note that $\mathcal{C}$ can even work consistently with the tuple of this form without performing the test mentioned below. Further note that for a fixed c_1 and encryption-count, there will be only one such tuple in the list L_G.)
- Retrieve the tuples of the form $\langle c_1, w_2, c_2, \texttt{encryption-count}, h_2\rangle$ from the list L_H. Consider that there are $\hat{n}_H$ such tuples. Choose the corresponding (w_{2j}, h_{2j}) values, for $j = 1$ to $\hat{n}_H$. (Even in this case, if the retrieved tuple is of the form $\langle c_1, -, c_2, \texttt{encryption-count}, h_2\rangle$, the tuple was generated by $\mathcal{C}$ during an encryption oracle query in phase II. $\mathcal{C}$ can even work consistently with the tuple of this form without performing the test mentioned below. This is because for a fixed c_1, c_2 and encryption-count there will be only one tuple of this form available in the list L_H.)
- For $i = 1$ to $\hat{n}_G$
  - For $j = 1$ to $\hat{n}_H$
    - * Check whether $w_{2j} \stackrel{?}{=} c_1^{z_1}/w_{1i}^{z_2}$.

  * If the check holds for some index $\hat{\imath}$ and $\hat{\jmath}$, choose the corresponding $h_{1\hat{\imath}}$ and $h_{2\hat{\jmath}}$ and return $w_{1\hat{\imath}}^{u^{*-1}}$ as the solution to the CDH problem instance. If the check does not hold for any tuple then reject the ciphertext c.
- Compute $m' = c_3 \oplus h_{2\hat{\jmath}}$.
- Retrieve m from the tuple of the form $\langle c_1, w_{1\hat{\imath}}, m, \texttt{encryption-count}, h_{1\hat{\imath}} \rangle$ available in list L_G.
- If $(m = m')$, then compute $u' = c_2 \oplus h_{1\hat{\imath}}$, else reject the ciphertext c.
- Check whether a tuple of the form $\langle k, u \rangle$, where $k = c_1^{u'^{-1}}$ and $u = u'$ appears in list L_F. If it appears accept m' and return it as the message corresponding to c.
- If any of the required tuples did not appear in the lists L_F, L_G or L_H return $\perp$.

Guess: At the end of ***Phase II***, $\mathcal{A}$ produces a bit β' to $\mathcal{C}$, but $\mathcal{C}$ ignores the response and performs the following to output the solution for the CDH problem instance.

- Retrieves the tuples of the form $\langle c_1^*, w_1, m, \texttt{encryption-count} \rangle$ from the list L_G and checks whether a tuple of the form $\langle c_1^*, w_2, c_2^*, \texttt{encryption-count}, h_2 \rangle$ is available in list L_H. If a tuple of this form exists in the list L_H, $\mathcal{C}$ checks whether $w_2 \stackrel{?}{=} c_1^{*z_1}/w_1^{*z_2}$. If the check holds, compute $g' = w_1^{*u^{*-1}}$ as the solution to the CDH problem.
- Alternatively, retrieves the tuples of the form $\langle c_1^*, w_2, c_2^*, \texttt{encryption-count} \rangle$ and checks whether a tuple of the form $\langle c_1^*, w_1, m_\beta, \texttt{encryption-count}, h_1 \rangle$ is available in list L_G. If a tuple of this form exists in the list L_G, $\mathcal{C}$ checks whether $w_2 \stackrel{?}{=} c_1^{*z_1}/w_1^{*z_2}$. If the check holds, compute $g' = w_1^{*u^{*-1}} = g^{ab}$ as the solution to the CDH problem. (The correctness is given in the full version of the paper.)

Thus, $\mathcal{C}$ obtains the solution to the CDH problem with almost the same advantage of $\mathcal{A}$ in the IND-CCA game. The argument is similar to Theorem 1. ■

5 Comparison with Existing Schemes

In this section, we compare the new stateful public key encryption scheme $(\mathcal{N} - \mathcal{SPKE}_1)$, proposed in section 3 with the existing schemes related to them respectively. The legends are E - Exponentiation, B - Bilinear Pairing, H - Hash computation, $|\mathbb{G}|$ - Cardinality of the group $\mathbb{G}$, $||\mathbb{G}|| = log|\mathbb{G}|$ - Size of one group element, MAC - MAC Computation, $|MAC|$ - Size of a MAC value, $|R|$ - Size of a random string usually λ, $CBDH$ - Computational Bilinear Diffie Hellman Problem, $GBDH$ - Gap Bilinear Diffie Hellman Problem, GDH - Gap Diffie Hellman Problem and SDH - Strong Diffie Hellman Problem.

This table summarizes the computation complexity and ciphertext overhead of the stateful public key encryption schemes by Bellare et al. (BKS_{st} [5]), Baek et al. (BCZ_{st} [4]), $\mathcal{N} - \mathcal{SPKE}_1$ and $\mathcal{N} - \mathcal{SPKE}_2$. Here, μ is the size of the

Table 1. Stateful Public Key Encryption Schemes with Short Ciphertext

Scheme	Encryption Cost	Decryption Cost	Ciphertext Expansion	Assumption	Ciphertext Verifiability
BKS_{st} [5]	$1H + 1MAC$	$1E + 1H + 1MAC$	$\|\|\mathbb{G}\|\| + \|MAC\| + \|R\|$	GDH	YES
BCZ_{st} [4]	$2H$	$1E + 2H$	$\|\|\mathbb{G}\|\| + \lambda$	GDH	NO
$\mathcal{N} - \mathcal{SPKE}_1$	$2H$	$2E + 2H$	$\|\|\mathbb{G}\|\| + \lambda + \mu$	SDH	YES
$\mathcal{N} - \mathcal{SPKE}_2$	$2H$	$3E + 2H$	$\|\|\mathbb{G}\|\| + \lambda + \mu$	CDH	YES

index used in our scheme. To ensure ciphertext verifiability, it is possible to append 80-bits of known value (usually 80-bits of 0's) to the plaintext while encrypting and checking whether decryption produces those 80-bits at the end of the message. If this technique is used in the BCZ_{st} scheme, the ciphertext expansion will be $||\mathbb{G}||+\lambda+$'80−bits'. However, in the new schemes $\mathcal{N} - \mathcal{SPKE}_1$ and $\mathcal{N} - \mathcal{SPKE}_2$, the size of the `encryption-count` ($\mu$), is upper bound by 20-bits and hence can take a value starting from 1−bit, which is a considerable reduction for resource constrained devices like sensors, PDAs and mobile devices. The ciphertext overhead is also smaller than that of the BKS_{st} scheme, that offers ciphertext verifiability.

6 Conclusion

Two new stateful public key encryption schemes with ciphertext verifiability were proposed and the security of these schemes were supported by a formal proof. Our first stateful public key encryption scheme is proved to be secure assuming the SDH problem and the second assuming the CDH problem. However, the ciphertext overhead of both the schemes turns out to be the same. We have proved both the schemes in the random oracle model. An interesting open issue that can be looked at is designing a public key encryption scheme which offers compact ciphertext (ciphertext overhead of one group element) with ciphertext verifiability.

Acknowledgement. We would like to thank the anonymous referees of IWSEC-2012 for their insightful remarks, which helped in improving our paper greatly. Special thanks goes to Prof. Rui Zhang for shephard heading our paper and helping in improving the paper.

References

1. Abdalla, M., Bellare, M., Rogaway, P.: The Oracle Diffie-Hellman Assumptions and an Analysis of DHIES. In: Naccache, D. (ed.) CT-RSA 2001. LNCS, vol. 2020, pp. 143–158. Springer, Heidelberg (2001)

2. Abe, M., Gennaro, R., Kurosawa, K., Shoup, V.: Tag-KEM/DEM: A New Framework for Hybrid Encryption and A New Analysis of Kurosawa-Desmedt KEM. In: Cramer, R. (ed.) EUROCRYPT 2005. LNCS, vol. 3494, pp. 128–146. Springer, Heidelberg (2005)
3. Abe, M., Kiltz, E., Okamoto, T.: Compact CCA-Secure Encryption for Messages of Arbitrary Length. In: Jarecki, S., Tsudik, G. (eds.) PKC 2009. LNCS, vol. 5443, pp. 377–392. Springer, Heidelberg (2009)
4. Baek, J., Chu, C.-K., Zhou, J.: On Shortening Ciphertexts: New Constructions for Compact Public Key and Stateful Encryption Schemes. In: Kiayias, A. (ed.) CT-RSA 2011. LNCS, vol. 6558, pp. 302–318. Springer, Heidelberg (2011)
5. Bellare, M., Kohno, T., Shoup, V.: Stateful public-key cryptosystems: how to encrypt with one 160-bit exponentiation. In: ACM Conference on Computer and Communications Security, pp. 380–389. ACM (2006)
6. Boyen, X.: Miniature CCA2 PK Encryption: Tight Security Without Redundancy. In: Kurosawa, K. (ed.) ASIACRYPT 2007. LNCS, vol. 4833, pp. 485–501. Springer, Heidelberg (2007)
7. Cash, D., Kiltz, E., Shoup, V.: The twin diffie-hellman problem and applications. Journal of Cryptology 22(4), 470–504 (2009)
8. Cramer, R., Shoup, V.: Design and analysis of practical public-key encryption schemes secure against chosen ciphertext attack. SIAM Journal on Computing 33(1), 167–226 (2003)
9. Fujisaki, E., Okamoto, T.: How to Enhance the Security of Public-Key Encryption at Minimum Cost. In: Imai, H., Zheng, Y. (eds.) PKC 1999. LNCS, vol. 1560, pp. 53–68. Springer, Heidelberg (1999)
10. Fujisaki, E., Okamoto, T.: Secure Integration of Asymmetric and Symmetric Encryption Schemes. In: Wiener, M. (ed.) CRYPTO 1999. LNCS, vol. 1666, pp. 537–554. Springer, Heidelberg (1999)
11. Halevi, S., Rogaway, P.: A Tweakable Enciphering Mode. In: Boneh, D. (ed.) CRYPTO 2003. LNCS, vol. 2729, pp. 482–499. Springer, Heidelberg (2003)
12. Kiltz, E., Malone-Lee, J.: A General Construction of IND-CCA2 Secure Public Key Encryption. In: Paterson, K.G. (ed.) Cryptography and Coding 2003. LNCS, vol. 2898, pp. 152–166. Springer, Heidelberg (2003)
13. Kurosawa, K., Matsuo, T.: How to Remove MAC from DHIES. In: Wang, H., Pieprzyk, J., Varadharajan, V. (eds.) ACISP 2004. LNCS, vol. 3108, pp. 236–247. Springer, Heidelberg (2004)
14. Okamoto, T., Pointcheval, D.: The Gap-Problems: A New Class of Problems for the Security of Cryptographic Schemes. In: Kim, K. (ed.) PKC 2001. LNCS, vol. 1992, pp. 104–118. Springer, Heidelberg (2001)
15. Pass, R., Shelat, A., Vaikuntanathan, V.: Relations Among Notions of Non-malleability for Encryption. In: Kurosawa, K. (ed.) ASIACRYPT 2007. LNCS, vol. 4833, pp. 519–535. Springer, Heidelberg (2007)

Structured Encryption for Conceptual Graphs*

Geong Sen Poh, Moesfa Soeheila Mohamad, and Muhammad Reza Z'aba

Cryptography Lab, ADAM, MIMOS Berhad
Technology Park Malaysia, 57000 Kuala Lumpur, Malaysia
{gspoh,soeheila.mohamad,reza.zaba}@mimos.my

Abstract. We investigate the problem of privately searching encrypted data that is structured in the form of knowledge. Our rationale in such an investigation lies on the potential emergence of knowledge-based search using natural language, which makes content searches more effective and is context-aware when compared with existing keyword searches. With knowledge-based search, indexes and databases will consist of data stored using knowledge representation techniques such as description logics and conceptual graphs. This leads naturally to the issue of how to privately search this data, especially when most existing searchable encryption schemes are keyword-based. We propose the first construction with CQA2-security for searching encrypted knowledge, where the knowledge is represented in a well-established formalism known as basic conceptual graphs. Our proposals are based on structured encryption schemes of Chase and Kamara [8].

1 Introduction

Most existing search techniques query data based on keywords, but searches based on natural language would be more effective in providing context-aware results from documents. One way to realise natural language searches is to represent the underlying data in a form of *knowledge* with knowledge retrieval capabilities, so that a computing device may process and *understand* them. Knowledge, in this case, is traditionally defined as *"justified true belief or true opinion combined with reason"*. Models to capture these beliefs is known as *knowledge representation and reasoning* [9,14]. One of the main representations is *conceptual graphs* (CGs) [9], in which a sentence in a document is structured in a graph format with the edges representing "relations" between the words. Query methods are defined for CGs using graph homomorphism. We discuss CG in more details in Section 3.1.

In this scenario the database contains documents represented as CGs. When such knowledge database is stored in the cloud, we would want it to be encrypted while at the same time searchable without the cloud provider being able to access the searched knowledge. There are many existing schemes for searching encrypted data but most of them are constructed to address keyword-based

* Work supported by MOSTI eScience Fund 01-03-04-SF0020.

G. Hanaoka and T. Yamauchi (Eds.): IWSEC 2012, LNCS 7631, pp. 105–122, 2012.

search on text documents only [1,2,4,5,10,12,16,18], except for the recent structured encryption schemes proposed by Chase and Kamara [8]. The schemes by Chase and Kamara generalise symmetric searchable encryption (SSE) to also work for arbitrarily structured data. Practical applications mentioned that can utilise their schemes include social network and labeled web graphs.

None of these schemes, however, examines searchable encryption on data represented as knowledge. We propose such schemes by adapting Chase and Kamara's schemes and taking CGs as the knowledge representation models. We note that CGs is a reasonable choice for knowledge representation, given its well-established nature as discussed in [9,19,20].

Our Results. In the following we summarize our contributions.

1. We introduce searchable encryption for data represented as knowledge. In particular, we extend applications of structured encryption [8] to include knowledge represented in CGs.
2. We propose a main construction called a *Message Query* (`MeQ`) scheme. It queries an encrypted document database and retrieves encrypted document matching the query. The query is a CG. In other words, given a phrase (or a sentence) structured as a CG as the query, the scheme returns multiple documents that contain the query, phrases and sentences related to the query. This is performed by having the phrases and sentences in the documents represented as CGs as well. We prove security of the scheme by utilising the CQA2-security definition and proof methods of structured encryption [8].
3. We describe the possibility of constructing other more flexible schemes, such as queries based on concepts (or group of neighbouring concepts) in CG.

2 Related Work

Symmetric Searchable Encryption (SSE) was first proposed by Song, Wagner and Perrig [18]. Their schemes contain encryption methods specifically designed to allow for encryptions and searches of words in a document. The queries can be performed via sequential scanning or indexes. The sequential scan is inefficient since the server needs to scan through all documents while the indexes is incomplete as discussed in Goh [12]. Due to this, Goh proposed a data structure formally known as secure indexes. The technique, which is based on Bloom filters, improves on search efficiency. Building on Goh's proposal, Chang and Mitzenmacher [7] suggests stronger security model based on their observations of information leakage in Goh's secure indexes. However, this comes with a tradeoff on computation efficiency. Improved security notions on symmetric searchable encryption schemes is then proposed by Curtmola *et al.* in [10]. The main contribution is the notion of non-adaptive and adaptive chosen-keyword attacks. Following from this, Chase and Kamara [8] proposed the generalisation of all the above constructs, in particular of secure indexes. Their proposal, called *structured encryption*, extends the setting of searchable encryption on keyword-based data to arbitrarily-structured data. Our proposals fall into this category and are based on structured encryption. In addition, recently Cao *et al.* [6] proposed a searchable encryption scheme for graph-based data. Their scheme is efficient and it allows computation of inner product in the encrypted domain but their scheme induces false positives and security claims are heuristic. We further note that works on oblivious RAMs first examined by Goldreich and Ostrovsky [13], with a recent proposal in [17], can also provide searchable encryption. However, these are not as practical as the above discussed constructs.

Public Key based Searchable Encryption was first proposed by Boneh, Di Crescenzo, Ostrovsky and Persiano [4]. It is known as *Public Key Encryption with keyword Search* (PEKS) and the constructions are based on bilinear maps and trapdoor permutation. This proposal was extended in [1], which further refine the consistency properties of PEKS and its relations to anonymous identity-based encryption (IBE). Schemes based on the concept known as Private Information Retrieval (PIR) was also proposed in [5]. This scheme provides full concealment of encrypted search, unlike the previous PEKS schemes that leak access patterns. Other schemes of interests include schemes for multi-user settings in [3] and wildcarded identity-based encryption [2] that can be used for wildcarded searchable encryption. Recently, fully homomorphic encryption [11] has become one of the main techniques to provide searchable encryption due to its capability to execute arbitrary operations on encrypted data.

3 Preliminaries

3.1 Conceptual Graphs

Conceptual graph as a knowledge representation model was proposed by Sowa in [19]. It is defined as a graph representation for logic, which is based on the semantic networks of Artificial Intelligence (AI) and existential graphs [20]. Chein and Mugnier [9] further enhanced Sowa's proposal by formalising the model as a family of formalisms. One of them is *basic conceptual graphs* (BGs), which is central to the construction of graph-based knowledge representation. It is common in the literature to just denote BGs as conceptual graphs (CGs). We follow this notation. From an application viewpoint, a sentence can be constructed using a CG. A text document can be represented by a set of CGs. Figure 1 shows a simple example of CG of the sentence *"A boy named Bob possesses a toy and he plays with the toy"*. Formally, CGs require a vocabulary, which serves as the basis for CGs [9].

Vocabulary, $\mathcal{V}$. A CG is constructed under two kinds of nodes, *concept* and *relation*. Concept nodes represent the entities in an application domain while relation nodes represent the relationships between these entities. The set of concepts is denoted as T_C and the set of relations as T_R. There are also items known as *individual markers*, $\mathcal{I}$. For example, *Boy* is an entity of a concept type while *Bob* is an individual marker to the concept *Boy*, as shown in Figure 1.

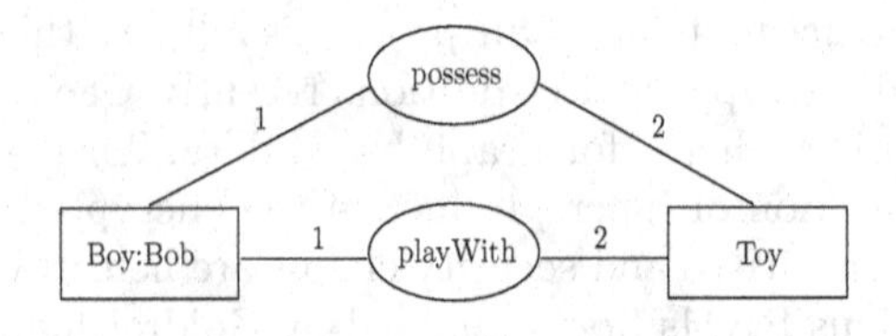

Fig. 1. An Example: A CG

There is also a *generic marker* $*$, which denotes an unspecified entity. For example, *Boy:** denotes any *Boy*. The sets of concepts, relations, $\mathcal{I}$ and $\{*\}$ compose the *vocabulary*.

CG. Formally, a basic conceptual graph CG defined over $\mathcal{V} = (T_C, T_R, \mathcal{I})$ is a quadruple $\mathcal{G} = (\mathcal{C}, \mathcal{R}, \mathcal{E}, \zeta)$ satisfying the following conditions [9]:

- $(\mathcal{C}, \mathcal{R}, \mathcal{E})$ is a finite, undirected and bipartite multigraph called the underlying graph of $\mathcal{G}$, denoted as $graph(\mathcal{G})$. $\mathcal{C}$ is the set of concept nodes, $\mathcal{R}$ is the set of relation nodes, and $\mathcal{E}$ is the family of edges.
- ζ is a labeling function of the nodes and edges of $graph(\mathcal{G})$ that satisfies:
  - A concept node c is labeled by a pair $type(c), marker(c)$ where $type(c) \in T_C$ and $marker(c) \in \mathcal{I} \cup \{*\}$,
  - A relation node r is labeled by $type(r) \in T_R$,
  - The degree of a relation node r is equal to the arity of $type(r)$,
  - Edges incident to a relation node r are totally ordered and they are labeled from 1 to $arity(type(r))$.

In our scheme the CGs, sets of concepts, relations and individual markers may serve as keywords (or queries) to retrieve CGs and messages matching CGs.

CG Homomorphisms. Homomorphism is the fundamental notion for CG reasoning. Informally, we may say that it is a mechanism to compare two CGs, G and H, and returns whether they are "similar" or not. This represents the central mean of *querying* database that contains CGs. We note that while deciding whether a graph is homomorphic to another is NP-complete, there are practical homomorphism algorithms for CG based on backtrack algorithms [9], under certain rules and constraints of the underlying application domains. It is implemented in Cogitant [21], a software package for constructing and querying CGs. Figure 2 shows an example of a graph homomorphism from g_1 to g_2. In an application scenario, we envisage sentences in documents being represented as CGs. Graph homomorphism is then performed on these CGs, allowing a user to categorise CGs that are related (or homomorphic) to one another.

3.2 Structured Encryption Schemes

Proposed by Chase and Kamara [8], structured encryption schemes generalise keyword-based SSE schemes to work on arbitrarily structured data such as web

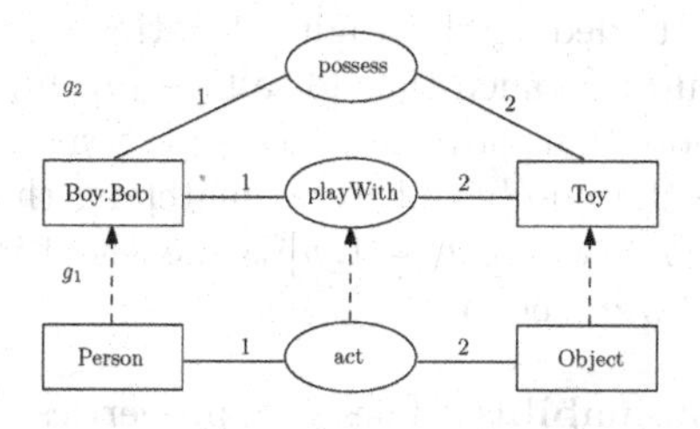

Fig. 2. An Example: A Homomorphism from g_1 to g_2

graph. Our constructions extend the applications of structured encryption to include knowledge represented in CGs. In the following we provide notation, building blocks and security definition as per defined by Chase and Kamara.

Notation. We denote the set of binary strings with length n as $\{0,1\}^n$ and the set of all finite binary strings as $\{0,1\}^*$. The set of integers $\{1,\ldots,n\}$ is denoted as $[n]$ and its power set is $\mathcal{P}[n]$. The empty set is $\emptyset$ or $\perp$. An algorithm $\mathcal{A}$ with an output x is denoted as $x \leftarrow \mathcal{A}$. We use $|S|$ to refer to the cardinality of a set S, and $|s|$ to refer to its bit length when s is a string. We further use $\mathcal{K}$ to denote the key space, $\mathcal{M}$ to denote the message space and $\mathcal{C}$ to denote the ciphertext space. Given $\mathbf{v}$ as a sequence of n elements, we denote v_i as its ith element.

Data Types. We consider a data type $\mathscr{T}$ in the form of sets, labels and dictionaries, which support query operations but not update operations. As in the original proposal of structured encryption, these data types have a single $\mathtt{Query}$ operation with a universe $\mathcal{U} = \{U_k\}_{k\in\mathbb{N}}$, where $\mathtt{Query}\colon \mathcal{U} \times \mathcal{Q} \rightarrow \mathcal{O}$, with $\mathcal{Q} = \{Q_k\}_{k\in\mathbb{N}}$ being the query space, $\mathcal{O} = \{O_k\}_{k\in\mathbb{N}}$ being the output space. It is also assumed that $\mathcal{U}$ is a *totally ordered set* and there is the element $\perp$ that denotes failure in $\mathcal{O}$. We remark that CGs are partially ordered. For example, with reference to Figure 2, given a $\leq$ relation in CGs, we have $boy \leq Person$, but $boy \nleq Object$. Due to this at first glance the structured encryption schemes may require fundamental changes since it operates under a totally ordered universe $\mathcal{U}$. However, as long as we restructure the representation of CGs such that the concepts and relations contained in the CGs are totally ordered, we can directly adopt and extend the schemes. One such technique is to build an index table (or labeling) "linking" all the related CGs, as what we propose in our constructions.

Symmetric Primitives. A CPA-secure symmetric encryption scheme $\Pi = (\mathtt{Gen}, \mathtt{Enc}, \mathtt{Dec})$ is required, where $\mathtt{Gen}$ is a probabilistic key generation algorithm, $\mathtt{Enc}$ a probabilistic encryption algorithm and $\mathtt{Dec}$ a deterministic decryption algorithm. Other primitives required include pseudo-random functions (PRF) and permutations (PRP). Formal definitions can be found in [15].

Induced Permutation. This permutation is performed in order to hide the locations of the items in a message sequence $\mathbf{m} = (m_1, \cdots, m_n)$ for $m_i \in \mathcal{M}$. It means given the locations of the items in the ciphertext sequence $\mathbf{c} = (c_1, \cdots, c_n)$

for $c_i \in C$, it is infeasible to deduce the original locations of the items in $\mathbf{m}$. We let π be the induced permutation such that for all $i \in [n]$, $m_i := \mathtt{Dec}(K, c_{\pi(i)})$, and π^{-1} as its inverse. We note that, however, access patterns are still leaked because for a server to retrieve the number of items matching the query $\{m_i : i \in I\}$, the server must be given I, where $I \subseteq [1, n]$ is the set of integer pointers to the data items in a message sequence $\mathbf{m}$.

Associativity and Chainability. These are properties that allow basic structured encryption schemes to be combined to construct more interesting schemes. A structured encryption scheme is said to be *associative* if the input message is defined as $\mathbf{M} = ((\mathbf{m}, \mathbf{v})) = ((m_1, v_1), \ldots, (m_n, v_n))$, where m_i is a message to be encrypted and v_i a semi-private data. A semi-private data is data that can be revealed given a matching query. In other words, the query operation in addition of returning the query results also returns the strings $(v_i)_{i \in I}$ related to the data items. *Chainability*, on the other hand, allows simpler structures to be "chained" to form a more complex structure using the associativity property. A possible chaining is to assign tokens on queries or encrypted message items of a simple structure as the semi-private data. These two properties are used to chain the basic label schemes in [8] to construct our main scheme.

Definition of Structured Encryption Schemes. An *associative symmetric structured encryption scheme* is a tuple of five polynomial-time algorithms $\Sigma = (\mathtt{Gen}, \mathtt{Enc}, \mathtt{Token}, \mathtt{Query}, \mathtt{Dec})$ where Gen is a probabilistic algorithm that generates a key K with input 1^k; Enc is a probabilistic algorithm that takes as input K, a data structure δ and a sequence of private and semi-private data $\mathbf{M}$ and outputs an encrypted data structure γ and a sequence of ciphertexts $\mathbf{c}$; Token is a (possibly probabilistic) algorithm that takes as input K and a query q and outputs a search token τ; Query is a deterministic algorithm that takes as input an encrypted data structure γ and a search token τ and outputs a set of pointers $J \subseteq [n]$ and a sequence of semi-private data $\mathbf{v}_I = (v_i)_{i \in I}$, where $I = \pi^{-1}[J]$; Dec is a deterministic algorithm that takes as input K and a ciphertext c_j and outputs a message m_j. Detailed and exact definition of the scheme can be found in [8].

4 Security Model

Our security model follows directly from that of a structured encryption scheme, where the aim is to provide confidentiality of stored data by preventing an adversary, which can be the storage provider, from reading the data. The adversary does have information on the access and query patterns. Specifically, the adversary has access to the following [8]:

- Encrypted data $(\gamma, \mathbf{c})$, where γ is the encrypted data structure containing indexes that map to the messages, while $\mathbf{c}$ is a sequence of ciphertexts.
- Tokens τ, where τ is an encrypted query used to retrieve the required items from the encrypted data structure.

- Query results $(J, \mathbf{v}_I)$, where J is a set of pointers to the messages and $\mathbf{v}_I$ the semi-private data.
- Query pattern $\text{QP}(q_t)$, where $q_t \in \mathbf{q}$ and $\mathbf{q}$, a non-empty sequence of queries, is a binary vector of length t with a value 1 at location i if $q_t = q_i$, and a value 0 otherwise. This allows an adversary to build a pattern of queries when queries are repeated, such as how frequent an identical query is made.
- Intersection pattern $\text{IP}(q_t)$, where $q_t \in \mathbf{q}$ and $\mathbf{q}$, a non-empty sequence of queries, is a sequence of length t with $f[I]$ at location t, where f is a fixed random permutation over $[n]$ and $I := \texttt{Query}(\delta, q_t)$. This means the access patterns are revealed when the same items are queried. However, the *exact* items are not revealed since every item in the message sequence $\mathbf{m}$ is permuted using the induced permutation π.

($\mathcal{L}_1$, $\mathcal{L}_2$)-security. A structured encryption scheme further defines two stateful leakage functions, $\mathcal{L}_1$ and $\mathcal{L}_2$. In general the $\mathcal{L}_1$ leakage function captures the leakage of size and length of the data items, that is, the information leaked by the encrypted data $(\gamma, \mathbf{c})$. On the other hand, $\mathcal{L}_2$ captures the leakage from the query and intersection patterns, by the token τ and query q. The actual form of leakage depends on the definition of $(\mathcal{L}_1, \mathcal{L}_2)$ of a concrete scheme.

Adaptive Chosen Queries Attack (CQA2) and CQA2-Security. Under this attack model the adversary is allowed to make a sequence of queries to the challenger. In return the adversary will be given the corresponding tokens. The adversary then makes the queries based on the tokens it obtained from all previous queries in such a way that it will be able to derive more information regarding the stored encrypted data. Formal definition is given in Appendix A.

5 Our Constructions

Two Approaches. There are two possible approaches in constructing structured encryption schemes for knowledge represented in CGs. The first approach is to pre-compute an index table as the data structure whereby all CGs homomorphic to a CG (which can be the query) is indexed. The table thus contains every CG linked to pointers pointing to other CGs homomorphic to it. In an application scenario, this means a user constructs CGs for all the documents to be stored, and performs graph homomorphism as described in Section 3.1 on these CGs to build the index table. We then construct various structured encryption schemes around the index table. The benefit of this approach is that query is efficient, without involving retrieval through graph homomorphism. The main limitation is the requirement for the user to pre-process all possible queries on his or her data. We note that similar pre-computation was also required for different data representation in the original structured encryption schemes.

The second approach is to perform graph homomorphism in the storage. This will give more flexibility to a user when constructing a query CG, whereby the query can be some new CGs not previously stored in the encrypted storage, yet allows the structured encryption scheme to retrieve related CGs and documents. A first thought would be to encrypt the concepts, relations and individual markers in a CG, but treat the CG's graph structure as a semi-private data so that it is possible to perform graph homomorphism between the query CG and the encrypted CGs in the storage. However this will not be secure since the specific graph structure of the query and the encrypted CGs are leaked, and this allows an adversary to distinguish between queries and between the returned results. To avoid such an issue, we would perform graph homomorphism in the encrypted domain through the underlying backtrack algorithm for graph homomorphism in CGs [9]. Our preliminary examination makes us to believe that it might not be possible to use an index-based scheme directly likes the first approach. Either a trusted third party must be involved or other approaches such as using fully homomorphic encryption schemes are required. This will lead to less efficient schemes compared to the first approach since graph homomorphisms in encrypted CGs have to be performed in real-time instead of pre-computed in the first approach, and with addition of trusted third party or using a fully homomorphic encryption scheme, computational workloads increase. In this paper we follow the first approach and reserve the second approach as our future work.

5.1 CG Query: $\texttt{CK-Label}_{\texttt{CGQ}}$

W first construct a basic structured encryption scheme for CG to CG queries. The aim is to allow a user to query and retrieve CGs from an encrypted CGs database. This can be achieved by adapting directly the structured encryption scheme for labeled data as proposed by Chase and Kamara [8]. The main difference is in the preparation of the data being queried. Here we term the scheme as $\texttt{CK-Label}_{\texttt{CGQ}}$. It will later be used to construct our main scheme (Section 5.2).

The scheme pre-processes a data structure known as a labeling δ_L. It is a data structure having a universe $\mathcal{U}$ containing the set of all binary relations between $[n]$ and the CGs. It supports a $\texttt{Search} : \mathcal{U} \times \mathcal{G} \rightarrow \mathcal{P}[n]$ operation with δ_L and $g \in \mathcal{G}$ as inputs and returns the set $\delta_L(g) = \{i \in [n] : (i, g) \in \delta_L\}$, where g denotes a CG and $\mathcal{G}$ the set of all possible CGs.

As a small hypothetical example, Figure 3 shows a query CG and the answers to the query, assuming the answers are retrieved from a set of CGs using CG homomorphism (where in practice CG can be represented in XML format and queried through implementation using Cogitant [21]). Given the query and answers, an index database can be prepared as shown in Table 1. Then the $\texttt{Search}$ operation for labeling $\delta_L(g_1)$, for example, will return $\{2, 3, 4\}$.

Table 1. An index database for query g_1 and the answers

Index	CGs	query answers
1	g_1	g_2,g_3,g_4
2	g_2	g_3
⋮	⋮	⋮
n	g_n	...

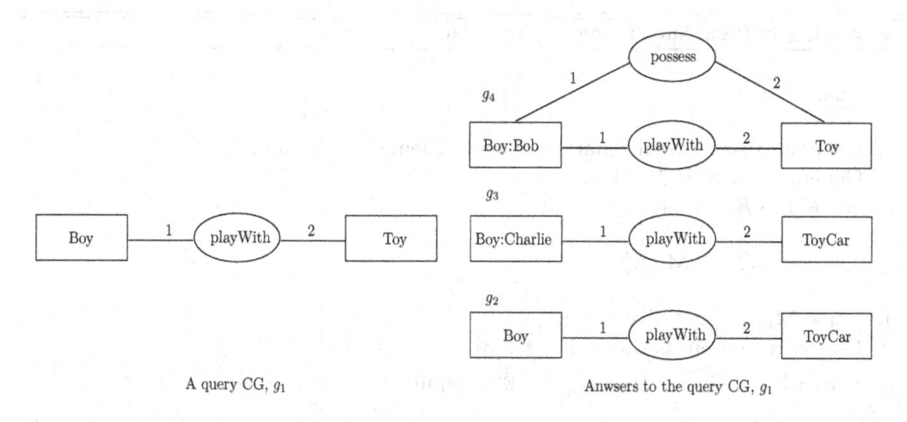

Fig. 3. An Example: A Query CG and the Possible Answers

The scheme also requires a data structure known as a dictionary T, which is a data structure constructed based on the content of δ_L. It contains pairs of (a, b), which are normally encrypted values, in such a way when given a, the value b can be retrieved efficiently. Given both δ_L and T, and let l, ω be integers, we define our message to be encrypted as $\mathbf{M}_G = (\mathbf{cg}, \mathbf{v})$ for $\mathbf{cg} = g_1, \ldots, g_n$, $|g_i| \leq l$ representing the sequence of CGs, and $\mathbf{v} = v_1, \ldots, v_n$, $|v_i| = \omega$ representing the sequence of semi-private data. We further denote g_q as the query CG. We also required a CPA-secure symmetric encryption scheme Π, and two PRFs $F : \{0,1\}^k \times \mathcal{G} \rightarrow \{0,1\}^{\max(\delta_L)\cdot(\log n+\omega)}$ and $H : \{0,1\}^k \times \mathcal{G} \rightarrow \{0,1\}^k$. The full algorithm is described in Figure 4.

As an illustration, we run the $\mathtt{CK\text{-}Label_{CGQ}}$ algorithm for $\delta_L(g_1) = \{2, 3, 4\}$ following Figure 3 and Table 1. In this case $\mathbf{M}_G = (\mathbf{cg}, \mathbf{v})$ for $\mathbf{cg} = (g_1, g_2, g_3, g_4)$ and $\mathbf{v} = (0, 0, 0, 0)$ since no semi-private data is required. We further assume the permutation $\pi = (2, 1, 4, 3)$. Executing the encryption function $\mathtt{Enc}$ produces $\langle (2,0), (1,0), (4,0), (3,0) \rangle \oplus F_{K_1}(g_1)$ and $H_{K_2}(g_1)$. $\mathbf{cg}$ is then permuted using π, resulting in $\mathbf{cg}^* = (g_2, g_1, g_4, g_3)$. Elements in $\mathbf{cg}^*$ are then padded so that all of them have the same length. Finally each elements in $\mathbf{cg}^*$ are encrypted as $\mathbf{c} = (c_1, c_2, c_3, c_4)$ using the symmetric encryption scheme Π. The encrypted structure γ is:

$$[\langle (2,0), (1,0), (4,0), (3,0) \rangle \oplus F_{K_1}(g_1), H_{K_2}(g_1)] \quad (1)$$

The resulting $(\gamma, \mathbf{c})$ can be queried for g_1 by executing $\mathtt{Token}$ and $\mathtt{Search}$. First, $\mathtt{Token}$ returns $\tau := (F_{K_1}(g_1), H_{K_2}(g_1))$ when g_1 is input as g_q. $\mathtt{Search}$ then uses $H_{K_2}(g_1)$ in τ as a search key to retrieve (1) from γ. Next $\mathtt{Search}$ XORs $\langle (2,0), (1,0), (4,0), (3,0) \rangle \oplus F_{K_1}(g_1)$ with $F_{K_1}(g_1)$, resulting in the output $J = (2, 1, 4, 3)$ and $\mathbf{v}_I = (0, 0, 0, 0)$. Using $J = (2, 1, 4, 3)$ as pointers, $\mathbf{c} = (c_1, c_2, c_3, c_4)$ is retrieved and the $\mathtt{Dec}$ algorithm decrypts $c_2 = g_1$, $c_1 = g_2$, $c_4 = g_3$ and $c_3 = g_4$.

CK-Label$_{\text{CGQ}}$ = (Gen, Enc, Token, Search, Dec)

$K \leftarrow$ Gen(1^k):

1. Generate two random binary sequence of length k, K_1 and K_2.
2. Generate $K_3 \leftarrow \Pi.\texttt{Gen}(1^k)$.
3. Set $K := (K_1, K_2, K_3)$.

$(\gamma, \mathbf{c}) \leftarrow$ Enc$(K, \delta_L, \mathbf{M}_G)$:

1. Parse $\mathbf{M}_G$ as **cg** and **v**.
2. Choose a random permutation $\pi : [n] \rightarrow [n]$.
3. For each $g \in \mathcal{G}$ such that $\delta_{L(g)} \neq \emptyset$, compute $F_{K_1}(g)$, $H_{K_2}(g)$, $\left\langle (\pi(i), v_i)_{i \in \delta_{L(g)}} \right\rangle$, and pad the strings $\left\langle (\pi(i), v_i)_{i \in \delta_{L(g)}} \right\rangle$ so that all of them have the same length. Then

$$\text{store } \left\langle (\pi(i), v_i)_{i \in \delta_{L(g)}} \right\rangle \oplus F_{K_1}(g) \text{ in } T \text{ with search key } H_{K_2}(g).$$

   where T denotes a dictionary.
4. Permute the elements in $\mathbf{cg}^*$ using π, where $\mathbf{cg}^*$ is the sequence that results from padding the elements of $\mathbf{cg}^*$ such that all of them have the same length.
5. For $1 \leq j \leq n$ compute $c_j \leftarrow \Pi.\texttt{Enc}(K_3, g_j^*)$.
6. Output $\gamma := T$ and $\mathbf{c} = (c_1, \ldots, c_n)$.

$\tau \leftarrow$ Token(K, g_q):

1. Output $\tau := (F_{K_1}(g_q), H_{K_2}(g_q))$

$(J, \mathbf{v}_I) :=$ Search(γ, τ):

1. Compute $\gamma(H_{K_2}(g_q)) \oplus F_{K_1}(g_q)$, where $\gamma(H_{K_2}(g_q))$ denotes the entry stored in γ with search key $H_{K_2}(g_q)$.
2. If $H_{K_2}(g_q)$ is not in γ then return $J = \emptyset$ and $\mathbf{v}_I = \perp$, else return $J = (j_1, \ldots, j_t)$ and $\mathbf{v}_I = (v_{i_1}, \ldots, v_{i_t})$.

$g_{\pi^{-1}(j)} :=$ Dec(K, c_j): output $g_{\pi^{-1}(j)} := \Pi.\texttt{Dec}(K_3, c_j)$.

Fig. 4. CG Query: CK-Label$_{\text{CGQ}}$ (following the labeled scheme in [8])

Assuming CK-Label$_{\text{CGQ}}$ is an exact instantiation of the labeled data scheme in [8], we say that CK-Label$_{\text{CGQ}}$ is $(\mathcal{L}_1, \mathcal{L}_2)$-secure under CQA2 following **Theorem 5.2** in [8].

5.2 CG-Message Query: MeQ

We now present a query to message (or document) structured encryption scheme by using CK-Label$_{\text{CGQ}}$ as the building block. The aim of the scheme is to extend the previous scheme to retrieve messages that contain the query CG and all CGs homomorphic to the query CG. The algorithmic description of the scheme is presented in Figure 5. Conceptually the scheme works on two encrypted data structures and chains them through the semi-private data to provide query to message retrieval. The data to be encrypted in this case consists of CGs (as in the previous scheme) with $\mathbf{M}_G = (\mathbf{cg}, \mathbf{v}^g)$, and messages containing CGs with $\mathbf{M}_M = (\mathbf{m}, \mathbf{v}^m)$. We thus need to pre-process two labelings, δ_L^g for CG to CGs structured encryption, and δ_L^m for CG to messages structured encryption. These labelings are used to produce the encrypted structures and the ciphertext sequences by using $\texttt{CK-Label}_{\texttt{CGQ}}$ separately for δ_L^g with $\mathbf{cg}$ and δ_L^m with $\mathbf{m}$. The resulting outputs are $(\gamma^g, \mathbf{c}^g)$ and $(\gamma^m, \mathbf{c}^m)$. Each of the constructions will provide query to retrieve CGs, and query to retrieve messages respectively. We denote the scheme for CG query as $\texttt{CK-Label}^g_{\texttt{CGQ}}$ and for message query as $\texttt{CK-Label}^m_{\texttt{CGQ}}$.

In the following we describe a simple example to illustrate the scheme. Using the same query and answers instance in Figure 3, we first assume, in addition to Table 1, there is a pre-processed index database that stores CG and messages containing CGs. Table 2 shows a hypothetical database.

Table 2. An index database for CGs and messages containing the CGs

Index	CGs	Messages
1	g_1	m_1,m_4,m_5
2	g_2	m_1, m_2
3	g_3	m_3
4	g_4	m_4
:	:	:
n	n	...

Given Table 2, the $\texttt{Search}$ operation for labeling $\delta_L^m(g_1)$, for example, will return $\{1,4,5\}$. Similarly $\delta_L^m(g_2)$ will return $\{1,2\}$ and so on. In order to execute the $\texttt{MeQ}$ scheme, we need both $\delta_L^g(g_1)$ (as presented in the previous section) and $\delta_L^m(g)$. We shall work on $\delta_L^g(g_1) = \{2,3,4\}$ and $\delta_L^m(g_1) = \{(1,4,5\}$ up to $L^m(g_4)$ $= \{4\}$. Also, we assume $\mathbf{M}_M = (\mathbf{m}, \mathbf{v}^m)$ for $\mathbf{m} = (m_1, m_2, m_3, m_4, m_5)$ and $\mathbf{v}^m = (0,0,0,0,0)$ as no semi-private data is required, while $\mathbf{M}_G = (\mathbf{cg}, \mathbf{v})$ for $\mathbf{cg} = (g_1, g_2, g_3, g_4)$ and $\mathbf{v}^g = (0,0,0,0)$. We further assume permutations $\pi_G = (2,1,4,3)$ for CG queries and $\pi_M = (2,5,1,4,3)$ for message queries.

The encryption $\texttt{Enc}$ consists of three stages. First with input $\delta_L^m(g_1) = \{1,4,5\}$ up to $L^m(g_4) = \{4\}$ and $\mathbf{m} = (m_1, m_2, m_3, m_4, m_5)$, $\texttt{CK-Label}^m_{\texttt{CGQ}}.\texttt{Enc}$ produces, with permutation π_M, γ^m as:

$$\begin{aligned}
&\langle((2,0),(4,0),(3,0))||pad\rangle \oplus F_{K_{1_1}}(g_1), H_{K_{1_2}}(g_1),\\
&\langle((2,0),(5,0))||pad\rangle \oplus F_{K_{1_1}}(g_2), H_{K_{1_2}}(g_2),\\
&\langle((1,0))||pad\rangle \oplus F_{K_{1_1}}(g_3), H_{K_{1_2}}(g_3),\\
&\langle((4,0))||pad\rangle \oplus F_{K_{1_1}}(g_4), H_{K_{1_2}}(g_4).
\end{aligned} \tag{2}$$

where $||$ denotes concatenation, K_{1_1}, K_{1_2} denotes the key for PRFs F and H respectively for $\texttt{CK-Label}^m_{\texttt{CGQ}}$, and *pad* denotes padding to the same length. $\mathbf{m}$ is also permuted using π_M resulting $\mathbf{m}^* = (m_2, m_5, m_1, m_4, m_3)$. Elements in $\mathbf{m}^*$ are padded so that all of them have the same length. Finally each elements in $\mathbf{m}^*$ are encrypted as $\mathbf{c}^m = (c_1^m, c_2^m, c_3^m, c_4^m, c_5^m)$.

In the second stage, search tokens $\tau_1^m = (F_{K_{1_1}}(g_1), H_{K_{1_2}}(g_1))$, ..., $\tau_4^m = (F_{K_{1_1}}(g_4), H_{K_{1_2}}(g_4))$ are generated by running $\texttt{CK-Label}^m_{\texttt{CGQ}}.\texttt{Token}$ on g_1 to g_4. These tokens are set as the semi-private data for $\mathbf{cg}$, as $\mathbf{v}^g = (\tau_1^m, \tau_2^m, \tau_3^m, \tau_4^m)$. Lastly, the third stage in the encryption involves computing the encrypted structure for $\mathbf{M}_G$ by executing $\texttt{CK-Label}^g_{\texttt{CGQ}}.\texttt{Enc}$ with permutation π_G to produce γ^g:

$$[\langle(2, \tau_1^m), (1, \tau_2^m), (4, \tau_3^m), (3, \tau_4^m)\rangle \oplus F_{K_{2_1}}(g_1), H_{K_{2_2}}(g_1)] \tag{3}$$

We note that for easier explanation we have only considered encryption for g_1 as in the previous section. The $\mathbf{cg}$ sequence is then padded and permuted, resulting in $\mathbf{cg}^*$ and each element in $\mathbf{cg}^*$ is then encrypted to generate a ciphertext sequence $\mathbf{c}^g = (c_1^g, c_2^g, c_3^g, c_4^g)$. The final output of the encryption function is $(\gamma^m, \gamma^g, \mathbf{c}^m, \mathbf{c}^g)$. The encrypted output can now be queried. Using g_1 as the query, $\texttt{Token}$ returns $\tau_1^g := (F_{K_{2_1}}(g_1), H_{K_{2_2}}(g_1))$. Given this token, $\texttt{Search}$ is conducted in two stages. Firstly, τ_1^g is used to search γ^g through $\texttt{CK-Label}^g_{\texttt{CGQ}}.\texttt{Search}$ by using $H_{K_{2_2}}(g_1)$ as a search key to retrieve (3). Next $\texttt{CK-Label}^g_{\texttt{CGQ}}.\texttt{Search}$ XORs (3) with $F_{K_{2_1}}(g_1)$ to retrieve the pointers $J = (2, 1, 4, 3)$ and semi-private data $\mathbf{v}_I = (\tau_1^m, \tau_2^m, \tau_3^m, \tau_4^m)$. The second stage involves $(\tau_1^m, \tau_2^m, \tau_3^m, \tau_4^m)$ as inputs to $\texttt{CK-Label}^m_{\texttt{CGQ}}.\texttt{Search}$ for retrieving all messages containing the CGs from γ^m. The searching and XORing follows the same steps as in $\texttt{CK-Label}^g_{\texttt{CGQ}}.\texttt{Search}$. For example, τ_1^m will allow for the retrieval of the pointers $(2, 4, 3)$. In the end $(J_j^m)_{j \in J^g} = ((2, 4, 3), (2, 5), (1), (4))$ are retrieved. These pointers point to $\mathbf{c}^m = (c_1^m, c_2^m, c_3^m, c_4^m, c_5^m)$, which are then decrypted as $c_2^m = m_1$, $c_5^m = m_2$, $c_1^m = m_3$, $c_4^m = m_4$ and $c_3^m = m_5$ using the $\texttt{CK-Label}^m_{\texttt{CGQ}}.\texttt{Dec}$ algorithm. The $\texttt{MeQ}$ scheme is effectively a scheme that chains two $\texttt{CK-Label}_{\texttt{CGQ}}$ constructions and therefore we consider its security through leak of information by the two constructions. Following the security notion for associative and chain-based construction for labeled web graph in [8], we say

Theorem 1. $\texttt{MeQ}$ *is* $(\mathcal{L}_1, \mathcal{L}_2)$*-secure under CQA2, if* $\texttt{CK-Label}^m_{\texttt{CGQ}}$ *is* $(\mathcal{L}_1^m, \mathcal{L}_2^m)$*-secure under CQA2 and if* $\texttt{CK-Label}^g_{\texttt{CGQ}}$ *is* $(\mathcal{L}_1^g, \mathcal{L}_2^g)$*-secure under CQA2, where* $\mathcal{L}_1(\delta_L^m, \delta_L^g, \mathbf{m}, \mathbf{cg}) = (\mathcal{L}_1^m(\delta_L^m, \mathbf{m}), \mathcal{L}_1^g(\delta_L^g, \mathbf{cg}))$ *and*

$$\mathcal{L}_2(\delta_L^m, \delta_L^g, g_q) = \left(\mathcal{L}_2^m(\delta_L^m, g_q), \mathcal{L}_2^g(\delta_L^g, g_q), (\mathcal{L}_2^m(\delta_L^m, g_i))_{i \in \delta^m_{L(g)}}\right).$$

In particular, let $|\delta_L|$ denote the number of query CGs such that $\delta_L(g)$ is nonempty and let $\max(\delta_L)$ be the size of the largest set $\delta_L(g)$, we have $\mathcal{L}_1(\delta_L^m, \delta_L^g, \mathbf{m}, \mathbf{cg}) = (|\delta_L|, max(\delta_L), n, l)$ since we can use padding so that $|\delta_L^m| = |\delta_L^g| = |\delta_L|$. Similarly we can arrive at the same conclusion for number of items n and length of the item l. We also have

$$\mathcal{L}_2(\delta_L^m, \delta_L^g, g_q) = \left(\mathrm{QP}(g_q), \mathrm{IP}(g_q), (\mathrm{QP}(g_i), \mathrm{IP}(g_i))_{i \in \delta^m_{L(g)}}, |\delta_L^m(g_q)|, |\delta_L^g(g_q)|\right).$$

This is by assumption that $\texttt{CK-Label}^m_{\texttt{CGQ}}$ and $\texttt{CK-Label}^g_{\texttt{CGQ}}$ are constructed from $\texttt{CK-Label}_{\texttt{CGQ}}$.

MeQ = (Gen, Enc, Token, Search, Dec)

Let $\texttt{CK-Label}^m_{\texttt{CGQ}}$ = (Gen, Enc, Token, Search, Dec) and $\texttt{CK-Label}^g_{\texttt{CGQ}}$ = (Gen, Enc, Token, Search, Dec) be associative structured encryption schemes for CGs.
$K \leftarrow \texttt{Gen}(1^k)$:

1. Generate $K_1 \leftarrow \texttt{CK-Label}^m_{\texttt{CGQ}}.\texttt{Gen}(1^k)$.
2. Generate $K_2 \leftarrow \texttt{CK-Label}^g_{\texttt{CGQ}}.\texttt{Gen}(1^k)$.
3. Set $K = (K_1, K_2)$.

$(\gamma^m, \gamma^g, \mathbf{c}^m, \mathbf{c}^g) \leftarrow \texttt{Enc}(K,\ \delta^m_L,\ \delta^g_L,\ \mathbf{M}_M,\ \mathbf{M}_G)$:

1. Compute $(\gamma^m, \mathbf{c}^m) \leftarrow \texttt{CK-Label}^m_{\texttt{CGQ}}.\texttt{Enc}(K_1, \delta^m_L, \mathbf{M}_M)$.
2. For $1 \leq i \leq n$,
   (a) compute $\tau^m_i \leftarrow \texttt{CK-Label}^m_{\texttt{CGQ}}.\texttt{Token}(K_1, g_i)$.
   (b) add τ^m_i to v^g_i, where v^g_i is the semi-private data of $\mathbf{M}_G = (\mathbf{cg}, \mathbf{v}^g)$, with δ^g_L the labeling generated from all the CGs in $\mathbf{cg}$.
3. Compute $(\gamma^g, \mathbf{c}^g) \leftarrow \texttt{CK-Label}^g_{\texttt{CGQ}}.\texttt{Enc}(K_2, \delta^g_L, \mathbf{M}_G)$.
4. Output $(\gamma^m, \gamma^g, \mathbf{c}^m, \mathbf{c}^g)$.

$\tau^g_q \leftarrow \texttt{Token}(K,\ g_q)$:

1. Compute $\tau^g_q \leftarrow \texttt{CK-Label}^g_{\texttt{CGQ}}.\texttt{Token}(K_2,\ g_q)$.
2. Output τ^g_q.

$(J^m_j)_{j \in J^g} := \texttt{Search}(\gamma^g,\ \gamma^m,\ \tau^g_q)$:

1. Compute $(J^g, \mathbf{v}^g_I) := \texttt{CK-Label}^g_{\texttt{CGQ}}.\texttt{Search}(\gamma^g,\ \tau^g_q)$.
2. Retrieve $(\tau^m_j)_j$ from $\mathbf{v}^g_I$.
3. For all $j \in J^g$, compute $J^m_j := \texttt{CK-Label}^m_{\texttt{CGQ}}.\texttt{Search}(\gamma^m,\ \tau^m_j)$.
4. Output $(J^m_j)_{j \in J^g}$.

$m_{\pi^{-1}(j)} := \texttt{Dec}(\mathrm{K},\ c^m_j)$:

1. Output $m_{\pi^{-1}(j)} := \texttt{CK-Label}^m_{\texttt{CGQ}}.\texttt{Dec}(K_1, c^m_j)$.

Fig. 5. Message Query - MeQ

Proof Sketch. By assumption there exists a simulator $\mathcal{S}_{\texttt{CGQ}}$ such that $\mathbf{Real}_{\texttt{CGQ},\mathcal{A}}(k)$ and $\mathbf{Ideal}_{\texttt{CGQ},\mathcal{A},\mathcal{S}_{\texttt{CGQ}}}(k)$ are indistinguishable. Given such a simulator, define the simulator $\mathcal{S}$ as follows:

1. It computes $(\gamma^m, \mathbf{c}^m) \leftarrow \mathcal{S}^m(\mathcal{L}^m_1)$ and $(\gamma^g, \mathbf{c}^g) \leftarrow \mathcal{S}^g(\mathcal{L}^g_1)$ using the information from $\mathcal{L}_1(\delta^m_L, \delta^g_L, \mathbf{m}, \mathbf{cg})$ and,
2. computes $\tau^m_j \leftarrow \mathcal{S}^m(\mathcal{L}^m_{2,j})$, $j \in [n]$ using the information from $\mathcal{L}_2(\delta^m_L, \delta^g_L, g_q)$,
3. outputs $(\tau^m_q) \leftarrow \mathcal{S}^m(\mathcal{L}^m_2)$ and $(\tau^g_q) \leftarrow \mathcal{S}^g(\mathcal{L}^g_2, \mathbf{v}^g)$ where $\mathbf{v}^g = (\tau^m_j)_{j \in [n]}$ using the information from $\mathcal{L}_2(\delta^m_L, \delta^g_L, g_q)$,

where $\mathcal{S}^m$ and $\mathcal{S}^g$ are simulators under CQA2-security by $\text{CK-Label}_{\texttt{CGQ}}$.

Given $\texttt{CK-Label}_{\texttt{CGQ}}$ secure under CQA2, we show that for all probabilistic polynomial time (PPT) adversary $\mathcal{A}$, the $\mathbf{Real}_{\Sigma,\mathcal{A}}(k)$ and $\mathbf{Ideal}_{\Sigma,\mathcal{A},\mathcal{S}}(k)$ experiments (Appendix A) is negligible by supposing the existence of a PPT adversary $\mathcal{A}$ that can differentiate the two experiments with non-negligible probability under simulation of both $\texttt{CK-Label}^{m}_{\texttt{CGQ}}$ and $\texttt{CK-Label}^{g}_{\texttt{CGQ}}$ schemes. The results follow directly that in such a case there exists a PPT adversary $\mathcal{B}$ that breaks the CQA2-security of $\texttt{CK-Label}_{\texttt{CGQ}}$. We show this by the following sequence of games, under similar arguments as in [8]:

Game$_0$: This represents the execution of the $\mathbf{Real}_{\Sigma,\mathcal{A}}(k)$ experiment. The challenger generates key $K = (K_1, K_2)$ and the adversary $\mathcal{A}$ generates (δ^m_L, δ^g_L, $\mathbf{M}_M$, $\mathbf{M}_G$). Next the challenger computes $\texttt{MeQ.Enc}(K, \delta^m_L, \delta^g_L, \mathbf{M}_M, \mathbf{M}_G)$ and gives the outputs $(\gamma^m, \gamma^g, \mathbf{c}^m, \mathbf{c}^g)$ to $\mathcal{A}$. The adversary $\mathcal{A}$ makes polynomially many adaptive queries and for each query g_q the challenger returns token $(\tau^m_q, \tau^g_q) \leftarrow \texttt{MeQ.Token}(K, g_q)$. Finally $\mathcal{A}$ outputs a bit b as the experiment result.

Game$_1$: In this game the call to $\texttt{CK-Label}^{g}_{\texttt{CGQ}}\texttt{.Enc}(K_2, \delta^g_L, \mathbf{M}_G)$ in Step 3 is replaced by calls to the simulator $\mathcal{S}^g(\mathcal{L}^g_1)$. The game begins with the challenger generating key $K = (K_1, K_2)$ and the adversary $\mathcal{A}$ generating (δ^m_L, δ^g_L, $\mathbf{M}_M$, $\mathbf{M}_G$). Given this generated data, the challenger computes $(\gamma^m, \mathbf{c}^m)$ $= \texttt{CK-Label}^{m}_{\texttt{CGQ}}\texttt{.Enc}(K_1, \delta^m_L, \mathbf{M}_M)$ and generates the semi-private data $\mathbf{v}_g$ in γ^g from $(\tau^m_i)_{i\in\delta^m_{L(g)}} = \texttt{CK-Label}^{m}_{\texttt{CGQ}}\texttt{.Token}(K_1, g_i)$. The simulator $\mathcal{S}^g$ is given $\mathcal{L}^g_1(\delta^g_L, \mathbf{c}^g)$ and generates $(\gamma^g, \mathbf{c}^g)$. The adversary $\mathcal{A}$ is given $(\gamma^m, \gamma^g, \mathbf{c}^m, \mathbf{c}^g)$ and makes polynomially many adaptive queries. For each query g_q, the challenger generates token (τ^m_q, τ^g_q) using the algorithm $\texttt{CK-Label}^{m}_{\texttt{CGQ}}\texttt{.Token}(K_1, g_q)$ and the simulator $\mathcal{S}^g(\mathcal{L}^g_2, \mathbf{v}^g)$. Token (τ^m_q, τ^g_q) is returned to $\mathcal{A}$. Finally $\mathcal{A}$ outputs a bit b as the experiment result.

We say that if there exists an adversary $\mathcal{A}$ that can distinguish **Game**$_0$ and **Game**$_1$ with non-negligible probability then there exists an adversary $\mathcal{B}$ that breaks the CQA2-security of $\texttt{CK-Label}^{g}_{\texttt{CGQ}}$. First we assume there exists such an adversary $\mathcal{A}$. We define $\mathcal{B}$ as the adversary who plays the $\mathbf{Real}(k)$ and $\mathbf{Ideal}(k)$ games while interacting with $\mathcal{A}$ to use the adaptive queries of $\mathcal{A}$:

1. $\mathcal{B}$ generates key $K_1 \leftarrow \texttt{CK-Label}^{m}_{\texttt{CGQ}}\texttt{.Gen}(1^k)$ and simulates $\mathcal{A}$.
2. Upon receiving (δ^m_L, δ^g_L, $\mathbf{M}_M$, $\mathbf{M}_G$) from $\mathcal{A}$, $\mathcal{B}$ passes this information to the challenger and receives $(\gamma^g, \mathbf{c}^g)$ from either the $\mathbf{Real}(k)$ or $\mathbf{Ideal}(k)$ game.
3. $\mathcal{B}$ gives the complete encrypted data $(\gamma^m, \gamma^g, \mathbf{c}^m, \mathbf{c}^g)$ to $\mathcal{A}$.
4. For each query g_q received from $\mathcal{A}$, $\mathcal{B}$ submits g_q to the challenger and obtains τ^g_q from either the $\mathbf{Real}(k)$ or $\mathbf{Ideal}(k)$ game.
5. $\mathcal{B}$ forwards the complete token (τ^m_q, τ^g_q) to $\mathcal{A}$.
6. $\mathcal{B}$ outputs the experiment result of $\mathcal{A}$.

Since $\mathcal{B}$ uses the adaptive queries of $\mathcal{A}$, which distinguishes between **Game**$_0$ and **Game**$_1$ with non-negligible probability, then the games $\mathbf{Real}(k)$ or $\mathbf{Ideal}(k)$ are also distinguishable. This breaks the CQA2-security of $\texttt{CK-Label}^{g}_{\texttt{CGQ}}$.

Game$_2$: This is the same as **Game**$_1$ except that we compute τ_i^m only when they are needed. Let $\mathcal{S}^g$ and $\mathcal{S}^m$ be as in **Game**$_1$. The challenger generates key $K = (K_1, K_2)$ and the adversary $\mathcal{A}$ generates $(\delta_L^m, \delta_L^g, \mathbf{M}_M, \mathbf{M}_G)$. The challenger computes $\texttt{MeQ.Enc}(K, \delta_L^m, \delta_L^g, \mathbf{M}_M, \mathbf{M}_G)$, except that it omits the computation of CK-Label$^m_{\texttt{CGQ}}$$\texttt{.Token}(K_1, g_i)$ in Step 2a. Next for each query g_q submitted by the adversary, the challenger computes (τ_q^m, τ_q^g) using $\texttt{CK-Label}^m_{\texttt{CGQ}}\texttt{.Token}(K_1, g_q)$ and $\mathcal{S}^g(\mathcal{L}_2^g, \mathbf{v}^g)$, where $\mathbf{v}^g = (\tau_i^m)_{i \in L^m(g)}$. The resulted tokens (τ_q^m, τ_q^g) are given to the adversary. The adversary outputs a bit b as the experiment result.

Game$_3$: This represents the simulation of **Game**$_2$ by replacing the outputs from $\texttt{CK-Label}^m_{\texttt{CGQ}}\texttt{.Enc}(K_1, \delta_L^m, \mathbf{M}_M)$ in Step 1 of the encryption algorithm with the simulation results from $\mathcal{S}^m(\mathcal{L}_1^m)$ and each token τ_i^m is replaced with the output from $\mathcal{S}^m(\mathcal{L}_2^m)$. Similarly the challenger generates key $K = (K_1, K_2)$ and the adversary $\mathcal{A}$ generates $(\delta_L^m, \delta_L^g, \mathbf{M}_M, \mathbf{M}_G)$. The challenger computes $\texttt{MeQ.Enc}(K, \delta_L^m, \delta_L^g, \mathbf{M}_M, \mathbf{M}_G)$, with the changes mentioned above and for each query g_q submitted by the adversary, the challenger computes (τ_q^m, τ_q^g) using $\mathcal{S}^m(\mathcal{L}_2^m)$ and $\mathcal{S}^g(\mathcal{L}_2^g, \mathbf{v}^g)$, where $\mathbf{v}^g = (\tau_i^m)_{i \in L^m(g)}$. Similarly in the end the adversary returns the experiment result.

As above with similar arguments for $\texttt{CK-Label}^g_{\texttt{CGQ}}$, by assuming there exists such an adversary $\mathcal{A}$, there exists an adversary $\mathcal{B}$ that breaks the CQA2-security of $\texttt{CK-Label}^g_{\texttt{CGQ}}$, and since $\mathcal{B}$ uses the adaptive queries of $\mathcal{A}$, which distinguishes between **Game**$_2$ and **Game**$_3$ with non-negligible probability, then the games **Real**(k) or **Ideal**(k) are also distinguishable. This breaks the CQA2-security of $\texttt{CK-Label}^m_{\texttt{CGQ}}$.

Game$_4$: This is the same as **Game**$_3$ except that both $\texttt{CK-Label}^g_{\texttt{CGQ}}$ and $\texttt{CK-Label}^m_{\texttt{CGQ}}$ are simulated, where $\mathcal{L}_1^m(\delta_L^m, \mathbf{m})$, $\mathcal{L}_1^g(\delta_L^g, \mathbf{cg})$, and for every query CG g_q, $\mathcal{L}_2^m(\delta_L^m, g_q)$, $\mathcal{L}_2^g(\delta_L^g, g_q)$ and $(\mathcal{L}_2^m(\delta_L^m, g_i))_{i \in \delta^m_{L(g)}}$ are provided by an oracle. In other words, we execute the **Ideal**(k) experiment with simulator $\mathcal{S}$.

By similar arguments, given $\mathcal{B}$ uses the adaptive queries of $\mathcal{A}$, **Game**$_3$ and **Game**$_4$ are distinguishable with non-negligible probability. This breaks the CQA2-security of $\texttt{CK-Label}^m_{\texttt{CGQ}}$.

Given these games, as $\texttt{CK-Label}_{\texttt{CGQ}}$ is secure under CQA2, and $\texttt{CK-Label}^g_{\texttt{CGQ}}$ and CK-Label$^m_{\texttt{CGQ}}$ are exact instantiation of $\texttt{CK-Label}_{\texttt{CGQ}}$, the $\texttt{MeQ}$ scheme is also secure under CQA2.

6 Other Constructions

In this section we discuss the possibility of constructing more flexible schemes to query CGs, extending the proposed `MeQ` scheme and the $\texttt{CK-Label}_{\texttt{CGQ}}$ scheme.

Keyword-CG Query. We first consider representing concepts in the CGs as the keywords and search the CGs based on keywords. While this approach seems to fall back to keyword search, concepts, as defined in Section 3.1, can be connected under the relationship of $\leq$ that can be represented as a tree. In such a case we can perform a neighbour search to first retrieve other keywords related to the query keyword. For example, a concept *Child* is at the higher level to the concept *Boy*, or $Boy \leq Child$, while an individual marker *Bob* is related to the concept *Boy*. Therefore when a *Boy* query is presented, we may retrieve all messages containing the words *Child* and *Bob*. By constructing such a scheme we can generalise the CK-Label$_{\texttt{CGQ}}$ scheme to return more general results. This can be achieved by first constructing an index table contains of a concept and other concepts related to it. The semi-private data in this case will contain search tokens of concepts for CGs. The search tokens thus chain the concepts to the related CGs, allowing the query keyword (which is a concept) to not just retrieve the related concepts but also CGs containing these concepts.

Keyword-CG-Message Query. Given the keyword-based scheme, we may combine it with the `MeQ` scheme to allow for a query of messages through a group of related keywords. In other words, given a keyword (i.e. a concept), the scheme first searches for other keywords related to the query keyword, and then the query and retrieved keywords are used to retrieve the related CGs, which in turns are used to retrieve the messages. In order to construct such a scheme, we define the semi-private data of the keyword (or concept) index table to contain the search tokens of concepts for CGs, and subsequently the semi-private data for the CGs' index table to contain the search tokens of CGs for the messages. The search tokens represent the two-level chaining from the keywords to the CGs and then to the messages, as opposed to the one chaining in the `MeQ` scheme and the Keyword-CG Query scheme.

7 Conclusions and Future Work

We propose structured encryption scheme for knowledge represented in conceptual graphs using the label scheme of Chase and Kamara [8]. As far as we know, this is the first structured encryption construction for knowledge-based database, in which one of the potential future applications being privacy-preserving natural language searches. Our next work following from this is to define and construct searchable encryption schemes for performing CGs graph homomorphisms in the encrypted form, which will allow for more flexible searches and reduce the required pre-processing in the current proposed scheme. It will also be interesting to examine schemes for knowledge represented in other representation models.

Acknowledgements. The authors thank Abdurashid Mamadolimov for fruitful discussions on conceptual graph and the anonymous referees for helpful comments.

References

1. Abdalla, M., Bellare, M., Catalano, D., Kiltz, E., Kohno, T., Lange, T., Malone-Lee, J., Neven, G., Paillier, P., Shi, H.: Searchable Encryption Revisited: Consistency Properties, Relation to Anonymous IBE, and Extensions. Journal of Cryptology 21, 350–391 (2008)

2. Abdalla, M., Birkett, J., Catalano, D., Dent, A., Malone-Lee, J., Neven, G., Schuldt, J., Smart, N.: Wildcarded Identity-Based Encryption. Journal of Cryptology, 1–41 (2010)
3. Bao, F., Deng, R.H., Ding, X., Yang, Y.: Private Query on Encrypted Data in Multi-user Settings. In: Chen, L., Mu, Y., Susilo, W. (eds.) ISPEC 2008. LNCS, vol. 4991, pp. 71–85. Springer, Heidelberg (2008)
4. Boneh, D., Di Crescenzo, G., Ostrovsky, R., Persiano, G.: Public Key Encryption with Keyword Search. In: Cachin, C., Camenisch, J.L. (eds.) EUROCRYPT 2004. LNCS, vol. 3027, pp. 506–522. Springer, Heidelberg (2004)
5. Boneh, D., Kushilevitz, E., Ostrovsky, R., Skeith III, W.E.: Public Key Encryption That Allows PIR Queries. In: Menezes, A. (ed.) CRYPTO 2007. LNCS, vol. 4622, pp. 50–67. Springer, Heidelberg (2007)
6. Cao, N., Yang, Z., Wang, C., Ren, K., Lou, W.: Privacy-Preserving Query over Encrypted Graph-Structured Data in Cloud Computing. In: 31st International Conference on Distributed Computing Systems (ICDCS 2011), pp. 393–402 (2011)
7. Chang, Y.-C., Mitzenmacher, M.: Privacy Preserving Keyword Searches on Remote Encrypted Data. In: Ioannidis, J., Keromytis, A.D., Yung, M. (eds.) ACNS 2005. LNCS, vol. 3531, pp. 442–455. Springer, Heidelberg (2005)
8. Chase, M., Kamara, S.: Structured Encryption and Controlled Disclosure. In: Abe, M. (ed.) ASIACRYPT 2010. LNCS, vol. 6477, pp. 577–594. Springer, Heidelberg (2010)
9. Chein, M., Mugnier, M.-L.: Graph-based Knowledge Representation: Computational Foundations of Conceptual Graphs. Advanced Information and Knowledge Processing Series. Springer-Verlag London Limited (2009)
10. Curtmola, R., Garay, J.A., Kamara, S., Ostrovsky, R.: Searchable Symmetric Encryption: Improved Definitions and Efficient Constructions. In: Juels, A., Wright, R.N., di Vimercati, S.D.C. (eds.) ACM Conference on Computer and Communications Security, CCS 2006, pp. 79–88. ACM (2006)
11. Gentry, C.: A Fully Homomorphic Encryption Scheme. PhD thesis, Stanford University (2009)
12. Goh, E.-J.: Secure indexes. Cryptology ePrint Archive, Report 2003/216 (2003), http://eprint.iacr.org/2003/216/
13. Goldreich, O., Ostrovsky, R.: Software Protection and Simulation on Oblivious RAMs. Journal of the ACM 43(3), 431–473 (1996)
14. Hilpinen, R.: Knowing that one knows and the classical definition of knowledge. Synthese 21, 109–132 (1970), doi:10.1007/BF00413541
15. Katz, J., Lindell, Y.: Introduction to Modern Cryptography. Chapman & Hall/CRC (2007)
16. Li, J., Wang, Q., Wang, C., Cao, N., Ren, K., Lou, W.: Fuzzy keyword search over encrypted data in cloud computing. In: INFOCOM 2010: Proceedings of the 29th Conference on Information Communications, pp. 441–445. IEEE Press (2010)
17. Shi, E., Chan, T.-H.H., Stefanov, E., Li, M.: Oblivious RAM with O((log N)3) Worst-Case Cost. In: Lee, D.H. (ed.) ASIACRYPT 2011. LNCS, vol. 7073, pp. 197–214. Springer, Heidelberg (2011)
18. Song, D.X., Wagner, D., Perrig, A.: Practical Techniques for Searches on Encrypted Data. In: SP 2000: Proceedings of the 2000 IEEE Symposium on Security and Privacy, p. 44. IEEE Computer Society (2000)
19. Sowa, J.F.: Conceptual Graphs for a Data Base Interface. IBM Journal of Research and Development 20, 336–357 (1976)

20. Sowa, J.F.: Conceptual Graphs. In: Handbook of Knowledge Representation, pp. 213–237 (2008)
21. LIRMM RCR team and LERIA ICLN team. Cogitant: A Conceptual Graph Library, `http://cogitant.sourceforge.net/`

A CQA2-Security

Definition 4.2 [8]. Given an associative private-key structured encryption scheme $\Sigma = (\texttt{Gen}, \texttt{Enc}, \texttt{Token}, \texttt{Query}, \texttt{Dec})$ for data type $\mathscr{T}$ that supports operation $\texttt{Query} : \mathcal{U} \times \mathcal{Q} \rightarrow \mathcal{P}[n]$ for $n \in \mathbb{N}$, $\mathcal{S}$ a simulator and $\mathcal{L}_1$ and $\mathcal{L}_2$ the stateful leakage functions, an adversary $\mathcal{A}$ performs two games:

$\mathbf{Real}_{\Sigma,\mathcal{A}}(k)$:
$\mathcal{A}$ generates a tuple $(\delta, \mathbf{M})$, where $\mathbf{M} = (\mathbf{m}, \mathbf{v})$ for $\mathbf{m} = (m_1, m_2, \ldots, m_n)$ and $\mathbf{v} = (v_1, v_2, \ldots, v_n)$. The challenger is given the tuple $(\delta, \mathbf{M})$ and runs $\texttt{Gen}(1^k)$ to generate a key K. Then the challenger runs $\texttt{Enc}(K, \delta, \mathbf{M})$ to output the encrypted data $(\gamma, \mathbf{c})$, where $\mathbf{c} = (c_1, c_2, \ldots, c_n)$. The encrypted data $(\gamma, \mathbf{c})$ is given to $\mathcal{A}$. Next $\mathcal{A}$ chooses a query q_0 and submit to the challenger and the challenger returns the corresponding token $\tau_0 = \texttt{Token}(K, q_0)$. For $t = 1, \ldots, p(k)$ where $p(.)$ is a polynomial, $\mathcal{A}$ chooses a query q_t based on observation of previous queries and the challenger returns the corresponding token $\tau_t = \texttt{Token}(K, q_t)$. After t many queries, $\mathcal{A}$ gives γ, $\mathbf{c}$, $(q_0, \ldots, q_{p(k)})$, $(\tau_0, \ldots, \tau_{p(k)})$ to distinguisher $\mathcal{D}$ and $\mathcal{D}(\gamma, \mathbf{c}, (q_0, \ldots, q_{p(k)}), (\tau_0, \ldots, \tau_{p(k)}))$ returns a bit b. Finally $\mathcal{A}$ outputs b.

$\mathbf{Ideal}_{\Sigma,\mathcal{A},\mathcal{S}}(k)$:
$\mathcal{A}$ generates a tuple $(\delta, \mathbf{M})$, where $\mathbf{M} = (\mathbf{m}, \mathbf{v})$ for $\mathbf{m} = (m_1, m_2, \ldots, m_n)$ and $\mathbf{v} = (v_1, v_2, \ldots, v_n)$. The simulator $\mathcal{S}$ is given $\mathcal{L}_1(\delta, \mathbf{M})$ and $\mathcal{S}$ generates encrypted data $(\gamma, \mathbf{c})$ and gives this to $\mathcal{A}$. Then $\mathcal{A}$ chooses a query q_0 and for this query $\mathcal{S}$ is given $(\mathcal{L}_2(\delta, q_0), \mathbf{v}_{I_0})$. $\mathcal{S}$ returns a token τ_0. For $t = 1, \ldots, p(k)$ where $p(.)$ is a polynomial, $\mathcal{A}$ chooses a query q_t based on observation of previous queries. $\mathcal{S}$ is given $(\mathcal{L}_2(\delta, q_t), \mathbf{v}_{I_t})$ and returns token τ_t. $\mathcal{A}$ gives γ, $\mathbf{c}$, $(q_0, \ldots, q_{p(k)})$, $(\tau_0, \ldots, \tau_{p(k)})$ to the distinguisher $\mathcal{D}$ and $\mathcal{D}(\gamma, \mathbf{c}, (q_0, \ldots, q_{p(k)}), (\tau_0, \ldots, \tau_{p(k)}))$ returns a bit b. Finally $\mathcal{A}$ outputs b.

Σ is $(\mathcal{L}_1, \mathcal{L}_2)$-secure under CQA2 if for all probabilistic polynomial-time adversaries $\mathcal{A}$, there exists a probabilistic polynomial-time simulator $\mathcal{S}$ such that

$$|\Pr[\mathbf{Real}_{\Sigma,\mathcal{A}}(k) = 1] - \Pr[\mathbf{Ideal}_{\Sigma,\mathcal{A},\mathcal{S}}(k)| \leq \texttt{negl}(k)$$

where $\texttt{negl}(k)$ is a negligible function.

Symmetric-Key Encryption Scheme with Multi-ciphertext Non-malleability

Akinori Kawachi, Hirotoshi Takebe, and Keisuke Tanaka

Department of Mathematical and Computing Sciences, Tokyo Institute of Technology, 2-12-1 Ookayama, Meguro-ku, Tokyo 152-8552, Japan

Abstract. A standard notion of non-malleability is that an adversary cannot forge a ciphertext c' from a single valid ciphertext c for which a plaintext m' of c' is meaningfully related to a plaintext m of c. The *multi-ciphertext non-malleability* is a stronger notion; an adversary is allowed to obtain multiple ciphertexts $c_1, c_2, \ldots$ in order to forge c'. We provide an efficient symmetric-key encryption scheme with an information-theoretic version of the multi-ciphertext non-malleability in this paper by using ℓ-wise almost independent permutations of Kaplan, Naor, and Reingold.

Keywords: symmetric-key encryption, information-theoretic security, non-malleability.

1 Introduction

Non-malleability is one of the most important security notions in modern cryptography and was introduced by Dolev, Dwork, and Naor [2]: Given a sample of ciphertext c, no adversary can generate another ciphertext c' of which a corresponding message m' is meaningfully related to the original message m of c. This notion has been studied extensively in a *computational* setting for security against computationally bounded adversaries. This notion is being extended to an *information-theoretic* setting for security against computationally unbounded adversaries. Hanaoka, Shikata, Hanaoka, and Imai [5] formalized the information-theoretic version of the non-malleability for the first time, and then McAven, Safavi-Naini, and Yung [9] extended the notion. See a comprehensive survey by Hanaoka [4] for more details of the information-theoretic non-malleability.

In the first formalization of the non-malleability, they considered a situation that an adversary is only given a single ciphertext c to generate a forged ciphertext c'. Considering general attacks of adversaries, it would be more natural for adversaries to deal with multiple ciphertexts $c_1, c_2, \ldots$ to forge another ciphertext c'. For example, Pass, Shelat, and Vaikuntanathan [10] considered several versions of non-malleability, including the model in which an adversary can obtain multiple ciphertexts, in a computational setting, and compare strength among the versions. Hereafter, we refer to this strong notion of the non-malleability as *multiple-ciphertext non-malleability*. Even in an information-theoretic setting, Kawachi, Portmann, and Tanaka [8] extended the original definition of Hanaoka et al. [5] to

G. Hanaoka and T. Yamauchi (Eds.): IWSEC 2012, LNCS 7631, pp. 123–137, 2012.

a multiple-ciphertext version of the non-malleability, and showed an equivalence between a naturally extended version of secrecy and the non-malleability.

While the multiple-ciphertext non-malleability has been discussed already even in the information-theoretic setting, there was no known scheme satisfying the information-theoretic multiple-ciphertext non-malleability so far. In this paper, we construct a symmetric-key encryption scheme satisfying the security notion.

In a single-ciphertext setting, namely, the original definition of the information-theoretic non-malleability, Hanaoka [4] provided a simple construction of a symmetric-key encryption scheme satisfying the single-ciphertext non-malleability from authentic codes. As pointed out in [8], his construction has a structure of pairwise independent hash functions, and they conjectured that ℓ-wise ones provide $(\ell-1)$-ciphertext non-malleability.

Hanaoka's construction in [4] is simple. The encryption function is defined as $c = am + b$ for a ciphertext $c \in \mathrm{GF}(2^n)$ and a message $m \in \mathrm{GF}(2^n)$, where $(a, b) \in \mathrm{GF}(2^n) \setminus \{0\} \times \mathrm{GF}(2^n)$ is a secret key. It is easy to extend this function to the following one with the ℓ-wise independence: $c = a_{\ell-1}m^{\ell-1} + a_{\ell-2}m^{\ell-2} + \cdots + a_1 m + a_0$ for a secret key $(a_0, ..., a_{\ell-1})$.

The ℓ-wise independence is indeed important for the non-malleability, but it is not enough. We need the invertibility with a secret key for decryption. In the naive extension mentioned above, we cannot uniquely decrypt the ciphertext c into the message m. We can uniquely decrypt only in the case of the function $c = am + b$ since it is a pairwise independent *permutation*.

In general, it is difficult to efficiently construct ℓ-wise independent permutations. Actually, pairwise and 3-wise independent permutations are only known so far [11,14,13] and there is no known efficient constructions of more than 3-wise ones.

However, by relaxing the notion of the independence, we can obtain a useful permutation for our purpose. There exist several constructions of ℓ-wise *almost* independent permutations (See a comprehensive survey in [7] for history of the constructions).

Recently, Kaplan, Naor, and Reingold [7] provided an efficient construction for a family of ℓ-wise almost independent permutations on a wide range of the parameter ℓ. They apply the derandomizing-composition method to the simple 3-bit permutations such as [3,6,1], and get a family of ℓ-wise almost independent permutations with a short description length. In our scheme, we directly make use of their construction to provide the multiple-ciphertext non-malleable encryption scheme. As a result, our scheme satisfies *approximate non-malleability* (already formalized in [8]), while Hanaoka's construction [4] satisfies *perfect non-malleability*. This relaxation does not hurt the security of our scheme significantly since we can make the gap from the perfect non-malleability arbitrarily small with reasonable overheads by the property of the almost independent permutations, as seen in the construction of the symmetric-key encryption.

We also observe that our scheme satisfies another security notion by the properties of the almost independent permutations. The multi-message secrecy is a security notion that even if we encrypt multiple different messages with a single secret-key no adversary can obtain information on these messages from corresponding ciphertexts. It is easy to see that the one-time pad has 1-message secrecy but not 2-message secrecy. Kawachi et al. [8] proved 2-message secrecy is equivalent to 1-ciphertext non-malleability, and thus Hanaoka's construction [4] provides not only a non-malleable encryption scheme but also 2-message secret one. They also showed a gap between approximate versions of the multiple-ciphertext non-malleability and multiple-message secrecy, and thus we cannot construct such a scheme in general directly from our scheme with multiple-ciphertext approximate non-malleability. However, the strong primitive, ℓ-wise almost independent permutations, can provide multiple-message approximate secrecy.

The remaining part of this paper is organized as follows. We give notation and definitions of the security notions in Section 2. In Section 3, we first review the result of the ℓ-wise almost independent permutations by Kaplan et al. [7], We then describe the symmetric-key encryption scheme, and prove it satisfies multiple-ciphertext approximate non-malleability and multiple-message approximate secrecy.

2 Definitions

We basically follow definitions given in [8] for notation and notions.

Notation. Calligraphic letters mean sets of some elements. Lowercase and uppercase letters mean elements and random variables, respectively. For a set $\mathcal{X}$, we denote $|\mathcal{X}|$ as the number of elements in $\mathcal{X}$. We denote by $P_X(x)$ the probability that the random variable X equals an element x, i.e., $P_X(x) = \Pr[X = x]$. Analogously, for two random variables X and Y, we denote by $P_{XY}(x, y)$ the probability associated with their joint probability, i.e., $P_{XY}(x, y) = \Pr[X = x \wedge Y = y]$ and by $P_{X|Y}(x|y)$ the conditional probability, i.e., $P_{X|Y}(x|y) = \Pr[X = x|Y = y]$. $X \cdot Y$ means a random variable according to the probability $P_{X \cdot Y}(x, y) := P_X(x)P_Y(y)$. Thus, it holds that $P_{X \cdot Y}(x, y) = P_{XY}(x, y)$ if and only if X and Y are independent.

For a set $\mathcal{X}$ and a random variable X distributed over $\mathcal{X}$, we define $X_{[\ell]}$ as a sequence of ℓ random variables $X_1, \ldots, X_\ell$ on sets $\mathcal{X}_1, \ldots, \mathcal{X}_\ell$ respectively, and for any $x \in \mathcal{X}_1 \times \cdots \times \mathcal{X}_\ell$, we denote by x_i the i-th element of ℓ-tuple $x = (x_1, \ldots, x_\ell)$. Furthermore, we define

$$\mathcal{X}_{\mathsf{diff}}^{\times \ell} := \{(x_1, \ldots, x_\ell) \in \mathcal{X}^{\times \ell} : \forall i, j \in \{1, \ldots, \ell\}, i \neq j \Rightarrow x_i \neq x_j\},$$

namely, a subset of $\mathcal{X}^{\times \ell}$ in which all the coordinates are different from the others.

Symmetric-key Encryption. A goal in this paper is to construct a symmetric-key encryption scheme (satisfying some security notions). We give the formal definition of the symmetric-key encryption scheme below.

Definition 1 (Symmetric-Key Encryption). *A symmetric-key encryption scheme consists of three algorithms* (Key-Generation, Encryption, Decryption). Key-Generation *picks a key k from a key set $\mathcal{K}$ according to a probability distribution $P_K(k)$ over $\mathcal{K}$.* Encryption *applies an encryption function f_k to a message*

m with a key k, and then outputs the ciphertext $c = f_k(m)$. Decryption *applies a decryption function f_k^{-1} to a ciphertext c, and then outputs $m = f_k^{-1}(c)$ if such a unique m exists, and $\perp$ otherwise.*

Throughout this paper, we denote messages, ciphertexts, and keys by random variables M, C, and K respectively. If C is determined by M with a key $k \in \mathcal{K}$, we write $C = f_k(M)$, or $C = f(M)$ simply.

Entropy and Statistical Distance. For discussions on information-theoretic security, we will use several variants of the Shannon entropy. The base of logarithms is 2 throughout this paper.

Definition 2 (Entropy). *For two random variables X over $\mathcal{X}$ and Y over $\mathcal{Y}$, we denote the entropy of X and the joint entropy of X and Y by*

$$H(X) := -\sum_{x\in\mathcal{X}} P_X(x)\log P_X(x) \quad \text{and}$$

$$H(XY) := -\sum_{x\in\mathcal{X},y\in\mathcal{Y}} P_{XY}(x,y)\log P_{XY}(x,y),$$

respectively. We also denote the entropy conditioned on $Y = y$ by

$$H(X|Y=y) := -\sum_{x\in\mathcal{X}} P_{X|Y}(x|y)\log P_{X|Y}(x|y),$$

and the conditional entropy of X on Y by

$$H(X|Y) := \sum_{y\in\mathcal{Y}} P_Y(y)H(X|Y=y) = H(XY) - H(Y),$$

respectively. In addition, we denote by

$$I(X;Y) := H(X) + H(Y) - H(XY)$$

the mutual information between X and Y.

Also, we will measure the distance between two random variables by variants of the statistical distance for definitions of security notions.

Definition 3 (Statistical Distance). *Let X, Y be random variables over $\mathcal{X}$. The* statistical distance *between X and Y is defined as*

$$d(X,Y) := \frac{1}{2}\sum_{x\in\mathcal{X}} |P_X(x) - P_Y(x)|.$$

For another random variable Z over $\mathcal{Z}$, a variant of the statistical distance we call the statistical distance *between X and Y* conditioned on $Z = z$ *for $z \in \mathcal{Z}$ is defined as*

$$d(X,Y|Z=z) := \frac{1}{2}\sum_{x\in\mathcal{X}} \left|P_{X|Z}(x|z) - P_{Y|Z}(x|z)\right|.$$

Further, we define the conditional *statistical distance as*

$$d(X,Y|Z) := \sum_{z \in \mathcal{Z}} P_Z(z) d(X,Y|Z=z).$$

The conditional statistical distance (called the expected variational distance in [8]) is an "average-case" version of the statistical distance conditioned on $Z = z$ in some sense. As we will seen later, the approximate non-malleability in [8] is defined by this average-case version. In contrast, the scheme we propose in this paper is based on the statistical distance conditioned on the "worst-case" message $z \in \mathcal{Z}$.

2.1 Security Notions

In this section, we define security notions of secrecy and non-malleability. These definitions have been already formalized in the literature such as [5,9,4,8]. As stated above, our formalization basically follows [8].

Secrecy. First, we define secrecy of encryption schemes. Since ciphertexts are sent over an insecure channel, an adversary can intercept them and try to get information on messages from them. Thus, an encryption scheme must satisfy (perfect) secrecy. We review the notion of information-theoretic secrecy defined by Shannon [15].

Definition 4 (Perfect Secrecy [15]). *We say that an encryption scheme satisfies perfect secrecy* (PS) *if for all the message random variables M on $\mathcal{X}$ independent from the key random variable K (i.e., $I(M;K) = 0$), it holds that*

$$H(M|C) = H(M), \text{ or } I(M;C) = 0. \tag{1}$$

We also define the approximate secrecy by relaxing the perfect one. In the approximate secrecy, a ciphertext is *almost independent* from the message. We formalize this notion via the statistical distance as in, e.g., [8].

Definition 5 (Approximate Secrecy [8]). *We say that an encryption scheme satisfies ϵ-secrecy* (ϵ-S) *if for all message random variables M on $\mathcal{X}$ independent from the key random variable K (i.e. $I(M;K) = 0$), it holds that*

$$d(MC, M \cdot C) \leq \epsilon, \tag{2}$$

where $C := f_K(M)$ is a random variable of the resulting ciphertext.

Note that this notion coincides with the perfect secrecy if $\epsilon = 0$. The above secrecy is naturally extended to the multiple-message secrecy, as defined in [8], which guarantees the approximate secrecy even if the same key is used for encryption repeatedly.

Definition 6 (ℓ-message Approximate Secrecy [8]). *We say that an encryption scheme satisfies ℓ-message ϵ-secrecy* (ϵ-S$^\ell$) *if for all ℓ-tuples of different* *message random variables* $M_{[\ell]}$ *on* $\mathcal{X}_{\mathrm{diff}}^{\times\ell}$ *independent from the key random variable* K *(i.e.* $I(M_{[\ell]}; K) = 0$*), it holds that*

$$d(M_{[\ell]}C_{[\ell]}, M_{[\ell]} \cdot C_{[\ell]}) \leq \epsilon, \tag{3}$$

where the C_i *are random variables of the resulting ciphertexts. If* $\epsilon = 0$*, we then say that the scheme satisfies* ℓ*-message perfect secrecy* (PS^ℓ).

Non-malleability. Second, we define non-malleability of encryption schemes. We first review an intuitive explanation in [8] on information-theoretic non-malleability. Informally, an encryption scheme is said to be non-malleable if any adversary cannot do meaningfully related modification of the ciphertext. In order to explain the notion of malleability, we consider the one-time pad as an example. If an adversary flips the first bit of a ciphertext c_1 of a message m_1 encrypted by the one-time pad, then the first bit of the resulting decrypted message m_2 are also flipped. Therefore the adversary can easily modify the resulting message m_2 even without knowing the original message m_1, and m_2 is meaningfully related to m_1; the first bit of m_1 is opposite to that of m_2. In this case, if given the original message m_1 after modifying the original c_1 to c_2, the adversary can easily get m_2. Thus it can be considered that (m_1, c_1, c_2) have more information about m_2 than (m_1, c_1). In the terminology of the entropy, we can express this fact as $H(M_2|M_1C_1C_2) < H(M_2|M_1C_1)$. Therefore, if the encryption scheme is non-malleable, it should satisfy $H(M_2|M_1C_1C_2) = H(M_2|M_1C_1)$, and equivalently $I(M_2; C_2|M_1C_1) = 0$. This formalization was proposed by Hanaoka et al. [5]. The criterion means given the original message M_1 and the original ciphertext C_1, the resulting message M_2 and the resulting ciphertext C_2 are independent.

For simplicity, we assume that the message and ciphertext sets have the same size.

Definition 7 (Perfect Non-malleability [5]). *We say that an encryption scheme satisfies perfect non-malleability* (PNM) *if for all message random variables* M_1 *on* $\mathcal{X}$ *independent from the key random variable* K *(i.e.* $I(M_1; K) = 0$*), and all ciphertext random variables* C_2 *on* $\mathcal{Y}$ *different from* C_1 *and independent from* K *given* M_1C_1 *(i.e.* $\Pr[C_1 = C_2] = 0$ *and* $I(C_2; K|M_1C_1) = 0$*), it holds that*

$$I(M_2; C_2|M_1C_1) = 0,$$

where $M_2 := f_K^{-1}(C_2)$ *is a message random variable obtained by decrypting* C_2 *with* K.

Notice that the condition $\Pr[C_1 \neq C_2] = 0$ is necessary for the definition of non-malleability. If it is not posed, an adversary can easily break the non-malleability by simply copying C_1 to C_2 since a trivial relation $M_1 = M_2$ holds then. So, we require the condition to exclude such a trivial attack.

In [8], the above non-malleability is extended to the following approximate version.

Definition 8 (Approximate Non-malleability [8]). *We say that an encryption scheme satisfies ϵ-non-malleability (ϵ-NM) if for all message random variable M_1 independent from the key random variable K, and all ciphertext random variables C_2 on $\mathcal{Y}$ different from C_1 and independent from K given M_1C_1 (i.e. $\Pr[C_1 = C_2] = 0$ and $I(C_2; K|M_1C_1) = 0$), it holds that*

$$d(M_2C_2, M_2 \cdot C_2|M_1C_1) \leq \epsilon,$$

where $M_2 := f_K^{-1}(C_2)$ is a message random variable obtained by decrypting C_2 with K.

Note that this coincides with the perfect non-malleability if $\epsilon = 0$.

We further extend the approximate non-malleability to a version that an adversary can get multiple ciphertexts for modification in a natural way. Specifically, we define ℓ-ciphertext approximate non-malleability which measures the independence of $M_{\ell+1}$ and $C_{\ell+1}$ when $M_{[\ell]}$ and $C_{[\ell]}$ are given.

Definition 9 (ℓ-ciphertext Approximate Non-malleability). *We say that an encryption scheme satisfies ℓ-ciphertext ϵ-non-malleability (ϵ-NM$^\ell$) if for all tuples of message random variables $M_1, \ldots, M_\ell \in \mathcal{X}_{\text{diff}}^{\times \ell}$ independent from the key random variable K, and all ciphertext random variables $C_{\ell+1}$ on $\mathcal{Y}$ different from C_i for all $i \in \{1, \ldots, \ell\}$ and independent from K given $M_{[\ell]}C_{[\ell]}$ (i.e. $\Pr[C_i = C_{\ell+1}] = 0$ for all $i \in \{1, \ldots, \ell\}$ and $I(C_{\ell+1}; K|M_{[\ell]}C_{[\ell]}) = 0$), it holds that*

$$d(M_{\ell+1}C_{\ell+1}, M_{\ell+1} \cdot C_{\ell+1}|M_{[\ell]}C_{[\ell]}) \leq \epsilon,$$

where $M_{\ell+1} := f_K^{-1}(C_{\ell+1})$ is a message random variable obtained by decrypting $C_{\ell+1}$ with K. If $\epsilon = 0$, we then call this notion ℓ-ciphertext perfect non-malleability (PNM$^\ell$).

2.2 A Variant of Non-malleability

In the above multi-ciphertext non-malleability, we took ℓ messages chosen by a sender as random variables $M_1, ..., M_\ell$ and bounded the statistical distance on average over the random variables. We now introduce a variant of the non-malleability on the "worst-case" messages. Our variant does not need to have message random variables for messages chosen by a sender, and thus, it only needs to deal with $M_{\ell+1}$ given by attack of an adversary as a message random variable, which simplifies the notion of the multi-ciphertext non-malleability.

Definition 10. *We say that an encryption scheme satisfies ℓ-ciphertext worst-case ϵ-non-malleability (ϵ-NM$_*^\ell$) if all tuples of messages $(m_1, \ldots, m_\ell)$ on $\mathcal{X}_{\text{diff}}^{\times \ell}$ which are independent from the key random variable K, all tuples of ciphertexts $(c_1, \ldots, c_\ell) \in \mathcal{Y}_{\text{diff}}^{\times \ell}$, and all ciphertext random variables $C_{\ell+1}$ on $\mathcal{Y} \setminus \{c_1, \ldots, c_\ell\}$, it holds that*

$$d(M_{\ell+1}C_{\ell+1}, M_{\ell+1} \cdot C_{\ell+1} | M_{[\ell]} = (m_1, ..., m_\ell), C_{[\ell]} = (c_1, ..., c_\ell))$$
$$= \frac{1}{2} \sum_{\substack{m_{\ell+1} \in \mathcal{X}', \\ c_{\ell+1} \in \mathcal{Y}'}} \left| P_{M_{\ell+1}C_{\ell+1}}(m_{\ell+1}, c_{\ell+1}) - P_{M_{\ell+1}}(m_{\ell+1}) P_{C_{\ell+1}}(c_{\ell+1}) \right|$$
$$\leq \epsilon,$$

where $\mathcal{X}' := \mathcal{X} \setminus \{m_1, \ldots, m_\ell\}$, $\mathcal{Y}' := \mathcal{Y} \setminus \{c_1, \ldots, c_\ell\}$, and $M_{\ell+1}$, which is defined on $\mathcal{X} \setminus \{m_1, \ldots, m_\ell\}$, is the message of $c_{\ell+1}$ under the key random variable K, i.e., $M_{\ell+1} := f_K^{-1}(c_{\ell+1})$.

While we formulate ϵ-NM$_*^\ell$ as the "worst-case" version of the non-malleability, we can prove that ϵ-NM$_*^\ell$ is equivalent to ϵ-NM$^\ell$, and thus, we only discuss the worst-case approximate non-malleability ϵ-NM$_*^\ell$ in the remaining part of this paper.

Proposition 1. *If a symmetric-key encryption scheme satisfies ϵ-NM$^\ell$, it also satisfies ϵ-NM$_*^\ell$, and vice versa.*

Proof. For simplicity, we consider only the case $\ell = 1$. This proof can be applied to the case $\ell > 1$ in a similar way.

First, we prove that ϵ-NM^1 implies ϵ-NM$_*^1$. This directly follows from the definitions by fixing the message random variable M_1 and the ciphertext random variable C_1 to the distributions that output any fixed message m_1 and ciphertext c_1 respectively with probability 1.

Second, we prove the converse direction. We define $\tilde{M_2} := (M_2 | M_1 = m_1, C_1 = c_1)$ and $\tilde{C_2} := (C_2 | M_1 = m_1, C_1 = c_1)$. We then have

$$d(M_2C_2, M_2 \cdot C_2 | M_1C_1)$$
$$= \sum_{m_1, c_1} P_{M_1C_1}(m_1, c_1) \cdot \frac{1}{2} \sum_{m_2, c_2} \left| P_{\tilde{M_2}\tilde{C_2}}(m_2, c_2) - P_{\tilde{M_2}}(m_2) P_{\tilde{C_2}}(c_2) \right|$$
$$= \sum_{m_1, c_1} P_{M_1C_1}(m_1, c_1) \cdot d(M_2C_2, M_2 \cdot C_2 | M_1 = m_1, C_1 = c_1) \leq \epsilon,$$

where the last inequality follows by the definition of ϵ-NM$_*^1$. □

3 Encryption Scheme and Its Security

In this section, we construct a symmetric-key encryption scheme based on ℓ-wise almost independent permutations. Moreover, we show that our scheme satisfies multi-message approximate secrecy and multi-ciphertext approximate non-malleability. While our scheme works well with any family of ℓ-wise (almost) independent permutations, we choose Kaplan, Naor, and Reingold's construction [7] since no explicit construction of ℓ-wise perfectly independent permutations is known in the case that $\ell \geq 4$ and their construction is the most efficient for general almost independent permutations.

We now formally define ℓ-wise ϵ-dependent permutations as follows.

Definition 11 (ℓ-wise ϵ-dependent permutation). *Let $\mathbf{F} = \{f : \mathcal{X} \to \mathcal{X}\}$ be a family of permutations and $\epsilon > 0$. The family $\mathbf{F}$ is ℓ-wise ϵ-dependent if for every ℓ-tuple of distinct elements $(x_1, \ldots, x_\ell) \in \mathcal{X}_{\text{diff}}^{\times \ell}$, the statistical distance between the distribution $(f(x_1), \ldots, f(x_\ell))$ where f is chosen uniformly at random and the uniform distribution on $\mathcal{X}_{\text{diff}}^{\times \ell}$ is at most ϵ. That is,*

$$\frac{1}{2} \sum_{(y_1,\ldots,y_\ell)\in\mathcal{X}_{\text{diff}}^{\times \ell}} \left| \Pr_f[f(x_1) = y_1 \wedge \cdots \wedge f(x_\ell) = y_\ell] - \frac{1}{\left|\mathcal{X}_{\text{diff}}^{\times \ell}\right|} \right| \leq \epsilon, \tag{4}$$

where f is chosen uniformly at random from $\mathbf{F}$. If $\epsilon = 0$, then the family $\mathbf{F}$ is said to be ℓ-wise independent.

As already mentioned, Kaplan et al. provided a general efficient construction of ℓ-wise ϵ-dependent permutations.

Theorem 1 ([7]). *There exists a family $\mathbf{F} = \{f : \{0,1\}^n \to \{0,1\}^n\}_n$ of ℓ-wise ϵ-dependent permutations such that every $f \in \mathbf{F}$ is representable by a binary string of length $O(n\ell + \log(\epsilon^{-1}))$, and there exist algorithms F and F^{-1} that run in polynomial time in n, ℓ and $\log(\epsilon^{-1})$, and $F(x) = f(x)$ and $F^{-1}(f(x)) = x$ for every $x \in \{0,1\}^n$.*

The construction of Kaplan et al. is based on a random composition of some simple permutations. Although it was shown that such a composition provides nice almost independent permutations [1], we then require a long seed to describe a fully random composition of the simple permutations. The main idea of their construction is to use the pseudorandom-walk generator, which was originally developed for derandomization of space-limited computation [12]. Derandomizing the random composition by the generator, they obtained a family of ℓ-wise ϵ-dependent permutations with a short description length.

Our scheme can be directly obtained from the above construction of Kaplan et al..

- Key-Generation: Let $\mathbf{F} = \{f_1, f_2, ...\}$ be the family of ℓ-wise ϵ-dependent permutations of Kaplan et al.. Sample $k \leftarrow \{1, \ldots, |\mathbf{F}|\}$ uniformly at random. The secret key is k.
- Encryption: For a message m, compute $c = f_k(m)$.
- Decryption: For a ciphertext c, compute $m = f_k^{-1}(c)$.

We drop the subscript k from f_k below if the key k is obvious. Note that the size of message set is equal to that of ciphertext set, i.e., $|\mathcal{X}| = |\mathcal{Y}|$, since f is a permutation.

One can immediately see that the key length of the scheme coincides with the description length $O(n\ell + \log(\epsilon^{-1}))$ of the permutation given in Theorem 1 from the construction. Therefore, the overhead of the key length is reasonable even if we set ϵ to be very small, for example, $\epsilon := 1/|\mathcal{X}|^c$ for a constant c.

3.1 Security Proofs

We next show that the above scheme satisfies ℓ-message approximate secrecy and $(\ell-1)$-ciphertext approximate non-malleability for $\ell \geq 2$.

ℓ-message approximate secrecy. First, we show that the scheme satisfies multi-message approximate secrecy.

Theorem 2. *The above scheme satisfies ℓ-message 2ϵ-secrecy.*

Proof. We consider only the case $\ell = 2$ for simplicity. This proof can be applied to the case $\ell > 2$ in a similar way.

The theorem can be proved from the following two claims.

Claim 1. *Let*

$$\alpha_1 := \max_{m_1,m_2} \left\{ \frac{1}{2} \sum_{c_1,c_2} \left| P_{C_1C_2|M_1M_2}(c_1,c_2|m_1,m_2) - \frac{1}{|\mathcal{Y}_{\mathrm{diff}}^{\times 2}|} \right| \right\},$$

$$\alpha_2 := \frac{1}{2} \sum_{c_1,c_2} \left| \frac{1}{|\mathcal{Y}_{\mathrm{diff}}^{\times 2}|} - P_{C_1C_2}(c_1,c_2) \right|.$$

For all $M_{[2]}$ on $\mathcal{X}_{\mathrm{diff}}^{\times 2}$, it holds $d(M_2C_2, M_2 \cdot C_2|M_1C_1) \leq \alpha_1 + \alpha_2$.

Claim 2. *We have $\alpha_2 \leq \alpha_1$.*

From these claims, we have $d(M_2C_2, M_2 \cdot C_2|M_1C_1) \leq 2\alpha_1$. By rewriting the probability $P_{C_1C_2|M_1M_2}(c_1,c_2|m_1,m_2)$ with the encryption function f, we have $P_{C_1C_2|M_1M_2}(c_1,c_2|m_1,m_2) = \Pr_f[f(m_1) = c_1 \wedge f(m_2) = c_2]$. Then, it immediately follows that $\alpha_1 \leq \epsilon$ from the definition of pairwise ϵ-dependent permutations, and thus, $d(M_2C_2, M_2 \cdot C_2|M_1C_1) \leq 2\alpha_1 \leq 2\epsilon$. We give the proofs of these claims as follows.

Proof (Claim 1). By the triangle inequality, we have

$$\begin{aligned}
&d(M_2C_2, M_2 \cdot C_2|M_1C_1) \\
&= \frac{1}{2} \sum_{m_1,m_2,c_1,c_2} |P_{M_1M_2C_1C_2}(m_1,m_2,c_1,c_2) - P_{M_1M_2}(m_1,m_2)P_{C_1C_2}(c_1,c_2)| \\
&= \frac{1}{2} \sum_{m_1,m_2} P_{M_1M_2}(m_1,m_2) \sum_{c_1,c_2} |P_{C_1C_2|M_1M_2}(c_1,c_2|m_1,m_2) - P_{C_1C_2}(c_1,c_2)| \\
&\leq \max_{m_1,m_2} \left\{ \frac{1}{2} \sum_{c_1,c_2} |P_{C_1C_2|M_1M_2}(c_1,c_2|m_1,m_2) - P_{C_1C_2}(c_1,c_2)| \right\} \\
&\leq \max_{m_1,m_2} \left\{ \frac{1}{2} \sum_{c_1,c_2} \left| P_{C_1C_2|M_1M_2}(c_1,c_2|m_1,m_2) - \frac{1}{|\mathcal{Y}_{\mathrm{diff}}^{\times 2}|} \right| \right\} \\
&\quad + \frac{1}{2} \sum_{c_1,c_2} \left| \frac{1}{|\mathcal{Y}_{\mathrm{diff}}^{\times 2}|} - P_{C_1C_2}(c_1,c_2) \right| = \alpha_1 + \alpha_2.
\end{aligned}$$

□

Proof (Claim 2). Note that

$$P_{C_1C_2}(c_1,c_2) = \sum_{m_1',m_2'} P_{M_1M_2}(m_1',m_2')P_{C_1C_2|M_1M_2}(c_1,c_2|m_1',m_2')$$

from the definition. We then obtain by the triangle inequality

$$\begin{aligned}
\alpha_2 &= \frac{1}{2}\sum_{c_1,c_2}\left|\frac{1}{|\mathcal{Y}_{\text{diff}}^{\times 2}|} - \sum_{m_1',m_2'} P_{M_1M_2}(m_1',m_2')P_{C_1C_2|M_1M_2}(c_1,c_2|m_1',m_2')\right| \\
&\le \frac{1}{2}\sum_{c_1,c_2}\sum_{m_1',m_2'} P_{M_1M_2}(m_1',m_2')\left|\frac{1}{|\mathcal{Y}_{\text{diff}}^{\times 2}|} - P_{C_1C_2|M_1M_2}(c_1,c_2|m_1',m_2')\right| \\
&\le \max_{m_1',m_2'}\frac{1}{2}\sum_{c_1,c_2}\left|\frac{1}{|\mathcal{Y}_{\text{diff}}^{\times 2}|} - P_{C_1C_2|M_1M_2}(c_1,c_2|m_1',m_2')\right| = \alpha_1.
\end{aligned}$$

□

This completes the proof of Theorem 2. □

$(\ell-1)$-ciphertext approximate non-malleability. Second, we show that the scheme satisfies multi-ciphertext approximate non-malleability.

Theorem 3. *Assume $|\mathcal{Y}| \ge 2$ and $0 \le \epsilon < 1/2|\mathcal{Y}|$. The above scheme satisfies $(\ell-1)$-ciphertext $O(|\mathcal{Y}|^{\ell-1})\epsilon$-non-malleability.*

Proof. By Proposition 1, it is sufficient to prove that the scheme satisfies $O(|\mathcal{Y}|^{\ell-1})\epsilon$-NM$_*^{\ell-1}$. As in the proof of Theorem 2, we consider only the case $\ell = 2$ for simplicity. This proof can be applied to the case $\ell > 2$ in a similar way.

We can take the same strategy as the proof of Theorem 2 at some technical parts. We consider the following two claims.

Claim 3. *Let*

$$\begin{aligned}
\beta_1(m_1,c_1,C_2) &:= \frac{1}{2}\sum_{m_2,c_2} P_{M_2}(m_2)\left|P_{C_2|M_2}(c_2|m_2) - \frac{1}{|\mathcal{Y}|-1}\right|, \\
\beta_2(m_1,c_1,C_2) &:= \frac{1}{2}\sum_{m_2,c_2} P_{M_2}(m_2)\left|\frac{1}{|\mathcal{Y}|-1} - P_{C_2}(c_2)\right|.
\end{aligned}$$

Then, it holds $d(M_2C_2, M_2\cdot C_2|M_1 = m_1, C_1 = c_1) \le \beta_1(m_1,c_1,C_2) + \beta_2(m_1,c_1,C_2)$ for all m_1, c_1, and C_2.

Claim 4. *We have $\beta_2(m_1,c_1,C_2) \le \beta_1(m_1,c_1,C_2)$ for all m_1, c_1, and C_2.*

We will give the proofs of the two claims later. From these claims, it is sufficient to bound $\beta_1(m_1,c_1,C_2)$ as in the proof of Theorem 2.

The most different part is to estimate a bound of $\beta_1(m_1,c_1,C_2)$ (The bound of α_1 was immediate from the definition of pairwise almost independent permutations). Assuming the two claims hold, we give the bound as the main lemma.

Lemma 1. *For all m_1, c_1, and C_2, it holds that*

$$\frac{1}{2}\sum_{m_2,c_2} P_{M_2}(m_2)\left|P_{C_2|M_2}(c_2|m_2) - \frac{1}{|\mathcal{Y}|-1}\right| < 4\,|\mathcal{Y}|\,\epsilon.$$

Since we have $\beta_1(m_1,c_1,C_2) \le 4|\mathcal{Y}|\epsilon$ from this lemma, it follows that $d(M_2C_2, M_2 \cdot C_2|M_1 = m_1, C_1 = c_1) \le 2\beta_1(m_1,c_1,C_2) \le 8|\mathcal{Y}|\epsilon$ for all m_1, c_1, and C_2.

Proof (Lemma 1). We define

$$\epsilon(c_1,c_2,m_1,m_2) := \frac{1}{2}\left(\Pr[f(m_1)=c_1]\cdot P_{C_2|M_2}(c_2|m_2) - \frac{1}{|\mathcal{Y}^{\times 2}_{\text{diff}}|}\right),$$

$$\delta(m_1,c_1) := \Pr[f(m_1)=c_1] - \frac{1}{|\mathcal{Y}|}.$$

Since $|\delta(m_1,c_1)| \le \epsilon < \frac{1}{2|\mathcal{Y}|}$ from the definition of pairwise ϵ-dependent permutations, it holds $\Pr[f(m_1)=c_1] \neq 0$. We therefore have

$$\begin{aligned}
&P_{C_2|M_2}(c_2|m_2) - \frac{1}{|\mathcal{Y}|-1}\\
&= \frac{1}{1+|\mathcal{Y}|\,\delta(m_1,c_1)}\left(-\frac{|\mathcal{Y}|\,\delta(m_1,c_1)}{|\mathcal{Y}|-1} + 2\,|\mathcal{Y}|\,\epsilon(c_1,c_2,m_1,m_2)\right).
\end{aligned}$$

Since $P_{C_2|M_2}(c_2|m_2) = \Pr_f[f(m_2) = c_2|f(m_1) = c_1 \wedge M_2 = m_2]$, we have $\sum_{c_1,c_2}|\epsilon(c_1,c_2,m_1,m_2)| \le \epsilon$ for all (m_1,m_2) from the definition of pairwise ϵ-dependent permutations. Also from the assumption that $|\mathcal{Y}| \ge 2$, we have

$$\begin{aligned}
&\left|P_{C_2|M_2}(c_2|m_2) - \frac{1}{|\mathcal{Y}|-1}\right|\\
&\le \frac{1}{1-|\mathcal{Y}||\delta(m_1,c_1)|}\left(\frac{|\mathcal{Y}||\delta(m_1,c_1)|}{|\mathcal{Y}|-1} + 2\,|\mathcal{Y}|\,|\epsilon(c_1,c_2,m_1,m_2)|\right)\\
&\le \frac{1}{1-|\mathcal{Y}|\,\epsilon}\left(2\,|\delta(m_1,c_1)| + 2\,|\mathcal{Y}|\,|\epsilon(c_1,c_2,m_1,m_2)|\right)\\
&\le (1+|\mathcal{Y}|\,\epsilon)\left(2\epsilon + 2\,|\mathcal{Y}|\,|\epsilon(c_1,c_2,m_1,m_2)|\right)\\
&\le 4\epsilon + 4\,|\mathcal{Y}|\,|\epsilon(c_1,c_2,m_1,m_2)|\,.
\end{aligned}$$

It then follows that

$$\begin{aligned}
&\frac{1}{2}\sum_{m_2,c_2} P_{M_2}(m_2)\left|P_{C_2|M_2}(c_2|m_2) - \frac{1}{|\mathcal{Y}|-1}\right|\\
&\le \frac{1}{2}\sum_{m_2,c_2} P_{M_2}(m_2)\cdot\left(4\epsilon + 4\,|\mathcal{Y}|\,|\epsilon(c_1,c_2,m_1,m_2)|\right)\\
&= \sum_{m_2} P_{M_2}(m_2)\sum_{c_2} 2\epsilon + \sum_{m_2} P_{M_2}(m_2)\sum_{c_1,c_2} 2\,|\mathcal{Y}|\,|\epsilon(c_1,c_2,m_1,m_2)|\\
&< 4\,|\mathcal{Y}|\,\epsilon.
\end{aligned}$$

This completes the proof of Lemma 1. □

Finally, we give the proofs of Claims 3 and 4.

Proof (Claim 3). For all m_1, c_1, and C_2, we have by the triangle inequality

$$\begin{aligned}
&d(M_2C_2, M_2 \cdot C_2 | M_1 = m_1, C_1 = c_1) \\
&= \frac{1}{2} \sum_{m_2, c_2} |P_{M_2C_2}(m_2, c_2) - P_{M_2}(m_2)P_{C_2}(c_2)| \\
&= \frac{1}{2} \sum_{m_2, c_2} P_{M_2}(m_2) \left| P_{C_2|M_2}(c_2|m_2) - P_{C_2}(c_2) \right| \\
&\leq \frac{1}{2} \sum_{m_2, c_2} P_{M_2}(m_2) \left| P_{C_2|M_2}(c_2|m_2) - \frac{1}{|\mathcal{Y}| - 1} \right| \\
&\quad + \frac{1}{2} \sum_{m_2, c_2} P_{M_2}(m_2) \left| \frac{1}{|\mathcal{Y}| - 1} - P_{C_2}(c_2) \right| \\
&= \beta_1(m_1, c_1, C_2) + \beta_2(m_1, c_1, C_2).
\end{aligned}$$

□

Proof (Claim 4). Note that $P_{C_2}(c_2) = \sum_{m_2'} P_{M_2}(m_2')P_{C_2|M_2}(c_2|m_2')$. We then have

$$\begin{aligned}
&\frac{1}{2} \sum_{m_2, c_2} P_{M_2}(m_2) \left| \frac{1}{|\mathcal{Y}| - 1} - P_{C_2}(c_2) \right| \\
&= \frac{1}{2} \sum_{m_2, c_2} P_{M_2}(m_2) \left| \frac{1}{|\mathcal{Y}| - 1} - \sum_{m_2'} P_{M_2}(m_2')P_{C_2|M_2}(c_2|m_2') \right| \\
&= \frac{1}{2} \sum_{m_2, c_2} P_{M_2}(m_2) \left| \sum_{m_2'} \left\{ P_{M_2}(m_2') \left(\frac{1}{|\mathcal{Y}| - 1} - P_{C_2|M_2}(c_2|m_2') \right) \right\} \right| \\
&\leq \frac{1}{2} \sum_{m_2, c_2} P_{M_2}(m_2) \left(\sum_{m_2'} P_{M_2}(m_2') \left| \frac{1}{|\mathcal{Y}| - 1} - P_{C_2|M_2}(c_2|m_2') \right| \right) \\
&= \frac{1}{2} \sum_{m_2', c_2} P_{M_2}(m_2') \left| \frac{1}{|\mathcal{Y}| - 1} - P_{C_2|M_2}(c_2|m_2') \right|.
\end{aligned}$$

□

This completes the proof of Theorem 3. □

4 Concluding Remarks

We have constructed a symmetric-key encryption scheme satisfying ℓ-message 2ϵ-approximate secrecy and $(\ell - 1)$-ciphertext $O(|\mathcal{Y}|^{\ell-1})\epsilon$-approximate non-malleability from a family of ℓ-wise ϵ-dependent permutations. In order to achieve ℓ-ciphertext ϵ'-approximate non-malleability from Kaplan, Naor, and Reingold's construction, one can easily see that the bit length of the keys is $O(n\ell + \log(\epsilon')^{-1})$ from Theorem 1.

Kawachi et al. [8] proved the matching lower bound of key length for 1-ciphertext perfect non-malleability and consequently Hanaoka's construction [4] is optimal on the key length. A major open problem is to extend their lower bound to the general non-malleability, namely, whether the key length given from our result is optimal for the multi-ciphertext approximate non-malleability.

Acknowledgments. We are grateful to the anonymous reviewers of IWSEC 2012 for a suggestion on the equivalence of two notions of non-malleability (Proposition 1) and other helpful comments. AK and KT were partially supported by the Ministry of Education, Science, Sports and Culture, Grant-in-Aid for Scientific Research (A) No.24240001. HT and KT were partially supported by a grant of I-System Co. Ltd., NTT Secure Platform Laboratories, and JSPS Global COE program "Computationism as Foundation for the Sciences".

References

1. Brodsky, A., Hoory, S.: Simple permutations mix even better. Random Struct. Algorithms 32(3), 274–289 (2008)
2. Dolev, D., Dwork, C., Naor, M.: Nonmalleable cryptography. SIAM J. Comput. 30(2), 391–437 (2000)
3. Gowers, W.T.: An almost m-wise independent random permutation of the cube. Combinatorics, Probability & Computing 5, 119–130 (1996)
4. Hanaoka, G.: Some Information Theoretic Arguments for Encryption: Non-malleability and Chosen-Ciphertext Security (Invited Talk). In: Safavi-Naini, R. (ed.) ICITS 2008. LNCS, vol. 5155, pp. 223–231. Springer, Heidelberg (2008)
5. Hanaoka, G., Shikata, J., Hanaoka, Y., Imai, H.: Unconditionally secure anonymous encryption and group authentication. Comput. J. 49(3), 310–321 (2006)
6. Hoory, S., Magen, A., Myers, S., Rackoff, C.: Simple permutations mix well. Theor. Comput. Sci. 348(2-3), 251–261 (2005)
7. Kaplan, E., Naor, M., Reingold, O.: Derandomized constructions of k-wise (almost) independent permutations. Algorithmica 55(1), 113–133 (2009)
8. Kawachi, A., Portmann, C., Tanaka, K.: Characterization of the Relations between Information-Theoretic Non-malleability, Secrecy, and Authenticity. In: Fehr, S. (ed.) ICITS 2011. LNCS, vol. 6673, pp. 6–24. Springer, Heidelberg (2011)
9. McAven, L., Safavi-Naini, R., Yung, M.: Unconditionally Secure Encryption Under Strong Attacks. In: Wang, H., Pieprzyk, J., Varadharajan, V. (eds.) ACISP 2004. LNCS, vol. 3108, pp. 427–439. Springer, Heidelberg (2004)
10. Pass, R., Shelat, A., Vaikuntanathan, V.: Relations Among Notions of Non-malleability for Encryption. In: Kurosawa, K. (ed.) ASIACRYPT 2007. LNCS, vol. 4833, pp. 519–535. Springer, Heidelberg (2007)
11. Rees, E.G.: Notes on geometry. Springer (1983)

12. Reingold, O.: Undirected connectivity in log-space. J. ACM 55(4) (2008)
13. Russell, A., Wang, H.: How to fool an unbounded adversary with a short key. IEEE Transactions on Information Theory 52(3), 1130–1140 (2006)
14. Saks, M.E., Srinivasan, A., Zhou, S., Zuckerman, D.: Low discrepancy sets yield approximate min-wise independent permutation families. Inf. Process. Lett. 73(1-2), 29–32 (2000)
15. Shannon, C.: Communication theory of secrecy systems. Bell System Technical Journal 28(4), 656–715 (1949)

Slide Cryptanalysis of Lightweight Stream Cipher RAKAPOSHI

Takanori Isobe[1,3], Toshihiro Ohigashi[2], and Masakatu Morii[3]

[1] Sony Corporation,
1-7-1 Konan, Minato-ku, Tokyo 108-0075, Japan
Takanori.Isobe@jp.sony.com
[2] Hiroshima University,
1-4-2 Kagamiyama, Higashi-Hiroshima, Hiroshima 739-8511, Japan
ohigashi@hiroshima-u.ac.jp
[3] Kobe University,
1-1 Rokkoudai, Nada-ku, Kobe 657-8501, Japan
mmorii@kobe-u.ac.jp

Abstract. In this paper, we analyze a slide property of RAKAPOSHI stream cipher. To begin, we show that any Key-IV pair has a corresponding *slide* Key-IV pair that generates an n-bit shifted keystream with probability of 2^{-2n}. Then we exploit this property in order to develop a key recovery attack on RAKAPOSHI in the related key setting. Our attack is able to recover a 128-bit key with time complexity of 2^{41} and 2^{38} chosen IVs. The result reveals that RAKAPOSHI is vulnerable to the related key attack. After that, we consider a variant of the slide property, called *partial slide* property. It enables us to construct a method for speeding up the brute force attack by a factor of 2 in the single key setting. Finally, we consider a slide property of K2 v2.0 stream cipher, and discuss the possibility of an attack exploiting the slide property.

Keywords: stream cipher, slide attack, related-key attack, RAKAPOSHI, K2 v2.0, initialization process.

1 Introduction

In recent years, with the large deployment of low resource devices such as RFID tags and sensor nodes, the demand for security in resource-constrained environments has been dramatically increased. As a result, lightweight cryptography is attracting attention of the cryptographic community. In fact, a number of lightweight primitives are proposed, e.g., block ciphers : PRESENT [6], KATAN [8], LED [14] and Piccolo [23], and hash functions : Quark [3], PHOTON [13] and SPONGENT [5]. As for lightweight stream ciphers, Grain v1 [16], Trivium [7] and MICKY2.0 [4] are selected by eSTREAM project for hardware applications with highly restricted resources [12]. In spite of considerable efforts in a multi-years project, it is still debatable that design and analysis of stream ciphers are mature enough. Indeed, after the end of this project in 2008, F-FCSR-H [2], which is initially contained in the final portfolio, and the 128-bit version of Grain are broken [15,11].

G. Hanaoka and T. Yamauchi (Eds.): IWSEC 2012, LNCS 7631, pp. 138–155, 2012.

Due to these facts, Cid et al. proposed a lightweight stream cipher RAKAPOSHI [10] after the eSTREAM project. RAKAPOSHI is a stream cipher supporting a 128-bit key and a 192-bit IV, and employs a bit-oriented Dynamic Linear Feedback Shift Register. This structure is also adopted in K2 v2.0 [19], which is recently selected in ISO standard stream ciphers [1] and currently discussed for inclusion into CRYPTREC [17]. RAKAPOSHI is considered as a variant of the K2 v2.0 for the low-cost hardware implementation. Its performance properties in hardware are comparable to stream ciphers selected in eSTREAM, e.g., the circuit size of RAKAPOSHI is estimated as about 3K gate. In addition, RAKAPOSHI can provide a 128 bit security while Grain and Trivium have only a 80-bit security. Thus, designers claim that RAKAPOSHI can complement the eSTREAM portfolio, and increase the choice of secure lightweight stream ciphers. Although RAKAPOSHI is an attractive lightweight stream cipher having notable features of design and implementations, there exist only designers' self evaluations, i.e., no external cryptanalysis has been published so far.

Our Contributions. In this paper, we analyze a slide property of RAKAPOSHI stream cipher. This property mainly exploits a weakness of an initialization algorithm, and has been applied to Grain v1 stream cipher [20,9]. Though designers claims that RAKAPOSHI is secure against attacks based on the weakness of the initialization algorithm, we demonstrate that a slide cryptanalysis is applicable to RAKAPOSHI.

To begin, by exploiting the self-similarity of the initialization algorithm of RAKAPOSHI, we show that any Key-IV pair has a corresponding *slide* Key-IV pair that generates an n-bit shifted keystream with probability of 2^{-2n}. For $n = 1$, a Key-IV pair has a corresponding Key-IV pair that generates a only 1-bit shifted keystream with probability of 2^{-2}, which is greatly high probability compared with an ideal stream cipher that generates a random keystream by Key-IV pair. Then we utilize this property in order to construct a related-key attack on RAKAPOSHI. Our attack can recover a 128-bit key with time complexity of 2^{41} and 2^{38} chosen IVs. This result reveals that RAKAPOSHI is vulnerable to the related key attack based on the slide property.

After that, we consider a variant of the slide property, which is called *partial slide property*, that occurs with higher probability than the basic slide property. Using this variant of the slide property, we give a method for speeding up the brute force attack in the single key setting by a factor of 2.

Finally, we consider a slide property of K2 v2.0, and discuss the possibility of an attack exploiting this property. Then, we show that K2 v2.0 has enough immunity against slide-type attacks.

Outline of the Paper. This paper is organized as follows. Brief descriptions of RAKAPOSHI and K2 v2.0 are given in Section 2. In Section 3, we introduce a slide property of stream ciphers, and we analyze a slide property of RAKAPOSHI stream cipher in Section 4. Then, related-Key attacks and a method for a speeding up a keysearch on RAKAPOSHI are given in Section 5 and 6, respectively. Section 7 consider a slide property of K2 v2.0. Finally, I conclude in Section 8.

2 Target Stream Ciphers

In this section, we give brief descriptions of RAKAPOSHI and K2 v2.0 stream ciphers.

2.1 Description of RAKAPOSHI

RAKAPOSHI is a stream cipher supporting a 128-bit key and a 192-bit IV. At time t, RAKAPOSHI consists of a 128-bit Non-linear Feedback Shift Register (NFSR) : $A^t = \{a_t, a_{t+1}, \ldots, a_{t+127}\}$ ($a_t \in \{0,1\}$), a 192-bit Linear Feedback Shift Register (LFSR) : $B^t = \{b_t, b_{t+1}, \ldots, b_{t+191}\}$ ($b_t \in \{0,1\}$) and an 8-to-1 nonlinear function v (see Fig. 1). Since RAKAPOSHI employs the bit-oriented Dynamic Linear Feedback Shift Register (DLFSR), two bits of the register A are used for dynamically updating the feedback function of the register B.

The NFSR A^t and the LFSR B^t are updated as follows:

$$\begin{aligned}
a_{t+128} &= g(a_t, a_{t+6}, a_{t+7}, a_{t+11}, a_{t+16}, a_{t+28}, a_{t+36}, a_{t+45}, a_{t+55}, a_{t+62}) \\
&= 1 \oplus a_t \oplus a_{t+6} \oplus a_{t+7} \oplus a_{t+11} \oplus a_{t+16} \oplus a_{t+28} \oplus a_{t+36} \oplus a_{t+45} \\
&\quad \oplus a_{t+55} \oplus a_{t+62} \oplus a_{t+7}a_{t+45} \oplus a_{t+11}a_{t+55} \oplus a_{t+7}a_{t+28} \\
&\quad \oplus a_{t+28}a_{t+55} \oplus a_{t+6}a_{t+45}a_{t+62} \oplus a_{t+6}a_{t+11}a_{t+62}, \\
b_{t+192} &= f(b_t, b_{t+14}, b_{t+37}, b_{t+41}, b_{t+49}, b_{t+51}, b_{t+93}, b_{t+107}, b_{t+120}, b_{t+134}, \\
&\quad b_{t+136}, b_{t+155}, b_{t+158}, b_{t+176}, c_0, c_1) \\
&= b_t \oplus b_{t+14} \oplus b_{t+37} \oplus b_{t+41} \oplus b_{t+49} \oplus b_{t+51} \oplus b_{t+93} \\
&\quad \oplus \overline{c_0} \cdot \overline{c_1} \cdot b_{t+107} \oplus \overline{c_0} \cdot c_1 \cdot b_{t+120} \oplus c_0 \cdot \overline{c_1} \cdot b_{t+134} \oplus c_0 \cdot c_1 \cdot b_{t+136} \\
&\quad \oplus \overline{c_0} \cdot b_{t+155} \oplus c_0 \cdot b_{t+158} \oplus b_{t+176},
\end{aligned}$$

where $\oplus$ is a bit-wise XOR, $\overline{x}$ is complement of x, and c_0 and c_1 are a_{t+41} and a_{t+89}, respectively. The 8-to-1 nonlinear function v is expressed as

$$s_t = v(a_{t+67}, a_{t+127}, b_{t+23}, b_{t+53}, b_{t+77}, b_{t+81}, b_{t+103}, b_{t+128}).$$

The details of the function v is given in Appendix A.

Initialization Process. A 128-bit key $K = \{k_0, k_1, \ldots, k_{127}\}$ ($k_i \in \{0,1\}$) and an initialization vector $IV = \{iv_0, iv_1, \ldots, iv_{191}\}$ ($iv_i \in \{0,1\}$) are loaded into the register A and B as follows:

$$a_i = k_i \ (0 \leq i \leq 127), \quad b_i = iv_i \ (0 \leq i \leq 191).$$

The initialization process updates the state 448 times without the keystream generation. It consists of a stage 1 (320 cycles) and a stage 2 (128 cycles). The difference of these stages is that s_t is XORed with b_{t+192} in the stage 1 and a_{t+128} in the stage 2, respectively. After the initialization process, the state $S^{448} = (\{A^{448}, B^{448}\})$ is obtained.

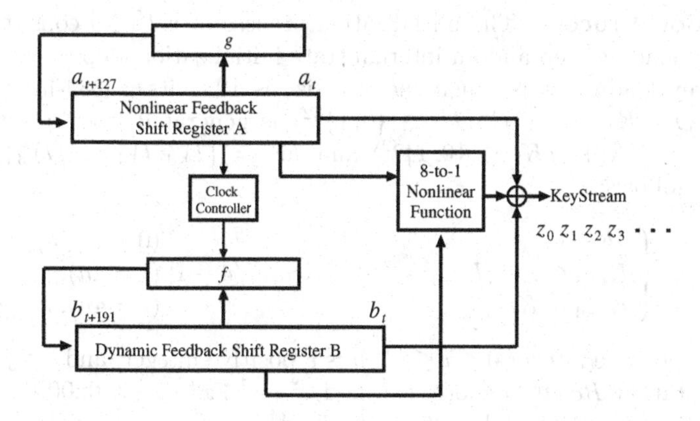

Fig. 1. RAKAPOSHI Stream Cipher

Keystream Generation. For $t \geq 448$, an internal state $S^t = (A^t, B^t)$ generates a keystream bit z_{t-448} such that $z_{t-448} = b_t \oplus a_t \oplus s_t$ with updating the internal state. Note that the fixed key and IV pair must be changed after 2^{64} keystream bits are generated.

2.2 Description of K2 v2.0

K2 v2.0 is a stream cipher supporting three key lengths: 128, 192, and 256 bits. The length of IV is 128 bits. K2 v2.0 consists of a 160-bit LFSR : $A^t = \{A_t, A_{t+1}, \ldots, A_{t+4}\}$ $(A_t \in \{0,1\}^{32})$, a 352-bit LFSR : $B^t = \{B_t, B_{t+1}, \ldots,$ $B_{t+10}\}$ $(B_t \in \{0,1\}^{32})$, and a non-linear function with four internal registers : $M^t = \{R1_t, R2_t, L1_t, L2_t\}$ $(R1_t, R2_t, L1_t, L2_t \in \{0,1\}^{32})$. Since K2 v2.0 also employs the DLFSR, the LFSR A and B are updated as follows:

$$A_{t+5} = \alpha_0 A_t \oplus A_{t+3}, \tag{1}$$
$$B_{t+11} = {\alpha_{12}}^{cl1_t} B_t \oplus B_{t+1} \oplus B_{t+6} \oplus {\alpha_3}^{cl2_t} B_{t+8}, \tag{2}$$

where ${\alpha_{12}}^{cl1_t} = {\alpha_1}^{cl1_t} + {\alpha_2}^{1-cl1_t} - 1$ and $\alpha_0, \alpha_1, \alpha_2, \alpha_3$ are 32-to-32 bit substitutions. Clock control bits $cl1_t$ and $cl2_t$ are described as,

$$cl1_t = A_{t+2}[30] \ \in \{0,1\},\ cl2_t = A_{t+2}[31] \ \in \{0,1\},$$

where $A_t[y]$ is the y-th bit of A_t. The non-linear function is expressed as

$$R1_{t+1} = Sub(L2_t \boxplus B_{t+9}),\ R2_{t+1} = Sub(R1_t), \tag{3}$$
$$L1_{t+1} = Sub(R2_t \boxplus B_{t+4}),\ L2_{t+1} = Sub(L1_t), \tag{4}$$

where $Sub(\cdot)$ is a 32-to-32 bit substitution and $\boxplus$ denotes 32-bit addition. For the details of the function, see [19].

Initialization Process. The initialization process of K2 v2.0 consists of two steps, a key loading step and an internal state initialization step.

In the key loading step, when the key size is 128 bits, a 384-bit extended key $K = \{K_0, K_1, \ldots, K_{11}\}$ ($K_i \in \{0,1\}^{32}$) is generated from a key $IK = \{IK_0, IK_1, \ldots, IK_3\}$ ($IK_i \in \{0,1\}^{32}$) and $IV = \{IV_0, IV_1, \ldots, IV_3\}$ ($IV_i \in \{0,1\}^{32}$) as follows:

$$K_i = \begin{cases} IK_i & (0 \le i \le 3), \\ K_{i-4} \oplus Sub(K_{i-1} \lll 8) \oplus Rcon[i/4-1] & (i = 4n), \\ K_{i-4} \oplus K_{i-1} & (i \ne 4n), \end{cases} \tag{5}$$

where i is a positive integer $0 \le i \le 11$, n is a positive integer, and $\lll j$ denotes j bits left rotation. $Rcon[j]$ denotes ($x^j \bmod x^8+x^4+x^3+x+1$, 0x00, 0x00, 0x00) and x is 0x02. For a 192-bit key, K is obtained as,

$$K_i = \begin{cases} IK_i & (0 \le i \le 5), \\ K_{i-6} \oplus Sub(K_{i-1} \lll 8) \oplus Rcon[i/6-1] & (i = 6), \\ K_{i-6} \oplus K_{i-1} & (7 \le i \le 11). \end{cases} \tag{6}$$

For a 256-bit key, K is obtained as,

$$K_i = \begin{cases} IK_i & (0 \le i \le 7), \\ K_{i-8} \oplus Sub(K_{i-1} \lll 8) \oplus Rcon[i/8-1] & (i = 8), \\ K_{i-8} \oplus K_{i-1} & (9 \le i \le 11). \end{cases} \tag{7}$$

Then, the internal state is initialized with K and IV as follows:

$$A_m = K_{4-m} \ (m = 0, 1, \ldots, 4),\ B_0 = K_{10},\ B_1 = K_{11},\ B_2 = IV_0,\ B_3 = IV_1,$$
$$B_4 = K_8,\ B_5 = K_9,\ B_6 = IV_2,\ B_7 = IV_3,\ B_8 = K_7,\ B_9 = K_5,\ B_{10} = K_6.$$
$$R1_0 = R2_0 = L1_0 = L2_0 = 0.$$

In the internal state initialization step, the internal state is updated 24 times ($t = 0, 1, \ldots, 23$) without the keystream generation. The internal state A_{t+5} and B_{t+11} are obtained as follows:

$$A_{t+5} = \alpha_0 A_t \oplus A_{t+3} \oplus z^L_{t-24}, \tag{8}$$
$$B_{t+11} = \alpha_{12}{}^{cl1_t} B_t \oplus B_{t+1} \oplus B_{t+6} \oplus \alpha_3^{cl2_t} B_{t+8} \oplus z^H_{t-24}. \tag{9}$$

At the same time, the internal register M_t is updated by eqs. (3) and (4).

Keystream Generation. For $t \ge 24$, an internal state $S^t = (A^t, B^t, M^t)$ generates a 64-bit keystream $z^{t-24} = \{z^L_{t-24}, z^H_{t-24}\}$ as follows:

$$z^L_{t-24} = (B_t \boxplus R2_t) \oplus R1_t \oplus A_{t+4},$$
$$z^H_{t-24} = (B_{t+10} \boxplus L2_t) \oplus L1_t \oplus A_t.$$

The fixed key and IV pair must be changed after 2^{64} keystream bits are generated.

3 Slide Property of Stream Cipher

If two different keys always convert same plaintexts into same ciphertexts, such a key pair is called equivalent key in terms of that these keys are functionally equivalent. Since stream ciphers additionally use IV for generating a keystream, equivalent Key-IV pairs can also be defined. Here, a ciphertext is obtained by XORing a plaintext with a keystream in stream ciphers. Thus, an equivalent Key-IV pair essentially means the pair generating same keystreams. The existence of these pairs indicates that effective (K, IV) space is smaller than the expected value which is the sum of the K and IV size. Also it may be exploited for a key recovery attack such as attacks in [24,18].

In stream ciphers, a variant of equivalent (K, IV) called *slide Key-IV pairs* is also defined in [20,9]. A slide Key-IV pair generates same keystream but $w \cdot n$-bit shifted, where w is the size of keystream z_t, e.g., RAKAPOSHI is $w = 1$ and K2 v2.0 is $w = 64$. Though the existence of this pair does not directly affect the effective (K, IV) space unlike the case of equivalent Key-IV pairs, it has the following interesting property.

Let $(K'_{(n)}, IV'_{(n)})$ be $w \cdot n$-bit slide Key-IV pair of (K, IV). In other words, $(K'_{(n)}, IV'_{(n)})$ generates the $w \cdot n$-bit shifted keystream with respected to that of (K, IV) such that $z'_t = z_{t+n}$ for $0 < t$. Suppose that plaintexts $P = \{p_0, p_1, \ldots, p_L\}$ and $P' = \{p'_0, p'_1, \ldots, p'_L\}$ are encrypted with (K, IV) and $(K'_{(n)}, IV'_{(n)})$, respectively. Then, ciphertexts $C = \{c_0, c_1, \ldots, c_L\}$ and $C' = \{c'_0, c'_1, \ldots, c'_L\}$ are as follows:

$$\begin{aligned} C &= \{c_0, c_1, \ldots, c_L\} = \{p_0 \oplus z_0, p_1 \oplus z_1, \ldots, p_L \oplus z_L\}, \\ C' &= \{c'_0, c'_1, \ldots, c'_L\} = \{p'_0 \oplus z'_0, p'_1 \oplus z'_1, \ldots, p'_L \oplus z'_L\} \\ &= \{p'_0 \oplus z_n, p'_1 \oplus z_{n+1}, \ldots, p'_L \oplus z_{n+L}\}. \end{aligned}$$

If an attacker can gets above ciphertexts which are generated from $w \cdot n$-bit slide Key-IV pairs, he can obtain information of plaintexts from only ciphertexts without knowledge of keys by XORing $w \cdot n$-bit shifted C to C' as follows:

$$\begin{aligned} c_{n+t} \oplus c'_t &= p_{n+t} \oplus z_{n+t} \oplus p'_t \oplus z_{n+t} \\ &= p_{n+t} \oplus p'_t. \end{aligned}$$

At first glance, this assumption seems to be very strong. However, it corresponds to the related-key and chosen IV setting for some classes of stream ciphers[1]. Beside, a slide Key-IV pair can be used not only for exposing plaintext information but also for related-key key recovery attacks [20,9]. Moreover, it may be able to utilize for a speed-up key search in the single key setting [9].

Therefore, the slide property is a very useful tool of analyses and evaluations of the security of stream ciphers.

4 Slide Property of RAKAPOSHI

In this section, we analyze a slide property of RAKAPHOSHI stream cipher.

[1] It highly depends on structures and algorithms of target stream ciphers.

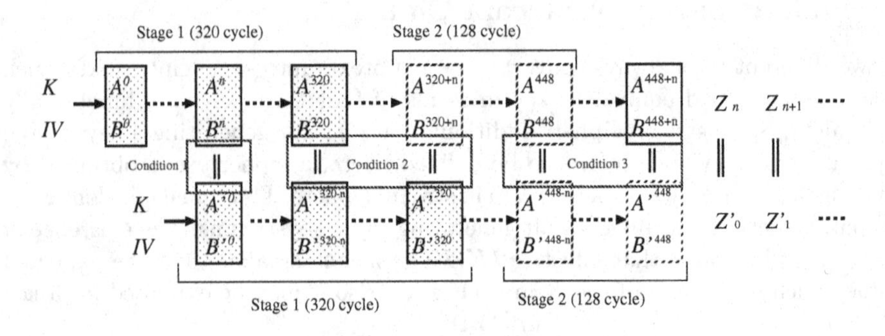

Fig. 2. n-bit slide pair of RAKAPOSHI

4.1 Conditions of Slide Pairs

For RAKAPOSHI, a (K, IV) pair has a corresponding n-bit *slide* pair $(K'_{(n)}, IV'_{(n)})$ that generates an n-bit shifted keystream for $0 \leq n < 320$, if these pairs satisfy following conditions:

$$\begin{aligned}
&\textbf{Condition 1}: \; S^n(=\{A^n, B^n\}) = S'^0(=\{A'^0, B'^0\}),\\
&\textbf{Condition 2}: \; s^{320+i} = 0 \;\; (0 \leq i < n),\\
&\textbf{Condition 3}: \; s^{448+i} = 0 \;\; (0 \leq i < n),
\end{aligned}$$

where S'^t is a state generated from $(K'_{(n)}, IV'_{(n)})$ at time t. Figure 2 illustrates these conditions for an n-bit slide pair.

Assume that the condition 1 holds, $S^{320}(=\{A^{320}, B^{320}\})$ and $S'^{320-n}(=\{A'^{320-n}, B'^{320-n}\})$ are identical, because S^n and S'^0 are updated in the same manner during the stage 1.

However, S^{320} and S'^{320-n} are updated by different update processes in the stage 1 and 2, respectively. As mentioned in Section 2.1, the difference of these stages is only usage of s_t, i.e., s_t is XORed with b_{t+192} in the stage 1 while it is XORed with a_{t+128} in the stage 2. When the condition 2 holds, the relation of $s^{320+i} = s'^{320-n+i} = 0$ is obtained for $0 \leq i < n$. It allows us to omit these differences of the stage 1 and 2, and then $S^{320+n}(=\{A^{320+n}, B^{320+n}\}) = S'^{320}(=\{A'^{320}, B'^{320}\})$. After that, since these states are updated in the same manner during the stage 2, $S^{448}(=\{A^{448}, B^{448}\})$ and $S'^{448-n}(=\{A'^{448-n}, B'^{448-n}\})$ are surely identical.

In the keystream generation, s_t is used for generating a keystream bits, and does not affect the state updating. Therefore, the condition 3 ensures that $S^{448+n}(=\{A^{448+n}, B^{448+n}\})$ and $S'^{448}(=\{A'^{448}, B'^{448}\})$ are identical. It means that (K', IV') produces n-bit sliding keystream with respect to (K, IV). In other words, the following equations holds, $z_i = z'_{i-n}$ $(n \leq i < 2^{64})$.

4.2 Evaluation

Let us estimate how many n -bit slide pair exist in RAKAPOSHI stream cipher. The condition 1 is expressed as

$$A^n = \{k_n, k_{n+1}, \ldots, k_{127}, x_0, \ldots, x_{n-1}\} = \{k'_0, k'_1, \ldots, k'_{127}\} = A'^0,$$
$$B^n = \{iv_n, iv_{n+1}, \ldots, iv_{191}, y_0, \ldots, y_{n-1}\} = \{iv'_0, iv'_1, \ldots, iv'_{191}\} = B'^0,$$

where $x_t = g(a_t, a_{t+6}, a_{t+7}, a_{t+11}, a_{t+16}, a_{t+28}, a_{t+36}, a_{t+46}, a_{t+55}, a_{t+62})$ and $y_t = f(b_t, b_{t+14}, b_{t+37}, b_{t+41}, b_{t+49}, b_{t+51}, b_{t+93}, b_{t+107}, b_{t+120}, b_{t+134}, b_{t+136}, b_{t+155}, b_{t+158}, b_{t+176}, a_{t+41}, a_{t+89}) \oplus s_t$. Since the state size and the sum of K and IV size are same, a (K, IV) pair surely has one pair of (K'_n, IV'_n) satisfying the condition 1 regardless of the value of n.

On the other hand, conditions 2 and 3 depend on the value of n. The probability that a (K, IV) pair satisfies the conditions 2 and 3 is $2^{-2n} (= 2^{-n} \times 2^{-n})$. Therefore, any (K, IV) pair theoretically has an n-bit slide pair $(K'_{(n)}, IV'_{(n)})$ that generates an n-bit shifted keystream with probability of 2^{-2n}. We have confirmed the correctness of this theoretical values by testing 2^{24} random chosen (K, IV) pairs for $n = 0, \ldots, 10$. Table 1 gives examples of 1 and 10 bits slide pairs. In addition, we can say that a (K, IV) pair having $(K'_{(n)}, IV'_{(n)})$ pairs also has $(K'_{(1)}, IV'_{(1)}) \ldots (K'_{(n-1)}, IV'_{(n-1)})$ pairs.

For $n = 1$, a (K, IV) pair has $(K'_{(1)}, IV'_{(1)})$ that generates a only 1-bit shifted keystream with probability of 2^{-2}, which is greatly high probability compared to an ideal stream cipher that generates a random keystream by (K, IV). If an attacker can access to stream ciphers using such a slide pair, it is easy to distinguish keystreams from random streams. Also, the ciphertext-only attack mentioned in Section 3 is feasible.

Table 1. Examples of slide pairs

1-bit slide pair			
K	IV	$K'_{(1)}$	$IV'_{(1)}$
4bdf973abdd66263$_x$ d4ef3bfb30609c57$_x$	49cba4aa656336eb$_x$ be0b3db8cc516480$_x$ 95b8910812c5c95b$_x$	97bf2e757bacc4c7$_x$ a9de77f660c138af$_x$	93974954cac66dd7$_x$ 7c167b7198a2c901$_x$ 2b712210258b92b7$_x$
keystream		keystream	
00100011001101010011010110011011$_2$ 10100010111111100111100100001000$_2$		01000110011010100110101100110111$_2$ 01000101111111001111001000010001$_2$	

10-bit slide pair			
K	IV	$K'_{(10)}$	$IV'_{(10)}$
d048119b66a37d84$_x$ d51287aef2f796d1$_x$	8b75aad54c32a2b6$_x$ f118a4764dd0560a$_x$ 88fc32827bc213cc$_x$	20466d9a8df61354$_x$ 4a1ebbcbde5b4738$_x$	d6ab5530ca8adbc4$_x$ 6291d93741582a23$_x$ f0ca09ef084f334b$_x$
keystream		keystream	
00100010100010100100010110010111$_2$ 10011100100100101001110011010111$_2$		00101001000101100101111001110010$_2$ 01001010011100110101110101001001$_2$	

5 Related-Key Attack on RAKAPOSHI

In this section, we exploit the slide property of RAKAPOSHI in order to construct a related-key attack on RAKAPOSHI. To begin, we give a method for determining a part of key bits by utilizing the 1-bit slide property. After that, we generalize it and propose a related-key attack on RAKAPOSHI based on the n-bit slide property.

5.1 Related-Key Attacks Using 1-Bit Slide Pair

Define the related key $K^*_{(1)}$ of this attack as [2]

$$K^*_{(1)} = \{k^*_0, k^*_1, \ldots, k^*_{127}\} = \{k_1, k_2, \ldots, k_{127}, x_0\}$$

where

$$\begin{aligned} x_0 &= g(a_0, a_6, a_7, a_{11}, a_{16}, a_{28}, a_{36}, a_{46}, a_{55}, a_{62}), \\ &= g(k_0, k_6, k_7, k_{11}, k_{16}, k_{28}, k_{36}, k_{46}, k_{55}, k_{62}). \end{aligned}$$

Since x_0 includes only key bits and does not depends on the value of IV, a related key K^* is determined if K is given. In the related key setting, an attacker knows that a pair of $(K, K^*_{(1)})$ holds this relation, though actual values of those are unknown.

This attack uses chosen IV pairs $(IV,\ IV^*_{(1)})$ satisfying following relation,

$$\begin{aligned} IV &= \{iv_1, iv_2, \ldots, iv_{191}, y_0\} \\ &= \{iv^*_0, iv^*_1, \ldots, iv^*_{191}\} = IV^*_{(1)}, \end{aligned}$$

where

$$\begin{aligned} y_0 &= f(b_0, b_{14}, b_{37}, b_{41}, b_{49}, b_{51}, b_{93}, b_{107}, b_{120}, b_{134}, b_{136}, b_{155}, \\ &\quad b_{158}, b_{176}, a_{41}, a_{89}) \oplus v(a_{67}, a_{127}, b_{23}, b_{53}, b_{77}, b_{81}, b_{103}, b_{128}) \\ &= f(iv_0, iv_{14}, iv_{37}, iv_{41}, iv_{49}, iv_{51}, iv_{93}, iv_{107}, iv_{120}, iv_{134}, iv_{136}, iv_{155}, \\ &\quad iv_{158}, iv_{176}, k_{41}, k_{89}) \oplus v(k_{67}, k_{127}, iv_{23}, iv_{53}, iv_{77}, iv_{81}, iv_{103}, iv_{128}). \end{aligned}$$

In the chosen-IV setting, an attacker is able to choose the values of IV freely. Given IV, we can determined $IV^*_{(1)}$ except iv^*_{191} $(= y_0)$, because y_0 includes four key bits, $\{k_{41}, k_{89}, k_{67}, k_{127}\}$, which are secret values even if in the related-key setting.

If the value of iv^*_{191} is correctly guessed, (K, IV) and $(K^*_{(1)}, IV^*_{(1)})$ satisfy the condition 1 regarding the 1-bit slide pair. Then, $(K^*_{(1)}, IV^*_{(1)})$ generates a 1-bit shifted keystream of (K, IV) with probability of 2^{-2}. Since the probability that iv^*_{191} is correctly guessed is 2^{-1}, we expect to obtain one 1-bit sliding keystream pair after testing 2^3 $(IV, IV^*_{(1)})$ pairs. Once such a $(IV, IV^*_{(1)})$ pair is found,

[2] This type of related keys has been utilized in attacks of Grain family [20,9].

we can confirm that $iv^*_{191}(=y_0)$ is correctly guessed. Then, a 1-bit equation of y_0, which includes 4 bit key bits of $\{k_{41}, k_{89}, k_{67}, k_{127}\}$, is obtained. Using four equations, $\{k_{41}, k_{89}, k_{67}, k_{127}\}$ can be determined with high probability.

The details of the attack procedure are given as follows:

1. Choose one pair of $(IV,\ IV^*_{(1)})$, where iv^*_{191} is guessed.
2. Obtain two keystreams of (K, IV) and $(K^*_{(1)}, IV^*_{(1)})$.
3. If these keystreams is the 1-bit sliding pair, then store the 1-bit equation of $\{k_{41}, k_{89}, k_{67}, k_{127}\}$ corresponding to iv^*_{191}.
4. Repeat step 1-3 until 4 equations are obtained.
5. Determine the key bits of $\{k_{41}, k_{89}, k_{67}, k_{127}\}$ by using four equations.
6. Obtain other 124 bits of the key in the brute force manner.

One equation can be obtained with probability of 2^{-3}. Thus, it is expected to repeat step 1-4 in 2^5 $(= 4 \times 2^3)$ times. The time complexity of the step 1-4 is 2^6 $(= 2 \times 2^5)$ initialization process, and the number of required chosen IVs is 2^6. In the step 5, we search $\{k_{41}, k_{89}, k_{67}, k_{127}\}$ by checking obtained equations. The time complexity of the step 5 is estimated as about less than 2^4 initialization process even if all these values are tested by four equations. Therefore, the whole time complexity is estimated as $2^{124}(\approx 2^4 + 2^6 + 2^{124})$ initialization process. This related key attack recovers a key with time complexity of 2^{124}, 2^6 chosen IVs and one related key.

5.2 Related-Key Attacks Using n-Bit Slide Pair

We extend the attack exploiting the 1-bit slide pair to an attack based on the n-bit slide pair. The related key $K^*_{(n)}$ and chosen IV pair are defined as,

$$
\begin{aligned}
K^*_{(n)} &= \{k^*_0, k^*_1, \ldots, k^*_{127}\} = \{k_n, k_{n+1}, \ldots, k_{127}, x_0, \ldots, x_{n-1}\}, \\
IV &= \{iv_n, iv_{n+1}, \ldots, iv_{191}, y_0, \ldots, y_{n-1}\} \\
&= \{iv^*_0, iv^*_1, \ldots, iv^*_{191}\} = IV^*_{(n)},
\end{aligned}
$$

assuming that the values of n is less than 127. Table 2 shows involved key bits of each y_t for $0 \le t \le 10$.

If the values of $\{y_0, \ldots, y_{n-1}\}$ are correctly guessed with probability of 2^{-n}, (K^*, IV^*) generate an n-bit sliding keystream of (K, IV) with probability of 2^{-2n}. Once we find such pairs, n equations regarding each value of $\{y_0, \ldots, y_{n-1}\}$ are obtained. If y_t includes m bits of the key, m independent equations of y_t are needed for determining m bits of the key.

As an example, let us consider the attack using a 4-bit slide pair. $\{y_0, \ldots, y_3\}$ includes 4, 13, 13 and 13 key bits, respectively, and in total these involve independent 41 key bits. If 13 independent equations regarding each y are obtained[3], we can determine key bits included in each equation. It implies that this attack requires 13 pairs of (IV, IV^*) causing a 4-bit sliding keystream. These pairs are

[3] For y_0, 4 independent equations are enough.

Table 2. Included key bits in each y_t

y_t	Included key bits
y_0	41, 67, 89, 127
y_1	42, 68, 90, (0, 6, 7, 11, 16, 28, 36, 45, 55, 62)
y_2	43, 69, 91, (1, 7, 8, 12, 17, 29, 37, 46, 56, 63)
y_3	44, 70, 92, (2, 8, 9, 13, 18, 30, 38, 47, 57, 64)
y_4	45, 71, 93, (3, 9, 10, 14, 19, 31, 39, 48, 58, 65)
y_5	46, 72, 94, (4, 10, 11, 15, 20, 32, 40, 49, 59, 66)
y_6	47, 73, 95, (5, 11, 12, 16, 21, 33, 41, 50, 60, 67)
y_7	48, 74, 96, (6, 12, 13, 17, 22, 34, 42, 51, 61, 68)
y_8	49, 75, 97, (7, 13, 14, 18, 23, 35, 43, 52, 62, 69)
y_9	50, 76, 98, (8, 14, 15, 19, 24, 36, 44, 53, 63, 70)
y_{10}	51, 77, 99, (9, 15, 16, 20, 25, 37, 45, 54, 64, 71)

obtained with time complexity of 2^{17} $(= 13 \times 2 \times 2^{3 \cdot 4})$ and $2^{17}(= 13 \times 2 \times 2^{3 \cdot 4})$ chosen IVs. Then, 41 bits of the key can be determined with complexity of $2^{15}(= 4 + 2^{13} + 2^{13} + 2^{13})$ by exhaustively checking obtained equations. Therefore, the whole time complexity is estimated as $2^{87}(= 2^{87}+2^{15}+2^{17})$ initialization process. This related attack recovers the key with time complexity of 2^{87} and 2^{17} chosen IV.

Using 11-bit slide pairs, each values of $\{y_0, \ldots, y_{11}\}$ includes 13 bits of the key except y_0 and in total these involve independent 88 key bits. 13 pairs of (IV, IV^*) causing a 10-bit sliding keystream are obtained with time complexity 2^{38} $(= 13 \times 2 \times 2^{3 \cdot 11})$ and $2^{38}(= 13 \times 2 \times 2^{3 \cdot 10})$ chosen IV. Then, in total 88 bits of the key are determined with complexity of $2^{17}(= 4 + 2^{13} \times 10)$ by exhaustively checking obtained equations. Therefore, the whole time complexity is estimated as $2^{41}(= 2^{40} + 2^{17} + 2^{38})$ initialization process. This related attack recovers the key with time complexity of 2^{41} and 2^{38} chosen IV.

Therefore, this result reveals that RAKAPOSHI is vulnerable to the related key attack.

6 Speed-Up Keysearch on RAKAPOSHI

In this section, we give a method for speeding up a keysearch in the single key setting. To begin, we consider a variant of the slide property, which is called *partial slide pair*. Then, this variant is utilized in order to construct a method for speeding up the brute force attack by a factor of 2.

6.1 Partial Slide Property of RAKAPOSHI

Recall that conditions 1-3 in Section 3.1 ensure that a (K, IV) pair has an n-bit slide pair $(K'_{(n)}, IV'_{(n)})$ that produces an n-bit sliding keystream of (K, IV). If the condition 3 does not holds, it is not ensured that $a_{448+128+i}$ and $a'_{448+128+i+n}$ are identical for $0 \leq i < n$, due to the difference of usage of s. However, these differences do no affect generations of $z_{n+1}(= z'_1), \ldots, z_{60}(= z'_{60-n})$. Thus, if only the conditions 1 and 2 holds, we can obtain the keystream pairs in which $\{z_{n+1}, \ldots, z_{60}\}$ and $\{z'_1, \ldots, z'_{60-n}\}$ are identical. We call such pair n-bit *partial slide pair*.

Therefore, a (K, IV) pair has an n-bit partial slide pair $(K''_{(n)}, IV''_{(n)})$ that generates an n-bit partial sliding keystream with probability of 2^{-n}, that occurs with higher probability than the basic slide property. For $n = 1$, a (K, IV) pair has a 1-bit partial slide pair $(K''_{(1)}, IV''_{(1)})$ with probability of 2^{-1}, where 59 bits of $\{z_2, \ldots, z_{60}\}$ and $\{z'_1, \ldots, z'_{59}\}$ are identical.

6.2 Speed-Up Keysearch Exploiting Partial Slide Pair

In order to improve the naive brute force attack, we exploit partial slide pairs. In particular, we utilize the observation that if the condition 2 regarding n-bit partial slide pairs holds, we can check n keys without recalculations of the initialization process.

Assume that an attacker aims to find K^{target} in the brute-force style search, i.e., test all keys with the keystream of $(K^{target}, IV^{target})$. Let us consider that a candidate pair of (K, IV) is set for the test. In the initialization process of (K, IV), if $s^{320+i} = 0$ (condition 2) holds for $0 \leq i < n$, then (K, IV) surely has $1, \ldots, n$ bit partial slide pairs such that $\{(K_1, IV_1), \ldots (K_n, IV_n)\} = \{(A_1, B_1), \ldots (A_n, B_n)\}$.

Then we can simultaneously verify n keys with only initialization call of (K, IV) by using additional keystreams of $\{(K^{target}, IV_1), \ldots (K^{target}, IV_n)\}$. Note that n bits of IV_n, namely $y_0, \ldots, y_n$, are uncontrollable and can not be fixed, while other $(192 - n)$ bits of IV_n is determined from IV^{target}. Thus, this attack requires a set of keystreams generated from $1 + 2^1 + 2^2 \ldots + 2^n$ chosen IVs.

The detailed algorithm is as follows:

1. Set $K = 0$ and $IV = IV^{target}$
2. Perform the initialization process and generate keystream bits $(z_0 \ldots z_{60})$.
3. For $t = 0$ to [smallest $0 \leq n < 60$ for which $s^{320+n+1} = 1$], check $(z_t \ldots z_{60})$ match the keystream of (K^{Target}, IV_t).
   - if matching, output $K^{target} = A^t$
4. Updating $K = A^{t+1}$, and Return step 2 only if $K \neq 0$.

As estimated in [9], this algorithm will eventually reach $K = 0$ again, because K is updated in the invertible way. Then, it is expected that this code check 2^{127} key values. The expected number of checked values of K in the step 3 for each loop of step 2-4 is 2 ($\approx 1 + 1 \cdot 1/4 + 2 \cdot 1/8 + \ldots$). Thus, the complexity of this algorithm is estimated as 2^{126} initialization processes of the step 2. If we can not find the target key, the algorithm can be repeated with different starting values which have a different cycle.

In order to estimate of the actual cost of the attack, we consider the case where $1-10$ bits partial slide pairs are used in this algorithm. Since the expected number of checked values of K in the step 3 is 2 ($\approx 1 + 1 \cdot 1/4 + 2 \cdot 1/8 + \ldots +$

$10 \cdot 1/1024$), time complexity for searching 2^{127} key values is estimate 2^{126}. After that, to cover all key space, we will check another cycle with same complexity. The while complexity is given as 2^{127} initialization processes. The number of the set of IV used for the attack is 2^{11} ($\approx 1 + 2 + 2^2 + \ldots + 2^{10}$).

Therefore, we can speed up keysearch by a factor two. In the Grain v1 attack [9], this type attack seems applicable in the case of that IV are all 1. Unlike the attack on Grain v1, our attack on RAKAPOSHI can be done for any IV while a set of chosen IVs are needed. This shows that RAKAPOSHI has a 127-bit security instead of 128 bits. However, this attack is a marginal improvement compared to the brute force attack. Thus, we do not claim this to be real attack based on algorithmic weakness.

7 Discussion : Slide Property of K2 v2.0

In this section, we analyze a slide property of K2 v2.0, and discuss the possibility of an attack exploiting the slide property.

7.1 Conditions of Slide pairs

For K2 v2.0, a $(IK'_{(n)}, IV'_{(n)})$ pair produces a $64n$-bit sliding keystream with respect to (IK, IV) if these pairs satisfy following conditions:

Condition 4: $S^n (= \{A^n, B^n, M^n\}) = S'^0 (= \{A'^0, B'^0, M'^0\})$,
Condition 5: $z^L_{24+i} = z^H_{24+i} = 0$, $(0 \leq i < n)$.

Suppose that the condition 4 holds, S^{24} $(= \{A^{24}, B^{24}, M^{24}\})$ and S'^{24-n} $(= \{A'^{24-n}, B'^{24-n}, M'^{24-n}\})$ are identical, because S^n and S'^0 are updated in same manner during the initialization process.

Though states of S^{24+i} and S'^{24-n+i} $(0 \leq i < n)$ are updated by using different update processes, the condition 3 ensures $S^{24+n} (= \{A^{24+n}, B^{24+n}, M^{24+n}\})$ and $S'^{24} (= \{A'^{24}, B'^{24}\})$ are identical for $(0 \leq i < n)$. Then, $(IK'_{(n)}, IV'_{(n)})$ produces $64n$-bit sliding keystream with respect to (IK, IV).

7.2 Analysis of 64-Bit Slide Pair

At first, we discuss a probability that a (IK, IV) pair has a 64-bit slide pair $(IK'_{(1)}, IV'_{(1)})$ that produces a 64-bit shifted keystream. According to the condition 4, the internal register must satisfy $M^1 = M'^0$. To achieve it, $R2'_0$ and $R2_1$ need to be identical, which is a partial condition of $M^1 = M'^0$. However, these values are always fixed as $R2'_0 = 0$ and $R2_1 = Sub(R1_0) = Sub(0) = \text{0x63636363}$. Thus, the condition 4 regarding 64-bit slide pair cannot be satisfied for any (K, IV). Table 3 shows a part of the values of input and output for function $Sub(\cdot)$. Therefore, there does not exist the 64-bit slide pair $(IK'_{(1)}, IV'_{(1)})$ for the 128, 192 and 256-bit key size.

Table 3. Relations between input x and output y of a function $y = Sub(x)$

y	x
0x00000000	0x52525252
0x52525252	0x48484848
0x63636363	0x00000000

7.3 Analysis of 128-Bit Slide Pair

Here, we discuss a probability of that a (IK, IV) pair has a 128-bit slide pair $(IK'_{(2)}, IV'_{(2)})$, which produces 128-bit shifted keystream.

Details of Condition of 128-bit Slide Pair : According the condition 4, the internal register must satisfy $M^2 = M'^0$ as follows:

$$\begin{aligned} R1_2(&= Sub(Sub(0) \boxplus K_6)) = R1'_0(=0),\\ R2_2(&= Sub(Sub(K_5)) = R2'_0(=0),\\ L1_2(&= Sub(0) \boxplus K_9) = L1'_0(=0),\\ L2_2(&= Sub(Sub(K_8)) = L2'_0(=0). \end{aligned}$$

From Table 3, above conditions are rewritten as follows:

$$K_5 = \text{0x48484848},\ K_6 = \text{0xEEEEEEEF}, \tag{10}$$

$$K_8 = \text{0x48484848},\ K_9 = \text{0xEEEEEEEF}. \tag{11}$$

In addition, the LFSR-A and LFSR-B must satisfy $A^2 = A'^0$ and $B^2 = B'^0$ as follows:

$$\begin{aligned} A^2 &= \{A_2, A_3, A_4, \alpha_0 A_0 \oplus A_3 \oplus z^L_{-24}, \alpha_0 A_1 \oplus A_4 \oplus z^L_{-23}\}\\ &= \{A'_0, A'_1, A'_2, A'_3, A'_4\} = A'^0,\\ B^2 &= \{B_2, B_3, B_4, B_5, B_6, B_7, B_8, B_9, B_{10},\\ &\quad \alpha_{12}{}^{cl1_0} B_0 \oplus B_1 \oplus B_6 \oplus \alpha_3^{cl2_0} B_8 \oplus z^H_{-24},\\ &\quad \alpha_{12}{}^{cl1_1} B_1 \oplus B_2 \oplus B_7 \oplus \alpha_3^{cl2_1} B_9 \oplus z^H_{-23}\}\\ &= \{B'_0, B'_1, B'_2, B'_3, B'_4, B'_5, B'_6, B'_7, B'_8, B'_9, B'_{10}\} = B'^0, \end{aligned}$$

where

$$\begin{aligned} z^L_{-24} &= (B_0 \boxplus R2_0) \oplus R1_0 \oplus A_4 = K_{10} \oplus K_0,\\ z^L_{-23} &= (B_1 \boxplus R2_1) \oplus R1_1 \oplus A_5\\ &= (K_{11} \boxplus Sub(0)) \oplus Sub(K_5) \oplus \alpha_0 K_4 \oplus K_1 \oplus K_{10} \oplus K_0,\\ z^H_{-24} &= (B_{10} \boxplus L2_0) \oplus L1_0 \oplus A_0 = K_6 \oplus K_4,\\ z^H_{-23} &= (B_{11} \boxplus L2_1) \oplus L1_1 \oplus A_1\\ &= ((\alpha_{12}{}^{cl1_0} K_{10} \oplus K_{11} \oplus IV_2 \oplus \alpha_3{}^{cl2_0} K_7 \oplus K_6 \oplus K_4) \boxplus Sub(0)) \oplus\\ &\quad Sub(K_8) \oplus K_3. \end{aligned}$$

Above conditions are rewritten as follows:

$$K'_0 = \alpha_0 K_3 \oplus (K_{11} \boxplus Sub(0)) \oplus Sub(K_5) \oplus \alpha_0 K_4 \oplus K_1 \oplus K_{10}, \quad (12)$$

$$K'_1 = \alpha_0 K_4 \oplus K_1 \oplus K_{10} \oplus K_0,\ K'_2 = K_0,\ K'_3 = K_1,\ K'_4 = K_2, \quad (13)$$

$$K'_5 = {\alpha_{12}}^{cl1_0} K_{10} \oplus K_{11} \oplus IV_2 \oplus {\alpha_3}^{cl2_0} K_7 \oplus K_6 \oplus K_4, \quad (14)$$

$$\begin{aligned} K'_6 = {} & {\alpha_{12}}^{cl1_1} K_{11} \oplus IV_0 \oplus IV_3 \oplus {\alpha_3}^{cl2_1} K_5 \oplus \\ & (({\alpha_{12}}^{cl1_0} K_{10} \oplus K_{11} \oplus IV_2 \oplus {\alpha_3}^{cl2_0} K_7 \oplus K_6 \oplus K_4) \boxplus Sub(0)) \oplus \\ & Sub(K_8) \oplus K_3, \end{aligned} \quad (15)$$

$$K'_7 = K_6, \quad (16)$$

$$K'_8 = IV_2,\ K'_9 = IV_3,\ K'_{10} = IV_0,\ K'_{11} = IV_1, \quad (17)$$

$$IV'_0 = K_8,\ IV'_1 = K_9,\ IV'_2 = K_7,\ IV'_3 = K_5. \quad (18)$$

According to the condition 5, the relation of $z_0^L = z_0^H = z_1^L = z_1^H = 0$ is given. The probability of that this condition holds is $1/2^{128}$.

128-bit key: For a 128-bit key, the relation of $K_9 = K_5 \oplus K_8$ always holds from eq. (5). This equation are not satisfied in conjunction with eqs. (10) and (11). Hence, there does not exist any 128-bit slide pair $(IK'_{(2)}, IV'_{(2)})$ for the 128-bit key size.

192-bit key: For a 192-bit key, from eqs. (6), (10), (11), and Table 3, four conditions of IK_0 = 0xBDBCBCBD, $IK_1 \oplus IK_2$ = 0xA6A6A6A7, IK_3 = 0xA6A6A6A7, and IK_5 = 0x48484848 are obtained. This 128-bit condition reduces a key space of IK to 2^{64} from 2^{192}.

Assume that (IK, IV) pair satisfy eqs. (10) and (11), there is one candidate of $(IK'_{(2)}, IV'_{(2)})$, which satisfy relations eqs. (12)–(14) and (18), because these are fully controlled by the values of $(IK'_{(2)}, IV'_{(2)})$. The remaining six conditions of eqs. (15)–(17) hold with probability of $1/2^{192}$. Therefore, a probability that a (IK, IV) pair satisfying eqs. (10) and (11) has $(IK'_{(2)}, IV'_{(2)})$ is $1/2^{320}$ $(=1/2^{192} \cdot 1/2^{128})$. Since the number of all candidates of (IK, IV) that satisfy eqs. (10) and (11) is 2^{192}, the expected number of slide Key-IV pairs for 192-bit key on all Key-IV space is $1/2^{128}$. It is negligibly-small. Therefore, it can be said that there does not exist any 128-bit slide pair $(IK'_{(2)}, IV'_{(2)})$ for the 192-bit key size.

256-bit key: For a 256-bit key, from eqs. (7), (10), (11), and Table 3, four conditions of $IK_0 \oplus Sub(IK_7 \lll 8)$ = 0x49484848, IK_1 = 0xA6A6A6A7, IK_5 = 0x48484848, and IK_6 = 0xEEEEEEEF are obtained. This 128-bit condition reduces a key space of IK to 2^{128} from 2^{256}. Assume that (IK, IV) pair satisfy eqs. (10) and (11), there is one candidate of $(IK'_{(2)}, IV'_{(2)})$, which satisfy relations eqs. (12)–(16) and (18). The remaining four relations (17) hold with probability of $1/2^{128}$. Therefore, a probability that the (IK, IV) pair satisfying eqs. (10) and (11) has $(IK'_{(2)}, IV'_{(2)})$ is $1/2^{256} (= 1/2^{128} \cdot 1/2^{128})$. Since the number of all candidates of (IK, IV) that satisfy eqs. (10) and (11) is 2^{256}, the expected value of number of slide Key-IV pairs for 256-bit key on all Key-IV space is only one. We think the number of this slide Key-IV pairs is not enough to execute key recovery attack. It is also negligibly-small. Therefore, it can be said that there does not exist any 128-bit slide pair $(IK'_{(2)}, IV'_{(2)})$ for the 256-bit key size.

For $64n$-bit slide pair $n \geq 3$, the probability of existence of it is obviously smaller than that of 128-bit slide pair. As a result, it seems to be difficult to construct attacks based on slide property to K2 v2.0 stream cipher in our evaluations.

8 Conclusion

This paper has investigated slide properties of RAKAPOSHI and K2 v2.0 stream ciphers.

Firstly, we have shown that for RAKAPOSHI, any Key-IV pair has a corresponding *slide* Key-IV pair that generates a n-bit shifted keystream with probability of 2^{-2n}. Then we exploited this property in order to construct the related-key attack on RAKAPOSHI. In this attack, we can recover a 128-bit key with time complexity of 2^{41} and 2^{38} chosen IVs. After that, we gave the variant of the slide property to construct the method for speeding up the brute force attack by a factor of 2 in the single key setting.

These results mainly exploit the self-similarity of the state update function of RAKAPOSHI. If the self-similarity is destroyed, this type attack can be avoided. For example, inserting a round constant or a counter value in each step is effective for preventing the attack presented in this paper.

Finally, we considered a slide property of K2 v2.0, and discuss the possibility of an attack exploiting the slide property. As a result, we have shown that K2 v2.0 has enough immunity against slide-type attacks. These are first evaluations with respect to slide properties of K2 v2.0. We believe that these results are meaningful for the accurate security evaluation of K2 v2.0.

Acknowledgments. The authors would like to thank to the anonymous referees for their fruitful comments and suggestions. This work was supported in part by Grant-in-Aid for Scientific Research (C) (KAKENHI 23560455) for Japan Society for the Promotion of Science.

References

1. ISO/IEC 18033-4. Amendment 1 - Information technology - security techniques - Encryption algorithms - Part 4: Stream ciphers, JTC 1/SC 27 (IT security tech.) (2011), `http://www.iso.org`
2. Arnault, F., Berger, T.P.: F-FCSR: Design of a New Class of Stream Ciphers. In: Gilbert, H., Handschuh, H. (eds.) FSE 2005. LNCS, vol. 3557, pp. 83–97. Springer, Heidelberg (2005)

3. Aumasson, J.-P., Henzen, L., Meier, W., Naya-Plasencia, M.: QUARK: A Lightweight Hash. In: Mangard, S., Standaert, F.-X. (eds.) CHES 2010. LNCS, vol. 6225, pp. 1–15. Springer, Heidelberg (2010)
4. Babbage, S., Dodd, M.: The MICKEY Stream Ciphers. In: Robshaw and Billet [22], pp. 191–209
5. Bogdanov, A., Knezevic, M., Leander, G., Toz, D., Varici, K., Verbauwhede, I.: SPONGENT: A Lightweight Hash Function. In: Preneel and Takagi [21], pp. 312–325
6. Bogdanov, A., Knudsen, L.R., Leander, G., Paar, C., Poschmann, A., Robshaw, M.J.B., Seurin, Y., Vikkelsoe, C.: PRESENT: An Ultra-Lightweight Block Cipher. In: Paillier, P., Verbauwhede, I. (eds.) CHES 2007. LNCS, vol. 4727, pp. 450–466. Springer, Heidelberg (2007)
7. De Cannière, C., Preneel, B.: Trivium. In: Robshaw and Billet [22], pp. 244–266
8. De Cannière, C., Dunkelman, O., Knežević, M.: KATAN and KTANTAN — A Family of Small and Efficient Hardware-Oriented Block Ciphers. In: Clavier, C., Gaj, K. (eds.) CHES 2009. LNCS, vol. 5747, pp. 272–288. Springer, Heidelberg (2009)
9. De Cannière, C., Küçük, Ö., Preneel, B.: Analysis of Grain's Initialization Algorithm. In: Vaudenay, S. (ed.) AFRICACRYPT 2008. LNCS, vol. 5023, pp. 276–289. Springer, Heidelberg (2008)
10. Cid, C., Kiyomoto, S., Kurihara, J.: The RAKAPOSHI Stream Cipher. In: Qing, S., Mitchell, C.J., Wang, G. (eds.) ICICS 2009. LNCS, vol. 5927, pp. 32–46. Springer, Heidelberg (2009)
11. Dinur, I., Güneysu, T., Paar, C., Shamir, A., Zimmermann, R.: An Experimentally Verified Attack on Full Grain-128 Using Dedicated Reconfigurable Hardware. In: Lee, D.H., Wang, X. (eds.) ASIACRYPT 2011. LNCS, vol. 7073, pp. 327–343. Springer, Heidelberg (2011)
12. The eSTREAM Project, `http://www.ecrypt.eu.org/stream`
13. Guo, J., Peyrin, T., Poschmann, A.: The PHOTON Family of Lightweight Hash Functions. In: Rogaway, P. (ed.) CRYPTO 2011. LNCS, vol. 6841, pp. 222–239. Springer, Heidelberg (2011)
14. Guo, J., Peyrin, T., Poschmann, A., Robshaw, M.J.B.: The LED Block Cipher. In: Preneel and Takagi [21], pp. 326–341
15. Hell, M., Johansson, T.: Breaking the Stream Ciphers F-FCSR-H and F-FCSR-16 in Real Time. J. Cryptology 24(3), 427–445 (2011)
16. Hell, M., Johansson, T., Maximov, A., Meier, W.: The Grain Family of Stream Ciphers. In: Robshaw and Billet [22], pp. 179–190
17. Imai, H., Yamagishi, A.: CRYPTREC (Japanese Cryptographic Algorithm Evaluation Project). In: van Tilborg, H.C.A., Jajodia, S. (eds.) Encyclopedia of Cryptography and Security, 2nd edn., pp. 285–288. Springer (2011)
18. Isobe, T., Ohigashi, T., Kuwakado, H., Morii, M.: A Chosen-IV Key Recovery Attack on Py and Pypy. IEICE Transactions 92-D(1), 32–40 (2009)
19. Kiyomoto, S., Tanaka, T., Sakurai, K.: K2: A Stream Cipher Algorithm using Dynamic Feedback Control. In: Hernando, J., Fernández-Medina, E., Malek, M. (eds.) SECRYPT, pp. 204–213. INSTICC Press (2007)
20. Lee, Y., Jeong, K., Sung, J., Hong, S.: Related-Key Chosen IV Attacks on Grain-v1 and Grain-128. In: Mu, Y., Susilo, W., Seberry, J. (eds.) ACISP 2008. LNCS, vol. 5107, pp. 321–335. Springer, Heidelberg (2008)
21. Preneel, B., Takagi, T. (eds.): CHES 2011. LNCS, vol. 6917. Springer, Heidelberg (2011)

22. Robshaw, M.J.B., Billet, O. (eds.): New Stream Cipher Designs. LNCS, vol. 4986. Springer, Heidelberg (2008)
23. Shibutani, K., Isobe, T., Hiwatari, H., Mitsuda, A., Akishita, T., Shirai, T.: Piccolo: An Ultra-Lightweight Blockcipher. In: Preneel and Takagi [21], pp. 342–357
24. Wu, H., Preneel, B.: Differential Cryptanalysis of the Stream Ciphers Py, Py6 and Pypy. In: Naor, M. (ed.) EUROCRYPT 2007. LNCS, vol. 4515, pp. 276–290. Springer, Heidelberg (2007)

Appendix

A Rakaposhi Non-Linear Function

The non-linear function v is given as follows.

$$v(x0, x1, x2, x3, x4, x5, x6, x7) =$$
$$x_0x_1x_2x_3x_4x_5x_6 + x_0x_1x_2x_3x_4x_5 + x_0x_1x_2x_3x_4x_6 + x_0x_1x_2x_3x_5x_6x_7 +$$
$$x_0x_1x_2x_3x_5x_6 + x_0x_1x_2x_3x_5x_7 + x_0x_1x_2x_3x_5 + x_0x_1x_2x_3x_6x_7 +$$
$$x_0x_1x_2x_4x_5x_6 + x_0x_1x_2x_4 + x_0x_1x_2x_5x_6 + x_0x_1x_2x_5x_7 + x_0x_1x_2x_7 +$$
$$x_0x_1x_2 + x_0x_1x_3x_4x_5x_6x_7 + x_0x_1x_3x_4x_5x_7 + x_0x_1x_3x_4x_5 + x_0x_1x_3x_4x_7 +$$
$$x_0x_1x_3x4 + x_0x_1x_3x_6 + x_0x_1x_4x_5x_6x_7 + x_0x_1x_4x_5x_6 + x_0x_1x_4x_5x_7 +$$
$$x_0x_1x_4x_6x_7 + x_0x_1x_4x_7 + x_0x_1x_5x_6x_7 + x_0x_1x_5x_6 + x_0x_1x_5 + x_0x_1x_6 +$$
$$x_0x1 + x_0x_2x_3x_4x_5x_6 + x_0x_2x_3x_4x_5x_7 + x_0x_2x_3x_4 + x_0x_2x_3x_5x_6x_7 +$$
$$x_0x_2x_3x_5x_6 + x_0x_2x_3x_5x_7 + x_0x_2x_3x_6 + x_0x_2x_4x_5x_6x_7 + x_0x_2x_5x_6 +$$
$$x_0x2x_5 + x_0x_2x_6x_7 + x_0x_2x_7 + x_0x_3x_4x_5x_6x_7 + x_0x_3x_4x_5x_6 + x_0x_3x_4x_5x_7 +$$
$$x_0x_3x_4x_5 + x_0x_3x_4x_7 + x_0x_3x_5x_6x_7 + x_0x_3x_5 + x_0x_3x_6 + x_0x_3 + x_0x_4x_5x_6 +$$
$$x_0x_4x_6x_7 + x_0x_5x_6 + x_0x_6 + x_0 + x_1x_2x_3x_4 + x_1x_2x_3x_5x_6 + x_1x_2x_3x_5x_7 +$$
$$x_1x_2x_3x_5 + x_1x_2x_3 + x_1x_2x_4x_5x_6 + x_1x_2x_4x_6 + x_1x_2x_4 + x_1x_2x_5 + x_1x_2 +$$
$$x_1x_3x_4x_5x_6x_7 + x_1x_3x_4x_5x_7 + x_1x_3x_4x_6x_7 + x_1x_3x_4x_6 + x_1x_3x_4 +$$
$$x_1x_3x_5x_6 + x_1x_3x_5 + x_1x_3x_6 + x_1x_3x_7 + x_1x_4x_5x_6x_7 + x_1x_4x_5x_7 +$$
$$x_1x_5x_6 + x_1x_5x_7 + x_1x_5 + x_1x_6x_7 + x_1x_6 + x_1 + x_2x_3x_4x_5x_6 + x_2x_3x_4x_5x_7 +$$
$$x_2x_3x_4x_5 + x_2x_3x_4x_6x_7 + x_2x_3x_4 + x_2x_3x_5x_7 + x_2x_3x_6x_7 + x_2x_3x_6 +$$
$$x_2x_4x_5x_6 + x_2x_4x_5x_7 + x_2x_4x_5 + x_2x_4x_6x_7 + x_2x_4x_6 + x_2x_4x_7 + x_2x_4 +$$
$$x_2x_5x_6x_7 + x_2x_6x_7 + x_2x_6 + x_2x_7 + x_3x_4x_5x_6x_7 + x_3x_4x_5 + x_3x_4x_6x_7 +$$
$$x_3x_4x_6 + x_3x_4x_7 + x_3x_5x_6x_7 + x_3x_6x_7 + x_3x_6 + x_3x_7 + x_4x_5x_6 + x_4x_5 +$$
$$x_5x_6x_7 + x_5x_6 + x_5 + x_6 + x_7.$$

Boomerang Distinguishers for Full HAS-160 Compression Function

Yu Sasaki[1], Lei Wang[2], Yasuhiro Takasaki[2], Kazuo Sakiyama[2], and Kazuo Ohta[2]

[1] NTT Secure Platform Laboratories, NTT Corporation, 3-9-11 Midori-cho, Musashino-shi, Tokyo 180-8585 Japan
sasaki.yu@lab.ntt.co.jp

[2] The University of Electro-Communications, 1-5-1 Choufugaoka, Choufu-shi, Tokyo, 182-8585 Japan
{takasaki,sakiyama,kazuo.ohta}@uec.ac.jp

Abstract. This paper studies a boomerang-attack-based distinguisher against full steps of the compression function of HAS-160, which is the hash function standard in Korea. The attack produces a second-order collision for the full steps of the compression function with a complexity of $2^{76.06}$, which is faster than the currently best-known generic attack with a complexity of 2^{80}. Previously Dunkelman *et al.* in 2009 applied a boomerang-based key-recovery attack on the internal block cipher of HAS-160. Because the goal of their attack is different from ours, the attack on the compression function has been reconstructed and optimized from scratch. As a result of the exhaustive search of the message difference, we found that the same message difference as theirs is the best choice for the first subcipher. We then propose some improvement to construct a differential characteristic from the message difference, which the probability of the characteristic increases from 2^{-47} to 2^{-44}. Thus our new characteristic also improves their key-recovery attack on the internal block cipher of HAS-160.

Keywords: HAS-160, hash function, 4-sum, second-order collision, boomerang attack.

1 Introduction

Hash functions are important cryptographic primitives. They are used for various purposes all over the world, so their security deserves to be carefully analyzed, especially since they are practically used. Hash functions are required to satisfy several fundamental properties such as preimage resistance, second-preimage resistance, and collision resistance. Recently, researchers have also investigated other weaker properties, e.g. distinguishers on the compression function and key recovery attacks on the internal block cipher.

HAS-160 is a hash function developed in Korea, and was standardized by the Korean government in 2000 [1]. The first cryptanalysis on HAS-160 was presented by Yun *et al.* [2] in 2005. They found that a collision for HAS-160 reduced to

G. Hanaoka and T. Yamauchi (Eds.): IWSEC 2012, LNCS 7631, pp. 156–169, 2012.

45 steps out of 80 steps could be generated in a very small complexity. This was improved by Cho *et al.* [3] in 2006, which reported that a collision attack could be theoretically applied until 53 steps. This was further improved by Mendel and Rijmen [4] in 2007, where a real collision until 53 steps was generated and a differential characteristic yielding a 59-step collision was reported. After that, a preimage attack on 52 steps was proposed by Sasaki and Aoki in 2008 [5], and this was extended up to 68 steps by Hong *et al.* [6] in 2009. In 2009, Dunkelman *et al.* [7] proposed another cryptanalysis, which was a key recovery attack against the internal block cipher of HAS-160 with the related-key rectangle approach. This recovers a secret-key with 2^{155} chosen plaintexts and $2^{377.5}$ computations by using 4 simple relations in the key. So far, no attack has been known for the full steps of the hash function, compression function, or internal block-cipher with a complexity below 2^{160}.

In this paper, boomerang type differential properties are discussed. The boomerang attack was first proposed by Wagner for analyzing block ciphers [8]. It divides the cipher $E(\cdot)$ into two subparts E_0 and E_1 such that $E(\cdot) = E_1 \circ E_0(\cdot)$. Let the probabilities of differential paths for E_0 and E_1 be p and q, respectively. The boomerang attack exploits the fact that a second-order differential property with a probability p^2q^2 exists for the entire cipher E. Aumasson *et al.* [9] applied the boomerang attack to the internal cipher of the hash function Skein. However, the goal of the attack is still recovering the secret key. After that, Biryukov *et al.* [10] and Lamberger and Mendel [11] independently applied this property on the compression function so as to mount distinguishers. Then, Sasaki [12] showed the application of the framework of [10,11] to the MD4-family (using the single-branch structure) consisting of up to 5 rounds. Recently, Sasaki and Wang [13] have applied the framework to double-branch hash functions RIPEMD-128 and RIPEMD-160.

Our Results

In this paper, we propose a boomerang-attack-based distinguisher against full steps of the compression function of HAS-160. In our attack, the property which is required to the differential characteristic is different compared to the previous related-key rectangle attack for the internal block-cipher [7]. In general, for the block cipher analysis, maximizing the probability of the entire characteristic is important. On the other hand, for the hash function analysis, the internal chaining variable values are known to the attacker, and moreover a part of them can even be chosen by the attacker. Hence, a part of the characteristic can be satisfied very easily with the message modification technique [14]. Therefore, it is important to locate the low probability part of the characteristic to the step positions where the message modification cannot be applied. Due to this fact, the attack on the compression function needs to be reconstructed from scratch. In this paper, we firstly search for message differences suitable for our attack. As a result of the search, for the first half of the characteristic, we choose the same message difference as the previous work [7]. For the last half of the characteristic, we use a new message difference. Secondly, the message difference is propagated to chaining variables *i.e.*, the differential characteristic is constructed. At this stage, we introduce some improvement into the first half of the differential characteristic. A new differential characteristic with higher probability by a factor of 2^3 is derived, which improves the data complexity of the previous key-recovery attack on the internal block-cipher [7] by a factor of 2^3. Finally, we show that a second-order collision can be found for the full steps of the compression function with a complexity of $2^{76.06}$ computations, which is faster than currently best-known generic attack with a complexity of 2^{80}. Note that, a *second-order collision* [15,11] on a function $F(\cdot)$ with n-bit outputs is a set of two non-zero difference and an input $\{\Delta, \nabla, Y\}$ satisfying $F(Y+\Delta+\nabla)-F(Y+\Delta)-F(Y+\nabla)+F(Y)=0$. The attack is implemented for a reduced-round version and an example of the second-order collision up to the last 75 steps is presented in Table 7. The summary of our attack results is given in Table 1. Note that the complexity of the second-order collision attack against the last 77 steps is smaller than the information theoretic bound, which is the query complexity for finding the same property against an ideal function. Hence, the last 77 steps of the compression function can be concluded as non-ideal.

We admit that the practical impact of our distinguisher to HAS-160 is not clear at the current stage, but our distinguisher leads to a better understanding of the security margin of HAS-160, and might inspire further extensive attacks in the future.

Table 1. Comparison of Attacks on HAS-160

Attack	Target	#Steps	Information Theoretic Bound	Complexity Time	Complexity Data	Reference
Collision	Hash	45	2^{80}	2^{12}		[2]
Collision	Hash	53	2^{80}	2^{55}		[3]
Collision	Hash	59	2^{80}	2^{55}		[4]
Preimage	Compress	52	2^{160}	2^{144}		[5]
Preimage	Hash	52	2^{160}	2^{153}		[5]
Preimage	Compress	$68^{\dagger}$	2^{160}	$2^{150.7}$		[6]
Preimage	Hash	$68^{\dagger}$	2^{160}	$2^{156.3}$		[6]
Key Recovery	Internal BC	80 (full)	2^{512}	$2^{377.5}$	2^{155}	[7]
4-sum	Compress	$75^{\dagger}$	2^{40}	$2^{33.83}$		Ours
4-sum	Compress	$77^{\dagger}$	$2^{53.3}$	$2^{51\ddagger}$		Ours
2nd-order Coll	Compress	$77^{\dagger}$	$2^{53.3}$	2^{51}		Ours
2nd-order Coll	Compress	80 (Full)	$2^{53.3}$	$2^{76.06\ddagger}$		Ours
Key Recovery	Internal BC	80 (full)	2^{512}	$2^{377.5}$	2^{152}	Ours

†: the attack target is from a middle step to the last step.
‡: the generic attack complexities to find 4-sums and second-order collisions are $2^{53.3}$ and 2^{80}, respectively.

Paper Outline

The organization of this paper is as follows. In Section 2, the specification of HAS-160 is described. In Section 3, related work is summarized. In Section 4, our distinguisher on the full step compression function is explained. Finally, we conclude this paper in Section 5.

2 Description of HAS-160

HAS-160 [1] is a hash function that produces 160-bit hash values. It adopts the Merkle-Damgård structure, and uses 160-bit (5-word) chaining variables and a 512-bit (16-word) message block to compute a compression function. First, an input message M is processed to be a multiple of 512 bits. Then, the padded message is separated into 512-bit message blocks $(M_0, M_1, \ldots, M_{N-1})$. Let $CF : \{0,1\}^{160} \times \{0,1\}^{512} \rightarrow \{0,1\}^{160}$ be the compression function of HAS-160. Let H_i be a 160-bit value and IV be the initial value defined in the specification. A hash value H_N is computed as follows. 1) IV is loaded into H_0, 2) Compute $H_{i+1} \leftarrow CF(H_i, M_i)$ for $i = 0, 1, \ldots, N-1$,

The compression function of HAS-160 iterates a step function 80 times to compute a hash value. Steps 0-19, 20-39, 40-59, and 60-79 are called the first, second, third, and fourth rounds, respectively.

M_i is divided into sixteen 32-bit message-words $m_0, \ldots, m_{15}$. The message expansion of HAS-160 is a permutation of 20 message words in each round, which consists of $m_0, \ldots, m_{15}$ and four additional messages $m_{16}, \ldots, m_{19}$ computed from $m_0, \ldots, m_{15}$. The computation of $m_{16}, \ldots, m_{19}$ is shown in Table 2. Let $X_0, X_1, \ldots, X_{79}$ be the message word used in each step. The message m_j assigned to each X_j is also shown in Table 2.

The output of the compression function H_{i+1} is computed as follows.

1. $p_0 \leftarrow H_i$.
2. $p_{j+1} \leftarrow R_j(p_j, X_j)$ for $j = 0, 1, \ldots, 79$,
3. Output $H_{i+1} = (p_{80} + H_i)$, where "+" denotes 32-bit word-wise addition.

Table 2. Message Expansion of HAS-160

Computation of m_{16} to m_{19} in each round

	Round 1	Round 2	Round 3	Round 4
m_{16}	$m[0,1,2,3]$	$m[3,6,9,12]$	$m[12,5,14,7]$	$m[7,2,13,8]$
m_{17}	$m[4,5,6,7]$	$m[15,2,5,8]$	$m[0,9,2,11]$	$m[3,14,9,4]$
m_{18}	$m[8,9,10,11]$	$m[11,14,1,4]$	$m[4,13,6,15]$	$m[15,10,5,0]$
m_{19}	$m[12,13,14,15]$	$m[7,10,13,0]$	$m[8,1,10,3]$	$m[11,6,1,12]$

$m[i,j,k,l]$ denotes $m_i \oplus m_j \oplus m_k \oplus m_l$.

Message order in each step

Round 1: $X_0, X_1, \ldots, X_{19}$	18	0	1	2	3	19	4	5	6	7	16	8	9	10	11	17	12	13	14	15
Round 2: $X_{20}, X_{21}, \ldots, X_{39}$	18	3	6	9	12	19	15	2	5	8	16	11	14	1	4	17	7	10	13	0
Round 3: $X_{40}, X_{41}, \ldots, X_{59}$	18	12	5	14	7	19	0	9	2	11	16	4	13	6	15	17	8	1	10	3
Round 4: $X_{60}, X_{61}, \ldots, X_{79}$	18	7	2	13	8	19	3	14	9	4	16	15	10	5	0	17	11	6	1	12

Table 3. Function f, Constant k, and Rotations $s1$ and $s2$ of HAS-160

Round	Function $f_j(X, Y, Z)$	Constant k_j	Rotation $s2_j$
Round 1	$(X \wedge Y) \vee (\neg X \wedge Z)$	0x00000000	10
Round 2	$Z \oplus Y \oplus Z$	0x5a827999	17
Round 3	$Y \oplus (X \vee \neg Z)$	0x6ed9eba1	25
Round 4	$X \oplus Y \oplus Z$	0x8f1bbcdc	30

Rotation $s1_j$

j mod 20	0	1	2	3	4	5	6	7	8	9	10	11	12	13	14	15	16	17	18	19
$s1_j$	5	11	7	15	6	13	8	14	7	12	9	11	8	15	6	12	9	14	5	13

R_j is the step function for Step j. Let a_j, b_j, c_j, d_j, e_j be 32-bit values that satisfy $p_j = (a_j \| b_j \| c_j \| d_j \| e_j)$. $R_j(p_j, X_j)$ computes p_{j+1} as follows:

$$\begin{cases} a_{j+1} = (a_j \lll s1_j) + f_j(b_j, c_j, d_j) + e_j + X_j + k_j, \\ b_{j+1} = a_j, \\ c_{j+1} = b_j \lll s2_j, \\ d_{j+1} = c_j, \\ e_{j+1} = d_j, \\ p_{j+1} = a_{j+1} \| b_{j+1} \| c_{j+1} \| d_{j+1} \| e_{j+1} \end{cases}$$

where f_j, k_j, and $\lll s2_j$ represent bitwise Boolean function, constant number, and $s2_j$-bit left rotation defined in each round, and $\lll s1_j$ represents $s1_j$-bit left rotation depending on the value of j mod 20. These values are shown in Table 3.

We show a figure of the step function in Fig. 1. Note that $R_j^{-1}(p_{j+1}, X_j)$ can be computed in the same complexity as that of R_j.

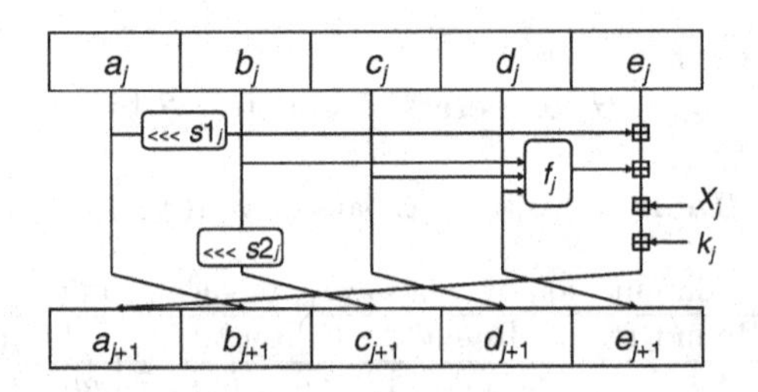

Fig. 1. Step function of HAS-160

3 Related work

3.1 4-sum and Second-order Collision

In this section, we explain two properties to be distinguished and query complexity to find each property against ideal primitives. There are two types of query complexity; information theoretic bound and generic attack complexity based on the current knowledge. The information theoretic bound only gives a bound. It does not imply that there would exist an attack with the same complexity as the bound. Therefore, discussing the generic attack complexity is also meaningful as well as the information theoretic bound.

A *4-sum* on a function $F(\cdot)$ with n-bit outputs is a set of four distinct inputs (Y_0, Y_1, Y_2, Y_3) satisfying

$$F(Y_0) \oplus F(Y_1) \oplus F(Y_2) \oplus F(Y_3) = 0.$$

If $F(\cdot)$ is an ideal compression function, it needs at least $2^{n/4}$ queries to find a 4-sum, where we mean by *ideal* that the output is uniformly distributed for each input. Therefore, if the 4-sum is obtained faster than $2^{n/4}$ computations, $F(\cdot)$ is regarded as non-ideal. On the other hand, apart from the information theoretic bound ($2^{n/4}$), the current best generic attack to find a 4-sum is a generalized birthday attack [16], which requires $2^{n/3}$ computations and $2^{n/3}$ memory. Hence, if 4-sums are generated with a complexity lower than $2^{n/3}$, $F(\cdot)$ is said to be weak because the same property cannot be detected on other functions with the current knowledge. Note that finding 4-sum quartets is interesting only if $F(\cdot)$ is a one-way function, and our attack target, HAS-160 compression function, is indeed a one-way function.

A *second-order collision* [15,11] on a function $F(\cdot)$ with n-bit outputs is a set of two non-zero difference and an input $\{\Delta, \nabla, Y\}$ satisfying

$$F(Y + \Delta + \nabla) - F(Y + \Delta) - F(Y + \nabla) + F(Y) = 0.$$

Second-order collision is a special form, in other words, a subset of the 4-sum, and can be viewed as limiting the form of input values on the 4-sum property. Previous work [15,11] showed that the information theoretic bound is $3 \cdot 2^{n/3}$ because the problem is essentially finding three parameters Δ, ∇, Y with an n-bit relation. On the other hand, the current best generic attack requires $2^{n/2}$. Similarly, if a second-order collision is obtained faster than $2^{n/3}$ computations, $F(\cdot)$ is regarded as non-ideal. Also if a second-order collision is obtained with a complexity lower than $2^{n/2}$, $F(\cdot)$ is said to be weak because the same property cannot be detected on other functions with the current knowledge.

3.2 Boomerang Attack on Internal Cipher of HAS-160

Dunkelman *et al.* analyzed the encryption mode of the HAS-160 compression function [7]. They applied the related-key rectangle attack for the internal block cipher, and recovered the secret key with 2^{155} chosen plaintexts and $2^{377.5}$ computations by using 4 simple relations in the key.

Our attack is related to [7] very closely because both attacks are based on the boomerang style attacks that build second-order differential characteristics on the attack target. In the following, we list differences of two researches in order to clarify contributions of this paper.

- For the secret-key recovery attack [7], the differential characteristic is constructed so that the probability for the entire characteristic can be optimized. On the other hand, for the attack on the compression function, the attack can choose the key values (messages) so that some part of the characteristic can be satisfied with a probability of 1. Hence, the differential characteristic needs to be reconstructed from scratch for attacking the compression function.
- In the encryption mode, the key size is 512 bits, and thus the key recovery attack can spend up to 2^{512} HAS-160 computations. On the other hand, all the properties discussed in our attack for the compression function will take at most 2^{80} HAS-160 computations, which is much smaller than the key-recovery attack.
- The first halves of the two attacks' differential characteristics share the same message differences. But we find a new differential characteristic with a higher probability, which also improves previous key-recovery attack on encryption mode.

4 Boomerang Distinguisher for Full HAS-160

To construct a boomerang distinguisher, we divide the internal cipher of HAS-160 into two subciphers denoted as E_0 and E_1. Since HAS-160 consists of four rounds, it seems natural to let E_0 consist of the first and second rounds and to let E_1 consist of the third and fourth rounds. Then, we adopt the start-from-the-middle approach to search for the second-order collision by following previous work [15,11,12].

More precisely, we start with fixing a quartet of the internal states between E_0 and E_1, and try to satisfy the differential characteristic of E_0 in backward direction and of E_1 in forward direction. Note that such a start-from-the-middle approach can also be seen in a series of rebound attacks, *e.g.*, [17,18]. The differential characteristic on E_0 is divided into *inside path* and *outside path*. The inside path refers to the part of the differential characteristic, which can be satisfied by the message modification technique. And the outside path refers to the remaining part. The same notations are also used for the differential characteristic on E_1. After the inside path is satisfied, we try to satisfy the outside paths on E_0 and E_1 probabilistically. The search for the outside paths is performed by using freedom degrees in the message words that do not appear in the inside path. Hence, once the inside path is satisfied, we never change the message words that relate to the inside path. Therefore, the attack complexity only depends on the search for the outside paths on E_0 and E_1. Let p and q be the probabilities of the outside paths on E_0 and E_1 respectively. Then, the complexity is written by $\frac{1}{p^2q^2}$. In order to minimize the complexity, the probabilities of the outside paths should be maximized. At the same time, the inside paths on E_0 and E_1 must not contradict. Otherwise, the boomerang attack cannot work.

4.1 Searching for the Message Differences

We mark each of $m_0, \ldots, m_{15}$ by a single bit: 1 stands for a non-zero difference and 0 for no difference. m_{16}, m_{17}, m_{18} and m_{19} in each round are marked by a single bit computed by XORing the mark bits of message words used to compute them. For example, in the first round, m_{16} is computed by XORing m_0, m_1, m_2 and m_3. The mark bit of m_{16} in the first round is computed by XORing the mark bits of these four words.

Adopting above approach, there are 2^{16} candidates for message differences. For E_0 (resp. E_1), we search for message differences, which locate at the very beginning stage in the first (resp. third) round, and at the very late stage in the second (resp. fourth) round. At the same time, we also pay attention to the absorption property of Boolean functions. Note that the Boolean function in the second round has no absorption property. Thus we intend to keep the inside path on E_0 short in order to avoid contradictions with the inside path on E_1 in advance. By exhaustively examining all the candidates, we decide to use the following message differences.

- On E_0: $\Delta m_0 =$ 0x80000000; $\Delta m_{10} =$ 0x80000000; $\Delta m_i = 0$ for $i \neq 0, 10$;
- On E_1: $\nabla m_6 =$ 0x80000000; $\nabla m_{12} =$ 0x80000000; $\nabla m_i = 0$ for $i \neq 6, 12$;

The reason of using the difference value 0x80000000 is to maximize the probability of the outside paths because a difference at MSB causes carries in fewer bits. A graphical view of the locations of the message differences is given in Table 4.

Table 4. Positions of Message Differences and Directions of Differential Propagations

round	message-word index for each round
1	⑱ ⓪ 1 2 3 19 4 5 6 7 ⑯ 8 9 ⑩ 11 17 12 13 14 15
	Outside path (OP) $\leftarrow \Delta$ | constant
2	18 3 6 9 12 19 15 2 5 8 16 11 14 1 4 17 7 ⑩ 13 ⓪
	constant | $\Delta \rightarrow$ IP
3	⑱ ⑫ 5 14 7 19 0 9 2 11 ⑯ 4 13 ⑥ 15 17 8 1 10 3
	Inside path (IP) $\leftarrow \nabla$ | constant
4	18 7 2 13 8 19 3 14 9 4 16 15 10 5 0 17 11 ⑥ 1 ⑫
	constant | $\nabla \rightarrow$ OP

4.2 Constructing Differential Characteristic

Our strategy of constructing the differential characteristic for E_0 is propagating the differences introduced by Δm_0 in backwards from internal states p_{14} to p_0 in the first round as the outside path, and Δm_{10} in forwards from p_{38} to p_{40} in the second round as the inside path. Similarly our strategy for E_1 is propagating the difference introduced by ∇m_6 in backwards from p_{54} to p_{40} in the third round as the inside path and ∇m_{12} in forwards from p_{78} to p_{80} in the fourth round as the outside path. An overview is given in Table 4.

For inside paths on E_0 and E_1, we should make sure that no contradiction occurs. A typical contradiction is that two inside paths set a joint internal state bit to different values at the same time. In our approach, we firstly fix the inside path on E_0 since f function in the second round has no absorption property and the inside path on E_0 covers only three steps. We also derive the conditions, which are $c_{40,16} = d_{40,16}$ and $b_{40,4} = a_{40,17}$. We secondly search for an inside path on E_1 backwards in the third round. Since the inside path on E_0 only has two conditions, we can easily get an inside path on E_1 not contradicting with it. The two inside paths we found are detailed in Tables 5 and 6.

For the outside path on E_0, we search for a differential characteristic with a high probability in order to lower the complexity of the attack. Firstly we simplify the step function by a linearization, mainly replacing the addition with XOR. At the same time we also consider the candidates with a limited number of bit carries caused by addition. Secondly launch a program to search for characteristics with low Hamming weight. The outside path on E_0 with a low Hamming weight is described in Table 5, which has in total 42 conditions. The outside path on E_1 covers only three steps, and thus can be easily constructed. An outside path on E_1 is given in Table 6.

Amplified Probability. Following previous works [11,12,13], we also consider the amplified probability of the outside paths, which is a sum of the probabilities of the multi-paths leading to a target property. For the outside path on E_0, we experimentally verified that the amplified probability of steps 3 to 1 is $2^{-19.06}$, and thus the amplified probability of the whole outside path on E_0 is $2^{-19.06-27*2} = 2^{-73.06}$. For outside path on E_1, we experimentally verified that its amplified probability is 2^{-1}.

4.3 Searching for a Second-Order Collision

We adopt a start-from-the-middle approach. Firstly choose a quartet of internal states at step 40 satisfying the differential characteristic. Secondly apply message modification technique to choose a corresponding quartet of message words used in the involved steps, which satisfy the inside paths. Finally exploit the freedom of the other undetermined message words to search for a quartet of messages which can satisfy the outside paths. A detailed search procedure is given below.

Initialization Phase. Set a random value to $p_{40}^{(1)}$ $(=a_{40}^{(1)}||b_{40}^{(1)}||c_{40}^{(1)}||d_{40}^{(1)}||e_{40}^{(1)})$, which satisfies the conditions in Tables 5 and 6. And compute $p_{40}^{(2)}$, $p_{40}^{(3)}$ and $p_{40}^{(4)}$ such that both $(p_{40}^{(1)}, p_{40}^{(2)})$ and $(p_{40}^{(3)}, p_{40}^{(4)})$ satisfy the difference at step 40 of the differential characteristic on E_0, and both $(p_{40}^{(1)}, p_{40}^{(3)})$ and $(p_{40}^{(2)}, p_{40}^{(4)})$ satisfy the difference at step 40 of the differential characteristic on E_1.

- $p_{40}^{(2)}$: $a_{40}^{(2)} = a_{40}^{(1)} \oplus$ 0x00020000, $b_{40}^{(2)} = b_{40}^{(1)} \oplus$ 0x00000010, $c_{40}^{(2)} = c_{40}^{(1)} \oplus$ 0x00010000, $d_{40}^{(2)} = d_{40}^{(1)}$, and $e_{40}^{(2)} = e_{40}^{(1)}$;
- $p_{40}^{(3)}$: $a_{40}^{(3)} = a_{40}^{(1)} \oplus$ 0x00102000, $b_{40}^{(3)} = b_{40}^{(1)} \oplus$ 0x02008000, $c_{40}^{(2)} = c_{40}^{(1)} \oplus$ 0x08102040, $d_{40}^{(3)} = d_{40}^{(1)} \oplus$ 0x80040110, and $e_{40}^{(3)} = e_{40}^{(1)} \oplus$ 0x08102040;
- $p_{40}^{(4)}$: $a_{40}^{(4)} = a_{40}^{(1)} \oplus$ 0x00122000, $b_{40}^{(4)} = b_{40}^{(1)} \oplus$ 0x00102010, $c_{40}^{(4)} = c_{40}^{(1)} \oplus$ 0x08112040, $d_{40}^{(4)} = d_{40}^{(1)} \oplus$ 0x80040110, and $e_{40}^{(4)} = e_{40}^{(1)} \oplus$ 0x08102040.

Table 5. Differential Characteristic on E_0

Outside Path			
Step j	Δp_j	ΔX_j	Conditions
1	$\Delta a_0 = 0x00002000$; $\Delta b_0 = 0x40000000$; $\Delta c_0 = 0x02040000$; $\Delta d_0 = 0x80000800$; $\Delta e_0 = 0x00040000$; $\Delta a_1 = 0x00000000$;	$0x80000000$;	$e_{0,18} \neq a_{0,13}$; $c_{0,30} = d_{0,30}$; $b_{0,18} = 0$; $b_{0,25} = 0$; $b_{0,31} = 0$; $b_{0,11} = 1$;
2	$\Delta a_2 = 0x00000800$;	$0x80000000$;	$c_{1,13} = d_{1,13}$; $b_{1,8} = 0$; $b_{1,18} = 1$; $b_{1,25} = 1$; $e_{1,11} = a_{2,11}$;
3	$\Delta a_3 = 0x06000000$;		$b_{2,23} = 0$; $b_{2,8} = 1$; $e_{2,18} \neq a_{2,11}$; $e_{2,25} \neq a_{3,25}$;
4	$\Delta a_4 = 0x00000000$;		$c_{3,11} = d_{3,11}$; $b_{3,23} = 0$; $e_{3,8} = a_{3,25}$;
5	$\Delta a_5 = 0x00800000$;		$b_{4,21} = 0$; $c_{4,25} = d_{4,25}$; $c_{4,26} = d_{4,26}$; $e_{4,23} = a_{5,23}$;
6	$\Delta a_6 = 0x00000000$;		$b_{5,3} = 0$; $b_{5,21} = 1$; $b_{5,4} = 1$; $b_{5,4} \neq a_{5,23}$;
7	$\Delta a_7 = 0x00200000$;		$c_{6,23} = d_{6,23}$; $b_{6,3} = 1$; $d_{6,4} = 1$; $e_{6,21} = a_{7,21}$;
8	$\Delta a_8 = 0x00000000$;		$b_{7,1} = 0$; $a_{7,21} = e_{7,3}$; $a_{7,21} = e_{7,4}$;
9	$\Delta a_9 = 0x00200000$;		$b_{8,1} = 1$; $c_{8,21} \neq d_{8,21}$; $(b_{8,21} \wedge c_{8,21}) \vee (\neg b_{8,21} \wedge d_{8,21}) = a_{9,21}$;
10	$\Delta a_{10} = 0x00000000$;		$b_{9,31} = 0$; $e_{9,5} \neq a_{9,21}$;
11	$\Delta a_{11} = 0x00000000$;	$0x80000000$;	$c_{10,21} = d_{10,21}$; $b_{10,31} = 1$;
12	$\Delta a_{12} = 0x00000000$;		$b_{11,31} = 1$;
13	$\Delta a_{13} = 0x00000000$;		$b_{12,31} = 1$;
14	$\Delta a_{14} = 0x00000000$;	$0x80000000$;	
Inside Path			
38	$\Delta p_{37} = 0$; $\Delta a_{38} = 0x80000000$;	$0x80000000$;	
39	$\Delta a_{39} = 0x00000010$;		$a_{38,31} = a_{37,31}$;
40	$\Delta a_{40} = 0x00020000$;	$0x80000000$;	$a_{39,4} = a_{40,17}$;

Inside Path Phase. Note that both $(p_{40}^{(1)}, p_{40}^{(2)})$ and $(p_{40}^{(3)}, p_{40}^{(4)})$ can satisfy the inside path on E_0 for any message word values. We mainly focus on the inside path on E_1. For $j = 41$ to 54 (except 51), we choose a random value for the internal state value $p_j^{(1)}$ but satisfies the conditions in Table 6, and then compute the corresponding value of the message word $X_j^{(1)}$. We then set $X_j^{(3)}$ be equal to $X_j^{(1)}$, and compute $p_j^{(3)}$. We check whether $p_j^{(3)}$ satisfies the conditions in Table 6. If not, the procedure is repeated with another random value for $p_j^{(1)}$. At step 51, m_{16} has been determined by $m_{12} \oplus m_5 \oplus m_{14} \oplus m_7$ after step 45. At this step, the search can go back to step 50, and re-choose another random value for $p_{50}^{(1)}$. It may be possible to further optimize the above procedure, but the complexity of the inside path phase is negligible compared with that of the outside path phase.

After the inside path phase, message words m_8, m_1, m_{10} and m_3 are not determined yet. However, these 4 words are restricted by a 32-bit condition because $m_{18} = m_8 \oplus m_1 \oplus m_{10} \oplus m_3$ in the third round is already fixed. Thus, 2^{96} freedom degrees are remaining for the outside paths, which is enough to satisfy them.

Table 6. Differential Characteristic on E_1

Step j	∇p_j	∇X_j	Conditions
	Inside Path		
41	$\nabla a_{40} = 0x00102000$; $\nabla b_{40} = 0x02008000$; $\nabla c_{40} = 0x08102040$; $\nabla d_{40} = 0x80040110$; $\nabla e_{40} = 0x08102040$; $\nabla a_{41} = 0x06000000$;	$0x80000000$;	$a_{40,20} \neq a_{41,25}$; $a_{40,20} = a_{41,26}$; $b_{40,18} = 0$; $c_{40,18} \oplus \neg d_{40,18} \neq a_{40,13}$; $d_{40,15} = 0$; $d_{40,25} = 0$; $b_{40,4} = 1$; $b_{40,8} = 1$; $b_{40,31} = 0$; $(b_{40,6} \vee \neg d_{40,6}) \oplus c_{40,6} \neq e_{40,6}$; $(b_{40,13} \vee \neg d_{40,13}) \oplus c_{40,13} \neq e_{40,13}$; $(b_{40,20} \vee \neg d_{40,20}) \oplus c_{40,20} \neq e_{40,20}$; $(b_{40,27} \vee \neg d_{40,27}) \oplus c_{40,27} \neq e_{40,27}$;
42	$\nabla a_{42} = 0x00102000$;	$0x80000000$;	$e_{41,4} = a_{41,25}$; $b_{41,20} \neq d_{41,20}$; $d_{41,13} = 1$; $b_{41,13} \oplus c_{41,13} = a_{42,13}$; $b_{41,6} = 1$; $b_{41,13} = 1$; $b_{41,27} = 1$; $(b_{41,8} \vee \neg d_{41,8}) \oplus c_{41,8} \neq e_{41,8}$; $(b_{41,18} \vee \neg d_{41,18}) \oplus c_{41,18} \neq e_{41,18}$;
43	$\nabla a_{43} = 0x00200000$;		$a_{42,13} \neq e_{42,20}$; $a_{42,20} \neq 42,27$; $d_{42,26} = 0$; $d_{42,25} = 1$; $b_{42,25} \oplus c_{42,25} = a_{43,25}$; $(b_{42,6} \vee \neg d_{42,6}) \oplus c_{42,6} \neq e_{42,6}$; $(b_{42,13} \vee \neg d_{42,13}) \oplus c_{42,13} \neq e_{42,13}$; $b_{42,8} = 1$; $b_{42,18} = 1$;
44	$\nabla a_{44} = 0x00102000$;		$a_{43,25} \neq e_{43,8}$; $d_{43,13} \neq b_{43,13}$; $d_{43,20} = 1$; $b_{43,20} \oplus c_{43,20} = a_{43,20}$; $b_{43,6} = 1$; $(b_{43,18} \vee \neg d_{43,18}) \oplus c_{43,18} \neq e_{43,18}$; $(b_{43,19} \vee \neg d_{43,19}) \oplus c_{43,19} \neq e_{43,19}$; $(b_{43,13} \vee \neg d_{43,13}) \oplus c_{43,13} = a_{44,13}$;
45	$\nabla a_{45} = 0x00200000$;		$d_{44,25} = 1$; $b_{44,18} = 1$; $b_{44,18} = 1$; $b_{44,19} = 0$; $b_{44,25} \oplus c_{44,25} \neq a_{44,20}$; $c_{44,19} \oplus \neg d_{44,19} \neq a_{44,13}$; $(b_{44,13} \vee \neg d_{44,13}) \oplus c_{44,13} \neq e_{44,13}$; $(b_{44,6} \vee \neg d_{44,6}) \oplus c_{44,6} \neq e_{44,6}$;
46	$\nabla a_{46} = 0x00000000$;		$b_{45,13} = d_{45,13}$; $d_{45,20} = 0$; $b_{45,6} = 0$; $c_{45,6} \oplus \neg d_{45,6} \neq a_{45,25}$; $(b_{45,18} \vee \neg d_{45,18}) \oplus c_{45,18} = e_{45,18}$; $(b_{45,18} \vee \neg d_{45,18}) \oplus c_{45,18} \neq e_{45,19}$;
47	$\nabla a_{47} = 0x00000000$;		$d_{46,25} = 0$; $b_{46,18} = 1$; $(b_{46,6} \vee \neg d_{46,6}) \oplus c_{46,6} \neq e_{46,6}$; $(b_{46,13} \vee \neg d_{46,13}) \oplus c_{46,13} \neq e_{46,13}$;
48	$\nabla a_{48} = 0x00000040$;		$b_{47,13} = 1$; $b_{47,6} = 0$; $c_{47,6} \oplus \neg d_{47,6} = a_{48,6}$; $(b_{47,18} \vee \neg d_{47,18}) \oplus c_{47,18} \neq e_{47,18}$;
49	$\nabla a_{49} = 0x00000040$;		$b_{48,18} = 1$; $e_{48,13} \neq a_{48,6}$; $e_{48,6} = a_{49,6}$;
50	$\nabla a_{50} = 0x00000000$;		$d_{49,6} = 0$; $a_{49,6} \neq e_{49,18}$;
51	$\nabla a_{51} = 0x00000000$;	$0x80000000$;	$d_{50,6} = 0$;
52	$\nabla a_{52} = 0x00000000$;		$b_{51,31} = 0$;
53	$\nabla a_{53} = 0x00000000$;		$b_{52,31} = 0$;
54	$\nabla a_{53} = 0x00000000$;	$0x80000000$;	
	Outside Path		
78	$\nabla p_{77} = 0$; $\nabla a_{78} = 0x80000000$;	$0x80000000$;	
79	$\nabla a_{79} = 0x00000010$;		$a_{78,31} = a_{79,4}$;
80	$\nabla a_{80} = 0x00020000$;	$0x80000000$;	$d_{79,31} = 1$; $a_{80,17} = a_{79,4}$;

Outside Path Phase. Randomly choose the values for message words $m_1^{(1)}$, $m_8^{(1)}$ and $m_{10}^{(1)}$, which determines the whole message quartet. Check whether the message quartet leads to a second-order collision on HAS-160. If not, repeat this procedure with another value for $m_1^{(1)}$, $m_8^{(1)}$ and $m_{10}^{(1)}$.

The Complexity. The outside path phase dominates the complexity. And in total $2^{73.06} \times 2^1$ quartets of messages need to be checked to produce a second-order collision. Thus the complexity is $2^{73.06+1+2} = 2^{76.06}$.

4.4 Summary of Distinguishers on HAS-160 Compression Function and Experiment Verification

This section summarizes the results of our boomerang-based attacks on HAS-160 with respect to the properties of 4-sum and second-order collision. Besides the full steps of HAS-160, we also evaluate step-reduced versions. In the following, t-step HAS-160 refers to the last t steps.

4-sum property. On 75-step HAS-160, a 4-sum can be obtained with a complexity of $2^{33.83}$, which is faster than $2^{40(=160/4)}$. Thus up to 75 steps, HAS-160 is non-ideal with respect to the notion of the 4-sum property. Note that up to 77 steps, a 4-sum can be obtained with a complexity of 2^{51}, which is faster than the generalized birthday attack, $2^{53.3(=160/3)}$.

2nd-order collision property. On 77-step HAS-160, a second-order collision is obtained with a complexity of 2^{51}, which is faster than $2^{53.3(=160/3)}$. Thus up to 77 steps, HAS-160 is non-ideal with respect to the notion of second-order collision property. As mentioned in Section 4.3, on full steps of HAS-160, it takes $2^{76.06}$ to produce a second-order collision, which is faster than the currently best-known generic attack with a complexity of 2^{80}.

In order to show the validity of our attack, we implemented the attack on the last 75 steps on a PC. We show a generated example of the 4-sum in Table 7.

4.5 Comparison with the Previous Characteristic [7]

Our attack target is a public function, which gives us the control of the internal state. Thus we select characteristics with short and simple sub-paths at the very beginning and at the very last steps of HAS-160. Differently from our attacks, Dunkelman *et al.* attacked the keyed block cipher of HAS-160 [7]. They selected characteristics with an overall minimum number of conditions. Regarding E_0, our characteristic shares the same message difference with theirs, but our characteristic has even fewer conditions, which is reduced to 44 from 47. Thus by adopting our characteristic on E_0, the data complexity of their related-key rectangle attack on the block cipher can be improved[1]. Let x be the number of input pairs with a specific difference Δ. It is known that the condition of x to form a rectangle quartet is written as follows:

$$x > 2^{n/2} \times \frac{1}{p} \times \frac{1}{q}, \tag{1}$$

where, n is the block-size of the cipher, which is 160 for the internal cipher of HAS-160. Therefore, the improvement by a factor of 2^3 for p results in the improvement of x by a factor of 2^3. Because the previous work [7] required 2^{155} chosen plaintexts, our attack requires 2^{152} chosen plaintexts.

[1] In rectangle attacks, the attack model is a chosen-plaintext attack, where attackers do not have to access the decryption oracle. This is different from boomerang attacks, where the attack model is an adoptively chosen-ciphertext attack.

Table 7. An Example of 4-sum on HAS-160 Reduced to the Last 75 steps

$p_5^{(1)}$	0x3c3fc642	0x7d021a93	0x189a5355	0xde513fb9	0x60a3b089
$M_i^{(1)}$	0x6f63d7e0	0xa931ea99	0xec9d5b8d	0xaa8a0aaa	0x1d2cc5ff
	0xda4ccf0f	0x9e2c11ba	0x9d14d81c	0x5fc94c41	0x30ee45ac
	0x8b5842b9	0xa0f14fa3	0x7bc50c4a	0x4fcf6a46	0x5101b564
	0xb702a1f8				
$p_{80}^{(1)}$	0x614dac1b	0xddf182b4	0x5f145d90	0x5ec72ad6	0x91bfb7ef
$p_5^{(2)}$	0x3c9fc642	0x7d221293	0x98ba5355	0x3ed13fb9	0xe023a87d
$M_i^{(2)}$	0xef63d7e0	0xa931ea99	0xec9d5b8d	0xaa8a0aaa	0x1d2cc5ff
	0xda4ccf0f	0x9e2c11ba	0x9d14d81c	0x5fc94c41	0x30ee45ac
	0x0b5842b9	0xa0f14fa3	0x7bc50c4a	0x4fcf6a46	0x5101b564
	0xb702a1f8				
$p_{80}^{(2)}$	0xa18577da	0xf58af24a	0x0c694920	0x9bb82689	0xe65d0b4c
$p_5^{(3)}$	0x5a259438	0xb9465d59	0x298dd564	0x640d2efa	0xfd396387
$M_i^{(3)}$	0x6f63d7e0	0xa931ea99	0xec9d5b8d	0xaa8a0aaa	0x1d2cc5ff
	0xda4ccf0f	0x1e2c11ba	0x9d14d81c	0x5fc94c41	0x30ee45ac
	0x8b5842b9	0xa0f14fa3	0xfbc50c4a	0x4fcf6a46	0x5101b564
	0xb702a1f8				
$p_{80}^{(3)}$	0x614fac1b	0xddf182c4	0x7f145d90	0x5ec72ad6	0x91bfb7ef
$p_5^{(4)}$	0x5a859438	0xb9665559	0xa9add564	0xc48d2efa	0x7cb95b7b
$M_i^{(4)}$	0xef63d7e0	0xa931ea99	0xec9d5b8d	0xaa8a0aaa	0x1d2cc5ff
	0xda4ccf0f	0x1e2c11ba	0x9d14d81c	0x5fc94c41	0x30ee45ac
	0x0b5842b9	0xa0f14fa3	0xfbc50c4a	0x4fcf6a46	0x5101b564
	0xb702a1f8				
$p_{80}^{(4)}$	0xa18777da	0xf58af25a	0x2c694920	0x9bb82689	0xe65d0b4c
4-sum	0x00000000	0x00000000	0x00000000	0x00000000	0x00000000

We emphasize that generating 4-sums with the current best generic attack requires $2^{53.3}$ computations and memory, which seem infeasible.

5 Conclusion

This paper has evaluated the security of HAS-160 compression function adopting boomerang attack framework. We successfully found a second-order collision attack faster than the currently best-known generic attack. While the impact of distinguishers might be unclear, our work has the contributions to a better understanding of the security margin of HAS-160, and hope that our distinguisher will lead to more powerful attacks on HAS-160 in the future.

References

1. Telecommunications Technology Association.: Hash Function Standard Part 2: Hash Function Algorithm Standard, HAS-160 (2000)
2. Yun, A., Sung, S.H., Park, S., Chang, D., Hong, S.H., Cho, H.-S.: Finding Collision on 45-Step HAS-160. In: Won, D.H., Kim, S. (eds.) ICISC 2005. LNCS, vol. 3935, pp. 146–155. Springer, Heidelberg (2006)
3. Cho, H.-S., Park, S., Sung, S.H., Yun, A.: Collision Search Attack for 53-Step HAS-160. In: Rhee, M.S., Lee, B. (eds.) ICISC 2006. LNCS, vol. 4296, pp. 286–295. Springer, Heidelberg (2006)

4. Mendel, F., Rijmen, V.: Colliding Message Pair for 53-Step HAS-160. In: Nam, K.-H., Rhee, G. (eds.) ICISC 2007. LNCS, vol. 4817, pp. 324–334. Springer, Heidelberg (2007)
5. Sasaki, Y., Aoki, K.: A Preimage Attack for 52-Step HAS-160. In: Lee, P.J., Cheon, J.H. (eds.) ICISC 2008. LNCS, vol. 5461, pp. 302–317. Springer, Heidelberg (2009)
6. Hong, D., Koo, B., Sasaki, Y.: Improved Preimage Attack for 68-Step HAS-160. In: Lee, D., Hong, S. (eds.) ICISC 2009. LNCS, vol. 5984, pp. 332–348. Springer, Heidelberg (2010)
7. Dunkelman, O., Fleischmann, E., Gorski, M., Lucks, S.: Related-Key Rectangle Attack of the Full HAS-160 Encryption Mode. In: Roy, B., Sendrier, N. (eds.) INDOCRYPT 2009. LNCS, vol. 5922, pp. 157–168. Springer, Heidelberg (2009)
8. Wagner, D.: The Boomerang Attack. In: Knudsen, L.R. (ed.) FSE 1999. LNCS, vol. 1636, pp. 156–170. Springer, Heidelberg (1999)
9. Aumasson, J.-P., Çalık, Ç., Meier, W., Özen, O., Phan, R.C.-W., Varıcı, K.: Improved Cryptanalysis of Skein. In: Matsui, M. (ed.) ASIACRYPT 2009. LNCS, vol. 5912, pp. 542–559. Springer, Heidelberg (2009); Extended version is available at Cryptology ePrint Archive: Report 2009/438
10. Biryukov, A., Nikolić, I., Roy, A.: Boomerang Attacks on BLAKE-32. In: Joux, A. (ed.) FSE 2011. LNCS, vol. 6733, pp. 218–237. Springer, Heidelberg (2011)
11. Lamberger, M., Mendel, F.: Higher-order differential attack on reduced SHA-256. Cryptology ePrint Archive, Report 2011/037 (2011), `http://eprint.iacr.org/2011/037`
12. Sasaki, Y.: Boomerang Distinguishers on MD4-Family: First Practical Results on Full 5-Pass HAVAL. In: Miri, A., Vaudenay, S. (eds.) SAC 2011. LNCS, vol. 7118, pp. 1–18. Springer, Heidelberg (2012)
13. Sasaki, Y., Wang, L.: 2-dimension sums: Distinguishers beyond three rounds of RIPEMD-128 and RIPEMD-160. Cryptology ePrint Archive, Report 2012/049 (2012), `http://eprint.iacr.org/2012/049`
14. Wang, X., Yu, H.: How to Break MD5 and Other Hash Functions. In: Cramer, R. (ed.) EUROCRYPT 2005. LNCS, vol. 3494, pp. 19–35. Springer, Heidelberg (2005)
15. Biryukov, A., Lamberger, M., Mendel, F., Nikolić, I.: Second-Order Differential Collisions for Reduced SHA-256. In: Lee, D.H., Wang, X. (eds.) ASIACRYPT 2011. LNCS, vol. 7073, pp. 270–287. Springer, Heidelberg (2011)
16. Wagner, D.: A Generalized Birthday Problem. In: Yung, M. (ed.) CRYPTO 2002. LNCS, vol. 2442, pp. 288–303. Springer, Heidelberg (2002)
17. Mendel, F., Rechberger, C., Schläffer, M., Thomsen, S.S.: The Rebound Attack: Cryptanalysis of Reduced Whirlpool and `Grøstl`. In: Dunkelman, O. (ed.) FSE 2009. LNCS, vol. 5665, pp. 260–276. Springer, Heidelberg (2009)
18. Mendel, F., Peyrin, T., Rechberger, C., Schläffer, M.: Improved Cryptanalysis of the Reduced `Grøstl` Compression Function, `ECHO` Permutation and AES Block Cipher. In: Jacobson Jr., M.J., Rijmen, V., Safavi-Naini, R. (eds.) SAC 2009. LNCS, vol. 5867, pp. 16–35. Springer, Heidelberg (2009)

Polynomial-Advantage Cryptanalysis of 3D Cipher and 3D-Based Hash Function

Lei Wang[1], Yu Sasaki[2], Kazuo Sakiyama[1], and Kazuo Ohta[1]

[1] The University of Electro-Communications,
1-5-1 Choufugaoka, Choufu-shi, Tokyo, 182-8585 Japan
{sakiyama,kazuo.ohta}@uec.ac.jp
[2] NTT Secure Platform Laboratories, NTT Corporation,
3-9-11 Midori-cho, Musashino-shi, Tokyo 180-8585 Japan
sasaki.yu@lab.ntt.co.jp

Abstract. This paper evaluates a block cipher mode, whose round functions of both the key schedule and the encryption process are independent of the round indexes. Previously related-key attack has been applied to such block cipher mode, and it can work no matter how many rounds are iterated in the cipher. This paper presents an accelerated key-recovery attack on this block cipher mode in the single-key setting. Similarly, our attack can also work no matter how many rounds are iterated in the cipher. More interestingly, the effectiveness of our attack, e.g. the relative advantage, increases with the number of rounds.

3D is a dedicated block cipher following the target mode. We apply the key-recovery attack to 3D cipher, and extend it to collision and preimage attacks on 3D-based hash functions. For a l-round instance of 3D (l is recommended as 22 by the designer), the complexity of recovering the secret key is $2^{512}/\sqrt{l/2}$ data, $2^{512}/\sqrt{l/2}$ offline computation, and $2^{512}/\sqrt{l/2}$ memory requirement. And the success probability is 0.63. Thus compared with the brute-force attack, the complexity is accelerated by a factor of $0.315 * \sqrt{l/2}$ in the sense of total computations (including both online and offline computations) under the same success probability 0.63. The total computations of finding collision and preimage on 3D-based hash functions are $2^{257}/l$ and $2^{513}/l$, namely accelerated by a factor of $l/2$ in the sense of total computations under the same success probability. Moreover, differently from the key-recovery attack, the collision and preimage attacks don't need to increase the memory requirement compared with the brute-force attack.

Finally we stress that all our attacks are polynomial-advantage attacks.

Keywords: 3D, key-recovery, collision, preimage, polynomial-advantage.

1 Introduction

Block cipher plays an important role in modern cryptography. It has been widely used for message encryptions and message authentications. Most block ciphers

G. Hanaoka and T. Yamauchi (Eds.): IWSEC 2012, LNCS 7631, pp. 170–181, 2012.

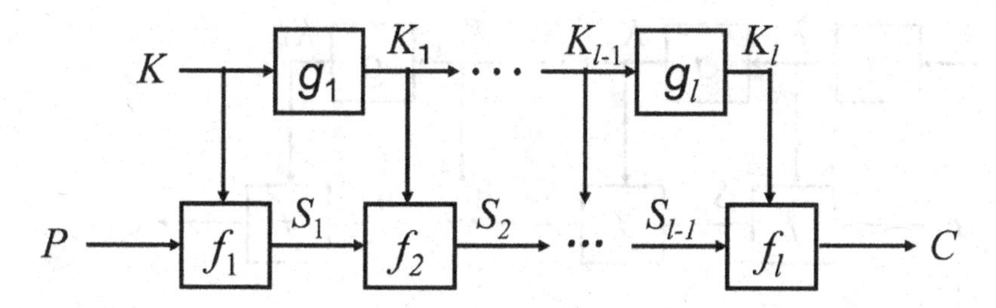

Fig. 1. Block Cipher Mode $E_K(\cdot)$

are constructed by a cascade of small round functions. See Fig. 1 for a graphical view. The number of rounds is l. The secret key K is expanded by a key schedule consisting of small round functions $g_1(\cdot)$, $g_2(\cdot)$, ..., and $g_l(\cdot)$. The expanded round keys are denoted as $(K_1, K_2, \ldots, K_l)$. The encryption process consists of small round functions $f_1(K_1, \cdot)$, $f_2(K_2, \cdot)$, ..., and $f_l(K_l, \cdot)$. All $f_i(K_i, \cdot)$s are permutations for decryption. Without loss of generality, we also assume that all $g_i(\cdot)$s are also permutations.

The effectiveness of most cryptanalysis techniques decreases with the increase of the number of rounds. So a simple countermeasure for modern block cipher is

enlarging the value of l to resist short-cut attacks.

Thus typically a cascaded block cipher relies its security on the sufficient number of rounds.

However, this countermeasure may not resist all the cryptanalysis techniques if the block cipher mode has some weak property. *Related-key* [1] and *slide attack* [4] are two attacks such that once they are applicable, the block cipher will be broken no matter how many rounds are used.[1] Particularly, related-key attack can work on a block cipher mode depicted in Fig. 2, whose round functions g_i and f_i are independent from round indexes, and thus denoted as g and f respectively. Similar related-key attack has also been applied to stream cipher [6]. Here we will omit the description of related attacks, and refer the details to [1]. Slide attack needs the block cipher mode has periodic round keys, e.g. identical round keys, so slide attack cannot be applied to the block cipher mode in Fig. 2 with an overwhelming probability.

Our Contributions

This paper presents an accelerated key-recovery attack on the block cipher mode in Fig. 2, which is in the single-key setting, and can always work no matter how many rounds are iterated.

[1] Here related attack is referred to as the approach on the block cipher mode described in Fig. 2 [1]. However, the related-key setting has been introduced to other attack approaches, such as differential attack, which are usually limited by the number of rounds.

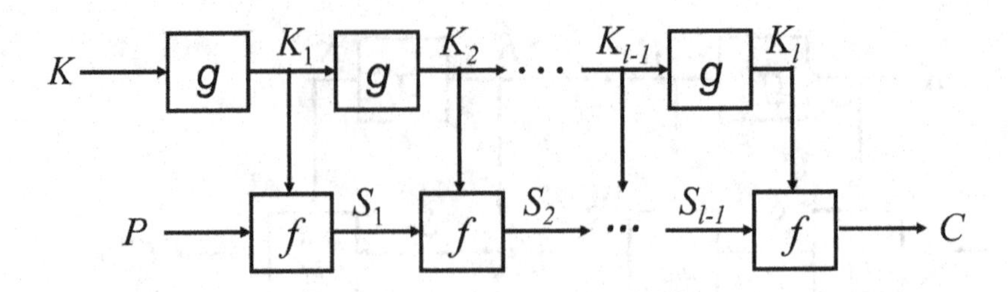

Fig. 2. Our Target Block Cipher Mode

Our attack is from the following observation. Select two random values x_0 and y_0 with suitable bit sizes. Then compute a sequence of (x_i, y_i) with $i = 1, 2, \ldots, t$ as below

$$x_i = g(x_{i-1}), \text{and } y_i = f(x_i, y_{i-1}).$$

And derive triples (x_{i-l}, y_{i-l}, y_i) with $i = l, l+1, \ldots, t$. There are in total $t - l + 1$ triples. A triple (x_{i-l}, y_{i-l}, y_l) implies that $E_{x_{i-l}}(y_{i-l}) = y_i$, which gives a plaintext-ciphertext pair encrypted by a key value x_{i-l}. If $t \gg l$, a triple (key, plaintext, ciphertext) is obtained with a complexity of one $g(\cdot)$ and one $f(\cdot, \cdot)$ computations on average. For the brute-force key-recovery attack, a plaintext-ciphertext pair encrypted by a guessed key value is obtained with a complexity of one execution of the entire block cipher, which consists of l computations of $g(\cdot)$ and l computations of $f(\cdot, \cdot)$. Thus we can accelerate the brute-force attack, which leads to a polynomial-advantage attack.

Finally the complexity of recovering the secret key is listed as below. Denote the bit sizes of the key and the block as k and n respectively.

- $k = n$ case: the complexity is $O(\frac{2^n}{\sqrt{l}})$;
- $k > n$ case: the complexity is $O(2^x + \frac{k}{n} \times 2^{k-n} + \frac{2^{k+n-x}}{l})$, where $\lceil log_2 \frac{k}{n} \rceil \leq x \leq n$,

where we stress the unit of complexity is one execution of the entire block cipher, which consists of l computations of $f(\cdot, \cdot)$ and l computations of $g(\cdot)$. Moreover, our attack can be transformed to accelerate collision and preimage attacks on a block-cipher-based hash function. And the complexity of our attack is $\frac{1}{l}$ of the complexity of the brute-force attacks. We have to point out that the advantage of our key-recovery attack is gained with a significant increase of memory requirement. On the other hand, collision and preimage attacks don't need to increase the memory requirement.

3D cipher [12], which has a block size 512 bits and a key size 512 bits, falls into the block cipher mode in Fig. 2. Thus our attack can be applied to 3D cipher. Moreover we extend the attack to collision and preimage attack on 3D-based hash functions. For a l-round instance of 3D (l is recommended as 22 by the designer), the complexity of recovering the secret key is $2^{512}/\sqrt{l/2}$ data, $2^{512}/\sqrt{l/2}$ offline computation, and $2^{512}/\sqrt{l/2}$ memory requirement. And the success probability is 0.63. Thus compared with the brute-force attack, the complexity is accelerated by a factor of $0.315 * \sqrt{l/2}$ in the sense of total computations (combining both online and offline computations) under the same success probability 0.63. The total computations of finding collision and preimage are $2^{257}/l$ and $2^{513}/l$, namely accelerated by a factor of $l/2$ in the sense of total computation under the same success probability.

Roadmap of the Paper

Section 2 describes notations and backgrounds. Section 3 illustrates our attack. Section 4 applies our attack to 3D and 3D-based hash functions. Section 5 concludes the paper.

2 Notations and Backgrounds

2.1 Notations

This section defines the notations used in this paper. See Fig. 2. Denote the block cipher as $E_K(\cdot)$, which consists of a key schedule and an encryption process. Denote the round function in the key schedule as $g(\cdot)$, and let $g(\cdot)$ be a permutation without loss of generality. Denote the round function in the encryption process as $f(\cdot, \cdot)$. Denote the secret key as K, and the i-th round key as K_i. Denote the key bit size as k. Denote plaintext as P, and ciphertext as C. Denote the output internal state value of $f(\cdot, \cdot)$ at i-th round as S_i. Denote the block bit size as n.

2.2 Cryptanalysis Techniques on Block Ciphers

The security of block ciphers are usually evaluated by how faster the key is recovered compared with the brute-force attack. The brute-force key search on $E_K(\cdot)$ is as below.

(1) Obtain a valid plaintext-ciphertext pair (P, C).
(2) for $K = 0$ to $2^k - 1$
(3) Compute $E_K(P)$ and match it to C. If there is a match, output K.

With the development of cryptanalysis techniques for block ciphers, many attack approaches are proposed, including differential cryptanalysis [3], linear cryptanalysis [18], truncated differential cryptanalysis [14], higher-order differential cryptanalysis [14,17], impossible differential cryptanalysis [2,15], boomerang attack [21], biclique attack [5] etc. Moreover, recently cryptanalysts also pay attentions to the security of block ciphers in the *known-key* model. Several known-key distinguishing attacks on block ciphers are proposed, including rebound attack [19] and super-sbox attack [10], etc. These short-cut attacks can be classified into two categories [20]: *exponential-advantage attack* and *polynomial-advantage attack*, compared with the brute-force attack.

Exponential-advantage attacks aim to reduce the number of repetitions, namely an acceleration of part (2). Most cryptanalysis techniques are exponential-advantage attacks. On the other hand, recently polynomial-advantage attack techniques also become popular, which aim to reduce the computational cost of one execution of $E_K(P)$, namely an acceleration of part (3). Biclique attack on AES [5] is a typical polynomial-advantage attack. Also in order to enlarge the number of the attacked rounds, recently cryptanalysts introduce polynomial-advantage techniques to traditional exponential-advantage attacks. Such attacks should be regarded as polynomial-advantage attacks because the polynomial-advantage attack part dominates the overall complexity.

Here we stress that our attack is a polynomial-advantage attack. People may suspect the significance of polynomial-advantage attacks considering that the gained complexity advantage is marginal. However, we stress that polynomial-advantage attacks contribute to a better understanding of the exact security bounds of block ciphers.

2.3 Cycles in a Permutation

We focus on $g(\cdot)$ in key schedule function of $E_K(\cdot)$, which is usually a permutation: $\{0,1\}^k \rightarrow \{0,1\}^k$. By cycle, we mean a set of k-bit distinct values $(x_1, x_2, \ldots, x_t)$ such that $x_i = g(x_{i-1})$ with $i = 2, 3, \ldots, t$ and $x_1 = g(x_t)$. Denote the value of t as the length of this cycle. For most permutations, there are several inside cycles with a Poisson distribution [11]. The expectation of the cycle length is 2^{k-1} [8].

We stress that the cycle distributions in the permutation $g(\cdot)$ will not influence the complexity of our attack, but may cause an increase of memory cost. For the details, refer to Section 3.

3 Our Attack on the Block Cipher Mode

This section describes our attack in detail on the block cipher mode in Fig. 2. For the simplicity of the description, we first focus on the case that $k = n$ and that there is only one cycle inside $g(\cdot)$ with a length 2^n. Later we will discuss the impact of the key length and that of the cycle distributions in $g(\cdot)$ in Sections 3.1 and 3.2 respectively.

Denote $f(g(x), y)$ by $f||g(x, y)$ for simplicity. The cipher can be regarded as iterating $f||g(\cdot,\cdot)$ l times on (K, P). Interestingly, select a random starting point (x_0, y_0), iteratively compute a long sequence of $(x_1, y_1) \leftarrow f||g(x_0, y_0)$, $(x_2, y_2) \leftarrow f||g(x_1, y_1)$, ..., and derive the triples (x_i, y_i, y_{i+l}). Such a triple (x_i, y_i, y_{i+l}) implies $y_{i+l} = E_{x_i}(y_i)$, which is a plaintext-ciphertext pair encrypted by a key x_i. On average, with one execution of $f||g(\cdot,\cdot)$, namely $\frac{1}{l}$ computation of the entire $E_K(\cdot)$, a plaintext-ciphertext pair of $E_{(\cdot)}(\cdot)$ is produced under a guessed key value at offline. Such a property leads to our attack, which is faster than the brute-force key search.

See Fig. 3 for an illustration. The main attack strategy is to recover the value of K by a match between the plaintext-ciphertext pairs encrypted by the real key K, which are obtained by queries at online phase, and the plaintext-ciphertext pairs encrypted by different guessed keys, which are computed at offline phase. If a match between (x_i, y_i, y_{i+l}) and (P_j, C_j) is found, namely (y_i, y_{i+l}) is equal to (P_j, C_j), it is with an overwhelming probability that x_i is equal to K. And for the negligible number of noisy matched pairs, we can easily erase them by a confirming computation using another queried plaintext-ciphertext pair. The main complexity advantage is gained from that a plaintext-ciphertext pair encrypted by a guessed key value is obtained with a complexity of $\frac{1}{l}$ on average.

Attack procedure. The attack consists of the following steps.

1. Query $\lceil \frac{2^n}{\sqrt{l}} \rceil$ different plaintexts to $E_K(\cdot)$, and store plaintext-ciphertext pairs in a table $\mathcal{T}$.
2. Select a random plaintext y_0 and a random key vaue x_0.
3. for $i = 1$ until l
   (a) Compute $(x_i, y_i) \leftarrow f||g(x_{i-1}, y_{i-1})$;
   (b) Match y_i to $\{C_j\}$ in $\mathcal{T}$.
   (c) If it is matched with a C_j, then check whether x_i is the round key K_l. And if x_i is equal to K_l, then compute the value of K and output it.
4. for $i = l + 1$ until $\lceil \sqrt{l} * 2^n \rceil$
   (a) Compute $(x_i, y_i) \leftarrow f||g(x_{i-1}, y_{i-1})$.
   (b) Match (y_{i-l}, y_i) to $\{(P_j, C_j)\}$ in $\mathcal{T}$.

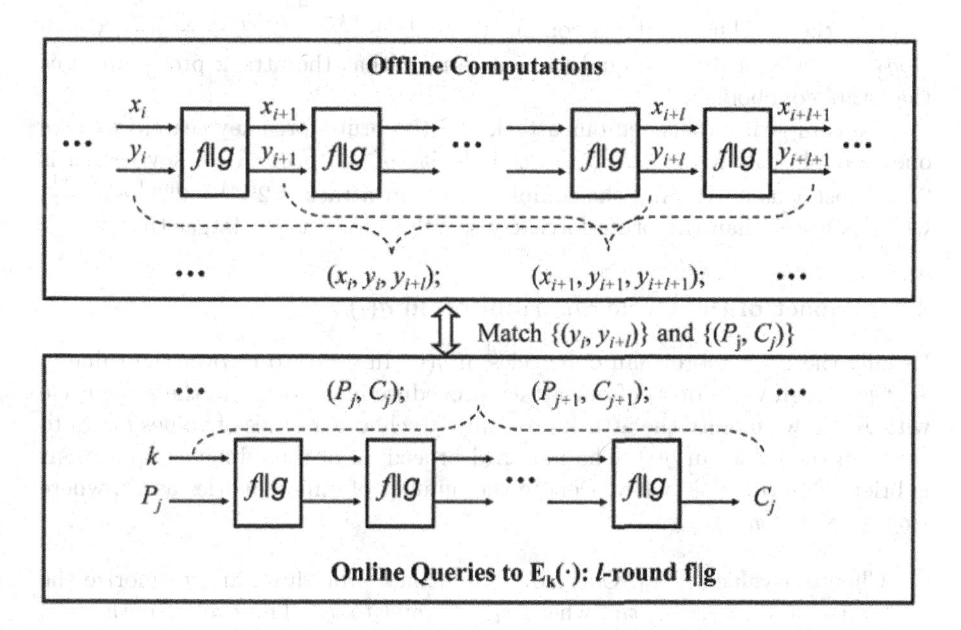

Fig. 3. Overview of Our Attack

(c) If it is matched with a (P_j, C_j), then check whether x_{i-l} is the real key K by matching $E_{x_{i-l}}(P_{j'})$ to $C_{j'}$, where $(P_{j'}, C_{j'})$ is another plaintext-ciphertext pair in $\mathcal{T}$. If x_{i-l} is the real key K, output it.

Complexity evaluation. The number of queries in total is $\lceil\frac{2^n}{\sqrt{l}}\rceil$. The offline computation is $\lceil\sqrt{l}*2^n*\frac{1}{l}\rceil$. The total complexity is $\frac{2*2^n}{\sqrt{l}}$. The memory is $\frac{2^n}{\sqrt{l}}$. Note at steps 3 and 4, we only need to memorize l pairs: (x_{i-l}, y_{i-l}), (x_{i-l+1}, y_{i-l+1}), ..., (x_i, y_i), which is negligible compared with the memory cost at step 1. Thus the dominant memory consumption is at step 1, which is $\frac{2^n}{\sqrt{l}}$.

Success evaluation. The success probability is $1-\frac{1}{e} \approx 0.63$ for the collision between the plaintext-ciphertext pairs queried at online phase and those computed at offline phase.

3.1 Impact of the Key Length

Some block cipher uses a key size larger than its block size. For the case $k > n$, our attack provides a variable tradeoff between data complexity and the offline complexity. Let the number of online queries be 2^x, where $x \leq n$. Note that the value of 2^x should be at least $\lceil\frac{k}{n}\rceil$ for identifying K. The offline complexity is $\frac{2^{k+n-x}}{l}$ for producing the expected collision with a probability $1-\frac{1}{e} \approx 0.63$. There are around 2^{k-n} noisy collisions, and it needs at most $\lceil\frac{k}{n}\rceil \times 2^{k-n}$ computations to erase them. Thus the total complexity is $2^x + \lceil\frac{k}{n}\rceil \times 2^{k-n} + \frac{2^{k+n-x}}{l}$, where $\lceil log_2\frac{k}{n}\rceil \leq x \leq n$. If x is equal to n, it implies that the attack procedure uses the entire codebook.

As a comparison between our attack and the brute-force key search, we give one example. Let k be $2n$. So the complexity of the brute-force key search is 2^{2n}. Select x as $n-1$. And the complexity of our attack is $2^{n-1} + 2^{n+1} + \frac{2^{2n+1}}{l}$, which is faster than the brute-force key search as long as l is larger than 2.

3.2 Impact of the Cycle Distribution in $g(\cdot)$

Usually there are more than one cycles in $g(\cdot)$. In order to recover the value of K, the chosen value of x_0 of our attack procedure must locate in the same cycle with K. So we modify the attack procedure to choose a series of values for x_0 to cover all the cycles in $g(\cdot)$. The modified procedure of the offline computations is briefly described as below. Denote the number of online queries as 2^x, where $\lceil log_2\frac{k}{n}\rceil \leq x \leq n$.

1. Choose a value for x_0. Carry out the attack procedure, and memorize the values of x_1, x_2, ..., x_{t_1}, where x_{t_1} is equal to x_0. The cycle length is t_1. For this cycle, iterate $f||g(\cdot,\cdot)$ $t_1 \times 2^{k-x}$ times at offline. If the value of K is recovered, output it.

2. Choose another value for x_0, which is not included in the cycle at step 1. Namely search K in another cycle of $g(\cdot)$. Denote the length of the second cycle as t_2. Iterate $f||g(\cdot,\cdot)$ $t_2 \times 2^{k-x}$ times at offline. If the value of K is recovered, output it.
3. Similarly with step 2, choose a series of values for x_0 until all the cycles in $g(\cdot)$ are covered.

The complexity of iterating $f||g(\cdot,\cdot)$ is $\frac{1}{l} \times 2^{k-x} \times (t_1+t_2+\cdots) = \frac{2^{k+n-x}}{l}$, which is the same with the case of only one cycle. So the total complexity remains the same, which is $2^x + \lceil \frac{k}{n} \rceil \times 2^{k-n} + \frac{2^{k+n-x}}{l}$. The success probability is $1 - \frac{1}{e} \approx 0.63$ due to the expected collision in the cycle where K locates. The memory may be increased, in the worst case 2^n, for storing the cycles in $g(\cdot)$.

4 Application on 3D Block Cipher and 3D-Based Compression Functions

4.1 3D Block Cipher

Nakahara Jr. proposed a block cipher 3D at CANS 2008 [12]. 3D follows the design framework of AES [7], but enlarges the block size and the key size. Both the block size and the key size of 3D are 512 bits. It seems an interesting research motivation: how to enhance AES considering the future development of computation power such that a 128-bit or even a 256-bit key becomes weak to resist the brute-force attack. Moreover, as also pointed out by Nakahara Jr. [12], such a AES-style block cipher with a large size is a suitable building block for hash functions, stream ciphers, etc. Here we briefly sketch the structure of 3D. For a completed specification, we refer to [12].

Encryption process. The i-th encryption round function is described as

$$\tau_i(\cdot) = \pi \circ \theta_{i \bmod 2} \circ \gamma \circ \kappa(\cdot).$$

And each transformation is detailed below.

- κ: bitwise XOR with a round key;
- γ: a byte-wise S-box transformation;
- θ_1, θ_2: two different byte-position shift transformations;
- π: a matrix multiplication transformation applied to columns of the state.

Similarly with AES, the last round function of encryption process does not include π operation but includes an extra key-whitening, which becomes $\kappa \circ \theta_{i \bmod 2+1} \circ \gamma \circ \kappa(\cdot)$.

Key schedule. The i-th round key is generated as[2]

$$\pi \circ \theta_{i \bmod 2+1} \circ \gamma \circ \kappa'(\cdot).$$

And new transformation is explained below.

[2] Encryption round function and key schedule round function use different γ. Here we omit the description.

- κ': bitwise XOR with a constant. The value of the constant depends on the number of rounds, namely the value of l, in order to resist the related-cipher attack [22]. We stress that κ' in each round of key schedule uses the same constant in a concrete 3D instance because the value of l is fixed.

The designer recommends that the number of rounds in 3D, namely l, is 22.

4.2 Previous Attacks on 3D

In [12], Nakahara Jr. proposed a key-recovery attack on 3D with 6 rounds. After that, he extended the number of the attacked round to 10 [13]. In [9], Dong *et al.* analyzed 3D in the known-key attack model, and found a distinguisher on 15-round 3D. Recently Koyama *et al.* proposed improved key-recovery attacks on 13-round 3D [16].

As a summary, all these previous attacks follow attack approaches including truncated differential cryptanalysis and impossible differential cryptanalysis, whose effectiveness decreases with the increase of the round numbers. So far, the best numbers of the attacked rounds on 3D are 13 in the secret-key attack model and 15 in the known-key attack model.

4.3 Application of Our Attack on 3D

This section applies our attack to 3D. Mainly we show the small functions $f(\cdot,\cdot)$ in the encryption process and $g(\cdot)$ in the key schedule, which are irrespective to the round indexes. And then the attack procedure in Section 3 can be easily applied.

$f(\cdot,\cdot)$ and $g(\cdot)$ in 3D. Only the byte position shift transformation θ is related to the round indexes. In the encryption process, θ_1 is used in the odd round, and θ_2 in the even round. In the key schedule, θ_1 is used in the even round, and θ_2 in the odd round. Regard two rounds in the encryption process as $f(\cdot,\cdot)$. More precisely, combine the first and second rounds, third and fourth rounds, and so on. $f(\cdot,\cdot)$ is irrespective to the round indexes. Similarly by regarding two rounds in the key schedule as $g(\cdot)$, $g(\cdot)$ is also irrespective to the round indexes.

Thus our attack can be applied. Note that the last round function in the encryption process is different from other round functions. It does not influence the applicability of our attack. We just need to compute a y'_{i+1} by an extra XOR during computing $(x_i, y_i) \leftarrow f||g(x_{i-1}, y_{i-1})$, and use $(x_{i-l/2}, y_{i-l/2}, y'_i)$ for the matching with online queried plaintext-ciphertext pairs. Since an extra XOR is negligible compared with $f||g(\cdot,\cdot)$, we will omit it. Finally the complexity is $\frac{2^{513}}{\sqrt{l/2}}$. The success probability is 0.63. And the memory requirement is $\frac{2^{512}}{\sqrt{l/2}}$.

Remark. For the recommended instance 22-round 3D, our attack is slightly better than the brute-force attack. We stress that our attack can be applied to all the instances of 3D. More interestingly, if l becomes larger, the effectiveness of our attack, e.g. the relative advantage compared with the brute-force attack, increases, while that most other short-cut attacks including truncated differential attacks, impossible differential attacks and so on decreases.

4.4 Collision and Preimage Attacks on 3D-Based Compression Function

Our attacks can also accelerate collision and preimage attacks on 3D-based compression functions. We use Davies-Meyer mode as an example. The compression function is $3D_m(h) \oplus h$, where the message block m is as the key, and the hash chaining value h as the plaintext. The collision attack procedure is as below. Without the loss of generality, let l be an even integer. For simplicity, we assume the last round function in the encryption process is the same with other round functions.

1. Select a random h_0 and m_0.
2. Initialize table $\mathcal{T}$ to empty.
3. for $i = 1$ to $l/2$, compute $(m_i, h_i) \leftarrow f||g(m_{i-1}, h_{i-1})$;
4. for $i = l/2 + 1$ to 2^{256},
   (a) Compute $(m_i, h_i) \leftarrow f||g(m_{i-1}, h_{i-1})$;
   (b) Compute $h_i \oplus h_{i-l/2}$;
   (c) Match it to stored triples in $\mathcal{T}$.
   (d) If it matches to z in a triple (x, y, z) in $\mathcal{T}$, output $(h_{i-l/2}, m_{i-l/2})$ and (x, y) as a collision.
   (e) Otherwise, store $(h_{i-l/2}, m_{i-l/2}, h_i \oplus h_{i-l/2})$ in $\mathcal{T}$.

The complexity is $\frac{2^{257}}{l}$ computation. And the success probability is the same with the brute-force birthday attack. Thus our attack is about $\frac{l}{2}$ times faster than the brute-force attack. Similarly we can launch a preimage attack. And the complexity is $\frac{2^{513}}{l}$, and is about $\frac{l}{2}$ times faster than the brute-force attack. The procedure is as below. Denote the target hash value as h.

1. Select a random h_0 and m_0.
2. Initialize table $\mathcal{T}$ to empty.
3. for $i = 1$ to $l/2$, compute $(m_i, h_i) \leftarrow f||g(m_{i-1}, h_{i-1})$ and store (m_i, h_i) to $\mathcal{T}$;
4. for $i = l/2 + 1$ to 2^{512},
   (a) Compute $(m_i, h_i) \leftarrow f||g(m_{i-1}, h_{i-1})$;
   (b) Compute $h_i \oplus h_{i-l/2}$;
   (c) Match it to h.
   (d) If it matches, output $(m_{i-l/2}, h_{i-l/2})$ as a preimage.
   (e) Otherwise, erase $(m_{i-l/2}, h_{i-l/2})$ from $\mathcal{T}$, and store (m_i, h_i) to $\mathcal{T}$.

Finally we point out that these attacks can also be regarded as a distinguishing attack on 3D in the chosen-key model.

5 Conclusion

This paper has proposed a polynomial-advantage attack, which is applicable to a block cipher mode whose round functions in both the encryption process and the key schedule are independent of the round indexes. We also applied the new attack to 3D and 3D-based hash functions.

References

1. Biham, E.: New Types of Cryptanalytic Attacks Using Related Keys. J. Cryptology 7(4), 229–246 (1994)
2. Biham, E., Biryukov, A., Shamir, A.: Cryptanalysis of Skipjack Reduced to 31 Rounds Using Impossible Differentials. In: Stern, J. (ed.) EUROCRYPT 1999. LNCS, vol. 1592, pp. 12–23. Springer, Heidelberg (1999)
3. Biham, E., Shamir, A.: Differential Cryptanalysis of DES-like Cryptosystems. In: Menezes, A., Vanstone, S.A. (eds.) CRYPTO 1990. LNCS, vol. 537, pp. 2–21. Springer, Heidelberg (1991)
4. Biryukov, A., Wagner, D.: Slide Attacks. In: Knudsen, L.R. (ed.) FSE 1999. LNCS, vol. 1636, pp. 245–259. Springer, Heidelberg (1999)
5. Bogdanov, A., Khovratovich, D., Rechberger, C.: Biclique Cryptanalysis of the Full AES. In: Lee, D.H., Wang, X. (eds.) ASIACRYPT 2011. LNCS, vol. 7073, pp. 344–371. Springer, Heidelberg (2011)
6. De Cannière, C., Küçük, Ö., Preneel, B.: Analysis of Grain's Initialization Algorithm. In: Vaudenay, S. (ed.) AFRICACRYPT 2008. LNCS, vol. 5023, pp. 276–289. Springer, Heidelberg (2008)
7. Daemen, J., Rijmen, V.: The Design of Rijndael: AES - The Advanced Encryption Standard. Springer (2002)
8. Davies, D.W., Parkin, G.I.P.: The Average Cycle Size of the Key Stream in Output Feedback Encipherment. In: Chaum, D., Rivest, R.L., Sherman, A.T. (eds.) CRYPTO, pp. 97–98. Plenum Press, New York (1982)
9. Dong, L., Wu, W., Wu, S., Zou, J.: Known-Key Distinguisher on Round-Reduced 3D Block Cipher. In: Jung, S., Yung, M. (eds.) WISA 2011. LNCS, vol. 7115, pp. 55–69. Springer, Heidelberg (2012)
10. Gilbert, H., Peyrin, T.: Super-Sbox Cryptanalysis: Improved Attacks for AES-Like Permutations. In: Hong, S., Iwata, T. (eds.) FSE 2010. LNCS, vol. 6147, pp. 365–383. Springer, Heidelberg (2010)
11. Granville, A.: Cycle lengths in a permutation are typically Poisson distributed. Electronic Journal of Combinatorics 13, R107 (2006)
12. Nakahara Jr., J.: 3D: A Three-Dimensional Block Cipher. In: Franklin, M.K., Hui, L.C.K., Wong, D.S. (eds.) CANS 2008. LNCS, vol. 5339, pp. 252–267. Springer, Heidelberg (2008)
13. Nakahara Jr., J.: New Impossible Differential and Known-Key Distinguishers for the 3D Cipher. In: Bao, F., Weng, J. (eds.) ISPEC 2011. LNCS, vol. 6672, pp. 208–221. Springer, Heidelberg (2011)
14. Knudsen, L.R.: Truncated and Higher Order Differentials. In: Preneel, B. (ed.) FSE 1994. LNCS, vol. 1008, pp. 196–211. Springer, Heidelberg (1995)
15. Knudsen, L.R.: DEAL- A 128-bit Block Cipher. Technical Report 151, Department of Informatics, University of Bergen, Beigen, Norway (1998)

16. Koyama, T., Wang, L., Sasaki, Y., Sakiyama, K., Ohta, K.: New Truncated Differential Cryptanalysis on 3D Block Cipher. In: Ryan, M.D., Smyth, B., Wang, G. (eds.) ISPEC 2012. LNCS, vol. 7232, pp. 109–125. Springer, Heidelberg (2012)
17. Lai, X.: High Order Derivatives and Differential Cryptanalysis. In: Communications and Cryptography, pp. 227–233 (1994)
18. Matsui, M.: Linear Cryptanalysis Method for DES Cipher. In: Helleseth, T. (ed.) EUROCRYPT 1993. LNCS, vol. 765, pp. 386–397. Springer, Heidelberg (1994)
19. Mendel, F., Rechberger, C., Schläffer, M., Thomsen, S.S.: The Rebound Attack: Cryptanalysis of Reduced Whirlpool and Grøstl. In: Dunkelman, O. (ed.) FSE 2009. LNCS, vol. 5665, pp. 260–276. Springer, Heidelberg (2009)
20. Shamir, A.: Dagstuhl Seminar Symmetric Cryptography (2012)
21. Wagner, D.: The Boomerang Attack. In: Knudsen, L.R. (ed.) FSE 1999. LNCS, vol. 1636, pp. 156–170. Springer, Heidelberg (1999)
22. Wu, H.: Related-Cipher Attacks. In: Deng, R.H., Qing, S., Bao, F., Zhou, J. (eds.) ICICS 2002. LNCS, vol. 2513, pp. 447–455. Springer, Heidelberg (2002)

Annihilators of Fast Discrete Fourier Spectra Attacks

Jingjing Wang[1,*], Kefei Chen[1,2,**], and Shixiong Zhu[3]

[1] Department of Computer Science and Engineering,
Shanghai Jiaotong University, Shanghai, China
{wangjingjing,kfchen}@sjtu.edu.cn
[2] Shanghai Key Laboratory of Scalable Computing and Systems, Shanghai, China
[3] Science and Technology on Communication Security Laboratory, Chengdu, China

Abstract. Spectra attacks proposed recently are more data efficient than algebraic attacks against stream cipher. They are also time-and-space efficient. A measurement of the security of a stream cipher against spectra attacks is spectral immunity, the lowest spectral weight of the annihilator of the key stream. We study both the annihilator and the spectral immunity. We obtain a necessary and sufficient condition for the existence of low spectral weight annihilator and find it is more difficult to decide the (non)existence of the low weight annihilator for spectra attacks than for algebraic attacks. We also give some basic properties of annihilators and find the probability of a periodic sequence to be the annihilator of another sequence of the same period is low. Finally we prove that the spectral immunity is upper bounded by half of the period of the key stream. As a result, to recover any key stream, the least amount of bits required by spectra attacks is at most half of its period.

Keywords: stream cipher, spectra attacks, spectral immunity, annihilator.

1 Introduction

Stream ciphers are popular for their efficiency in a wide range of applications including real-time encryptions and security applications for constrained environments.

Algebraic attacks have been successful on stream ciphers in recent years[1–4]. They recover key streams of a stream cipher by solving an overdefined algebraic equation system. They are efficient if low degree annihilators of Boolean functions are found[6]. Fast algebraic attacks[5] generalizes that; fast algebraic attacks are efficient if low degree relations of Boolean functions are found.

Fast discrete Fourier spectra attacks on stream ciphers[8] are algebraic attacks solving equations on spectra of key streams. They are parallel to fast algebraic

* Corresponding author.
** The author is supported by the National Natural Science Foundation of China (61133014, 60970111) and NLMC (9140C110201110C1102).

G. Hanaoka and T. Yamauchi (Eds.): IWSEC 2012, LNCS 7631, pp. 182–196, 2012.

attacks; they are efficient if low spectral weight relations of periodic sequences are found. But they can be more efficient than algebraic attacks and fast algebraic attacks, especially when the stream cipher uses an algebraic-immune Boolean function[12]. And they are generally applicable to any periodic sequence.

In fast discrete Fourier spectra attacks, the existence of low spectral weight relation of periodic sequences is required. A specialized case, the existence of low spectral weight sequence annihilator also fulfills the requirement. However, neither the sequence annihilator nor the more generalized relation has received extensive study.

The only few results are about the sequence annihilator. [8] proposes the concept of spectral immunity, which is the minimum spectral weight of annihilators of a periodic sequence; it also shows that the upper bound of spectral immunity will be greater than the smaller value between the weight of a periodic sequence and its complement. [10] generalizes the concept of spectral immunity and shows an upper bound of the spectral immunity in the algebraic immunity of the Boolean function when the underlying key stream is generated by a filter generator.

This paper also focuses on the sequence annihilator. We show that a sequence has a low spectral weight annihilator if and only if it allows a special matrix to be not of full column rank. We analyze the annihilator set with respect to a specific period of a sequence. We find that as the period associated with the annihilator set increases, the cardinality of the set grows but the ratio between that cardinality and the number of all sequences with that period approaches to zero. Finally, we prove that the spectral immunity of a periodic sequence is upper bounded by approximately half of its period. It is the first time that the upper bound of spectral immunity is expressed by one of the design parameters of a stream cipher.

The rest of the paper is organized as follows. Section 2 gives necessary definitions and notations for this paper. Section 3 shows how spectral attacks may exploit the properties of annihilators to gain efficiency. In Section 4 we provide a necessary and sufficient condition for a sequence to have low spectral weight annihilators. In Section 5 we discuss properties of annihilator sets. In Section 6 we give upper bounds of spectral immunity and show how that will affect the design criteria of a stream cipher. Section 6 concludes the paper.

2 Preliminaries

In this section we give some necessary definitions and notations about binary sequences and their discrete fourier transform over finite fields. See[9][11] for a thorough discussion.

For a positive integer T, suppose $T|2^n-1$ for some integer n. Let $\mathbf{s}$ be a binary sequence of period T and $s_0, s_1, \ldots, s_{T-1}$ be the terms of $\mathbf{s}$ in its first period. Let α be an element in $\mathbb{F}_{2^n}$ of order T. Then the discrete fourier transform of the sequence is defined by

$$S_r = \sum_{t=0}^{T-1} s_t \alpha^{-tr}, r = 0, 1, \ldots, T-1.$$

The result of the transform $S_0, S_1, \ldots, S_{T-1}$ is called the discrete fourier spectra of the sequence $\mathbf{s}$.

The inverse discrete fourier transform is

$$s_t = \sum_{r=0}^{T-1} S_r \alpha^{rt}, t = 0, 1, \ldots, T-1.$$

Let $\Gamma_2(T)$ be the set of the leaders of the cyclotomic coset modulo T (with respect to 2) and n_g be the size of the coset led by a leader $g \in \Gamma_2(T)$. If we partition the set of integers $\{0, 1, \ldots, T-1\}$ into the cyclotomic cosets modulo T the inverse discrete fourier transform is also

$$s_t = \sum_{g \in \Gamma_2(T)} \mathrm{Tr}_1^{n_g}(S_g \alpha^{gt}), t = 0, 1, \ldots, T-1$$

with $\mathrm{Tr}_1^y(x)$ being the trace function from $\mathbb{F}_{2^y}$ to $\mathbb{F}_2$. This inverse transform formula is also referred to as the trace representation of the sequence $\mathbf{s}$.

If a sequence $\mathbf{s}$ has w nonzero terms in one period we say its weight with respect to (its period) T is w. If a sequence $\mathbf{s}$ has v nonzero terms in its discrete fourier spectra we say its spectral weight is v. Let $l(\mathbf{s})$ be the linear complexity of $\mathbf{s}$. Then $l(\mathbf{s}) = v$. In the rest of the paper, we use the linear complexity $l(\mathbf{s})$ to refer to the spectral weight sometimes in order to be consistent with the symbol in [8].

Note that despite the dependency of the discrete fourier transform on the period of a sequence, no matter which one of the periods is used to do the transform the number of nonzero terms in the spectra of a sequence remains the same.

For the positive integer T, we denote the set of all sequences of period T by Ω_T. We define operations for sequences as termwise. In detail, if $\mathbf{s}$ is the sequence $s_0, s_1, \ldots$ and $\mathbf{z}$ is the sequence $z_0, z_1, \ldots$ then the sum $\mathbf{s} + \mathbf{z}$ is taken to be the sequence $s_0 + z_0, s_1 + z_1, \ldots$ and the product $\mathbf{s} \cdot \mathbf{z}$ is the sequence $s_0 \cdot z_0, s_1 \cdot z_1, \ldots$. Under these definitions of addition and multiplication, the set Ω_T is a ring. Its additive identity is the sequence $0, 0, \ldots$ denoted by $\mathbf{0}$ and its multiplicative identity is the sequence $1, 1, \ldots$ denoted by $\mathbf{1}$.

For the convenience of statement, we define the operator "concatenation" $||$ for vectors. Let $\boldsymbol{col_0}$ and $\boldsymbol{col_1}$ be two vectors. Let $\boldsymbol{col_0} = (col_{0,0}, col_{0,1}, \cdots, col_{0,n_0})^T$ and $\boldsymbol{col_1} = (col_{1,0}, col_{1,1}, \ldots, col_{1,n_1})^T$. Then $\boldsymbol{col_0}||\boldsymbol{col_1} = (col_{0,0}, col_{0,1}, \ldots, col_{0,n_0}, col_{1,0}, col_{1,1}, \ldots, col_{1,n_1})^T$. The concatenation of more than two vectors is defined similarly.

3 Fast Discrete Fourier Spectra Attacks: Revisited

In [8], fast discrete Fourier spectra attacks are described under the assumption that low spectral weight relations exist. The aim of this section is to show how the assumption of low spectral weight relation is related to that of low spectral weight annihilator and why this paper focuses on the latter in studying fast discrete Fourier spectra attacks. In the rest of the paper, the term "fast discrete Fourier spectra attacks" is shortened to "spectra attacks" for convenience.

Recall the assumption of low spectral weight relation in the spectra attack algorithm. Let $\mathbf{s}$ be the periodic sequence to be attacked; let $l(\cdot)$ the spectral weight of a sequence.

Assumption. Let $\mathbf{a}$ be the shifted sequence of $\mathbf{s}$. Assume that there exists two periodic sequences $\mathbf{c}, \mathbf{d}$ such that $\mathbf{ac} = \mathbf{d}$ and $l(\mathbf{c}) + l(\mathbf{d}) < l(\mathbf{a})$.

The attack algorithm makes use of this assumption as follows. Let β be the shift difference between sequences $\mathbf{a}$ and $\mathbf{s}$. Let $\mathbf{b}, \mathbf{u}$ be the shifted sequences of $\mathbf{c}, \mathbf{d}$ with the same shift difference β. Then $\mathbf{sb} = \mathbf{u}$. This is an equation of variable β; given the spectra of $\mathbf{a}$, $\mathbf{c}$ and $\mathbf{d}$, β can be solved and $\mathbf{s}$ will be recovered.

Naturally this assumption has two sub-assumptions:

$$\text{Sub-assumption S1. } \mathbf{ac} = \mathbf{d}, \mathbf{d} \neq \mathbf{0} \text{ and } l(\mathbf{c}) + l(\mathbf{d}) < l(\mathbf{a}),$$
$$\text{Sub-assumption S2. } \mathbf{ae} = \mathbf{0} \text{ and } l(\mathbf{e}) < l(\mathbf{a}),$$

where $\mathbf{c}, \mathbf{d}, \mathbf{e}$ are just some periodic sequences but $\mathbf{a}$ is the shifted sequence of $\mathbf{s}$. The complexity results of spectra attacks can be separated for attacks under the two disjoint assumptions.

Table 1. Complexity Results of Spectra Attacks under Disjoint Assumptions

	S1. $\mathbf{ac} = \mathbf{d} \neq \mathbf{0}$	S2. $\mathbf{ae} = \mathbf{0}$
data complexity	$l(\mathbf{c}) + l(\mathbf{d})$	$l(\mathbf{e})$
time complexity (pre-computation)	$O(l(\mathbf{d})[n(\log n)^2 + (\log(l(\mathbf{d})))^3 + \eta(n)^{\mathrm{a}}])$	0
time complexity (computation)	$O(l(\mathbf{c})\log(l(\mathbf{c}))\eta(l(\mathbf{c})) + 4\lvert\mathcal{N}_{\mathbf{c}}\rvert^{\mathrm{b}}\,\eta(n-1))$	$O(l(\mathbf{e})\log(l(\mathbf{e}))\eta(l(\mathbf{e})) + 4\lvert\mathcal{N}_{\mathbf{e}}\rvert\eta(n-1))$

[a] $\eta(n) = n \log_2 n \log_2 \log_2 n$.

[b] $\mathcal{N}_{\mathbf{c}}$ is the set of coset leaders such that the spectrum on that coset leader for the sequence $\mathbf{c}$ is nonzero.

The assumption of low spectra weight annihilator is actually Sub-assumption S2. Spectra attacks under S2 performs better than spectra attacks under S1 when $l(\mathbf{c}) \approx l(\mathbf{e})$, eliminating pre-computation and using fewer data bits. If we can find such a low spectral weight annihilator that $l(\mathbf{e}) \ll l(\mathbf{a})$, spectra attacks under Sub-assumption S2 will be efficient.

More importantly, as the low spectral weight annihilator can be constructed from a low spectral weight relation, we find that spectra attacks using the constructed low spectral weight annihilators require less data complexity than spectra attacks using the original low spectral weight relations.

Let $\mathbf{ac} = \mathbf{d}$ be a low spectral weight relation that satisfies Sub-assumption S1. Then as

$$(\mathbf{a} + \mathbf{1}) \cdot \mathbf{d} = \mathbf{a} \cdot \mathbf{ac} + \mathbf{ac} = \mathbf{0},$$

it produces an annihilator that satisfies Sub-assumption S2 except that the annihilator $\mathbf{d}$ is an annihilator of the sequence $\mathbf{a} + \mathbf{1}$. The attacker can use this annihilator to recover the sequence $\mathbf{s}+\mathbf{1}$ and then recover the sequence $\mathbf{s}$. Therefore both are feasible assumptions in recovering the sequence $\mathbf{s}$.

However their data complexity is different. The data complexity of spectra attacks under the constructed low spectral weight annihilator is $l(\mathbf{d})$ while that of spectra attacks under the original low spectral weight relation is $l(\mathbf{c}) + l(\mathbf{d})$. This lead to a fact that the least data complexity that spectra attacks could achieve must occur at when the low spectral annihilator is used as assumption.

Therefore the annihilator is important for spectra attacks to fulfill its assumption and to profile its data complexity. This paper is going to show some results of annihilator in both aspects.

4 Annihilator

This section discusses the concept of an annihilator and the condition for a sequence to have a low spectral weight annihilator. [8] mentions annihilators in its definition of spectral immunity but it does not use the term "annihilator". No formal definition of annihilator has been given yet. We must formulate one in order to investigate its properties.

Definition 1. *For a binary sequence* $\mathbf{s}$ *of period* T*, a binary sequence* $\mathbf{a} \neq \mathbf{0}$ *also of period* T *satisfying* $\mathbf{a} \cdot \mathbf{s} = \mathbf{0}$ *under termwise multiplication is called an annihilator of* $\mathbf{s}$.

The period is not necessary in the definition of an annihilator. As long as $\mathbf{a}$ and $\mathbf{s}$ are periodic sequences they always share some common period such as a common multiple of their minimal periods. Nevertheless the definition is more consistent with that of spectral immunity in [8] if the period is referred to. If a specific common period T is required for the sequence and the annihilator, the latter will be called an annihilator with respect to (the common period) T.

Let V be the spectral weight of annihilator reasonably low for performance of fast discrete fourier spectra attacks. In the rest of this section, the low spectral weight annihilator means the annihilator with spectral weight no greater than V.

The existence of such annihilator can be decided by a special matrix which is defined as follows. For any integer g, define a row vector

$$\boldsymbol{u}^T(g;t) = (\mathrm{Tr}_1^{n_g}(\alpha^{g\cdot 0} \cdot \alpha^{gt}), \mathrm{Tr}_1^{n_g}(\alpha^{g\cdot 1} \cdot \alpha^{gt})), \ldots, \mathrm{Tr}_1^{n_g}(\alpha^{g\cdot(n_g-1)} \cdot \alpha^{gt}).$$

And for a set of integers $G = \{g_0, g_1, \cdots, g_{m-1}\}$, where $m = |G|$, define a row vector

$$\boldsymbol{u}^T(G;t) = \boldsymbol{u}^T(g_0;t)||\boldsymbol{u}^T(g_1;t)||\cdots||\boldsymbol{u}^T(g_{m-1};t).$$

Then for this set of integers G, define a matrix

$$U(G;1_{\mathbf{s}}) = (\boldsymbol{u}(G;t_0), \boldsymbol{u}(G;t_1), \ldots, \boldsymbol{u}(G;t_{|1_{\mathbf{s}}|-1}))^T$$

where $1_{\mathbf{s}} = \{t_0, t_1, \ldots, t_{|1_{\mathbf{s}}|-1}\} = \{t|s_t = 1\}$. We find that the existence of low spectral weight annihilator is equivalent to the matrix $U(G;1_{\mathbf{s}})$ for some particular G to be not of full column rank.

Proposition 1. *Let* $\mathbf{s}$ *be a sequence of period* T. *Let* $\Gamma_2(T)$ *denote the set of all coset leaders of cyclotomic cosets modulo* T. *Then the sequence* $\mathbf{s}$ *has a low weight annihilator of period* T *if and only if a set of coset leaders* $G = \{g_0, g_1, \ldots, g_{m-1}\} \subseteq \Gamma_2(T)$ *exists such that* $v = \sum_{i=0}^{m-1} n_{g_i} \leq V$ *and that the rank of the matrix* $U(G;1_{\mathbf{s}})$ *is less than* v.

Proof. First consider the necessary condition. Let $\mathbf{s}$ have a low spectral weight annihilator of period T. Let $\mathbf{a} = a_0, a_1, \ldots$ be this annihilator. Then the spectral weight of $\mathbf{a}$ is no greater than V.

For the spectra of $\mathbf{a}$, $\{A_0, A_1, \ldots, A_{T-1}\}$, we have:

$$\begin{aligned}&\{A_0, A_1, \ldots, A_{T-1}\}\\ =& \bigcup_{g\in\Gamma_2(T)} \{A_{g\cdot 2^j} | 0 \leq j \leq n_g - 1\}\\ =& \bigcup_{g\in\Gamma_2(T)} \{(A_g)^{2^j} | 0 \leq j \leq n_g - 1\}\end{aligned}$$

where the first equality follows the definition of cyclotomic cosets and the second equality follows from the fact that $A_{g\cdot 2^j} = (A_g)^{2^j}$ [9]. Let G be the set of coset leaders $\{g|A_g \neq 0, g \in \Gamma_2(T)\}$. It follows that $(A_g)^{2^j} \neq 0$ for any $0 \leq j \leq n_g - 1$. Therefore the spectral weight of the annihilator $\mathbf{a}$ is: $v = \sum_{g\in G} n_g$ and it is no greater than V.

The trace representation of the annihilator $\mathbf{a}$ is:

$$\begin{aligned}a_t =& \sum_{g\in\Gamma_2(T)} \mathrm{Tr}_1^{n_g}(A_g\alpha^{gt})\\ =& \sum_{g\in G} \mathrm{Tr}_1^{n_g}(A_g\alpha^{gt})\end{aligned}$$

where α is an element in $\mathbb{F}_2^n$ of order T, n_g is the size of coset led by g. As $A_{g\cdot 2^j} = (A_g)^{2^j}$, $A_g = (A_g)^{2^{n_g}}$. then A_g is in $\mathbb{F}_2^{n_g}$. Since α^g is primitive in $\mathbb{F}_2^{n_g}$, A_g can be expressed as a linear combination over $\mathbb{F}_2$ of the basis $\alpha^{g\cdot 0}, \alpha^{g\cdot 1}, \ldots, \alpha^{g\cdot(n_g-1)}$ in $\mathbb{F}_2^{n_g}$: $A_g = \sum_{j=0}^{n_g-1} e_{g,j}\alpha^{g\cdot j}$ where the coefficients $\{e_{g,j}|0 \leq j \leq n_g - 1\}$ are elements in $\mathbb{F}_2$. Let $\boldsymbol{e}(g)$ be the vector $(e_{g,0}, e_{g,1}, \ldots, e_{g,n_g-1})^T$, then

$$
\begin{aligned}
A_g \alpha^{gt} =& (\sum_{j=0}^{n_g-1} e_{g_j} \alpha^{gj}) \cdot \alpha^{gt} \\
=& \sum_{j=0}^{n_g-1} e_{g_j} (\alpha^{gj} \alpha^{gt}) \\
=& \boldsymbol{u}^T(g;t) \cdot \boldsymbol{e}(g).
\end{aligned}
$$

Let $G = \{g_0, g_1, \ldots, g_{m-1}\}$ and let $\boldsymbol{e}(G)$ be the vector $\boldsymbol{e}(g_0)||\boldsymbol{e}(g_1)|| \ldots ||\boldsymbol{e}(g_{m-1})$. Then:

$$a_t = \boldsymbol{u}^T(G;t) \cdot \boldsymbol{e}(G).$$

Since $a_t = 0$ whenever $s_t = 1$, we have the following equation system for unknowns $\bigcup_{g \in G}\{e_{g,j} | 0 \leq j \leq n_g - 1\}$:

$$
\begin{cases}
\boldsymbol{u}^T(G;t_0) \cdot \boldsymbol{e}(G) = 0 \\
\boldsymbol{u}^T(G;t_1) \cdot \boldsymbol{e}(G) = 0 \\
\cdots \\
\boldsymbol{u}^T(G;t_{|1_{\mathbf{s}}|-1}) \cdot \boldsymbol{e}(G) = 0
\end{cases}
. \tag{1}
$$

Note if $\boldsymbol{e}(G) = \mathbf{0}$ then $\mathbf{a} = \mathbf{0}$. Therefore as the annihilator $\mathbf{a} \neq \mathbf{0}$, this equation system must have nonzero solutions. Since the system is equivalent to

$$U(G;1_{\mathbf{s}}) \cdot \boldsymbol{e}(G) = \mathbf{0},$$

the rank of the $v \times |1_{\mathbf{s}}|$ matrix $U(G;1_{\mathbf{s}})$ must be less than v.

On the other hand, if there exists a set of coset leaders G such that the rank of of $(U(G;1_{\mathbf{s}}))$ is less than v, then the equation system (1) will have nonzero solutions, which in turn gives an annihilator of the sequence $\mathbf{s}$ which has spectral weight no greater than V. □

There is a similar result for the annihilators of Boolean function in [7]. However, the necessary and sufficient condition for the Boolean function annihilators shows that the test of the rank of one matrix is sufficient to decide the the (non)existence of annihilators while that for the sequence annihilators requires much much more matrices to be considered. It shows that spectral attacks are more flexible than algebraic attacks (as one sequence has potentially much more annihilators) and designers may find more difficulties to defend spectral attacks (as the number of matrices to be tested is much greater).

5 Properties of Annihilator Set

This section defines the concept of annihilator set. It discusses the equivalence between two sub-assumptions of spectra attacks and also the possibility of a random periodic sequence being an annihilator for a specified sequence.

For a sequence $\mathbf{s}$ of period T, its annihilator is a sequence of period T of which the termwise product with $\mathbf{s}$ is $\mathbf{0}$. This concept of an annihilator implies that the annihilator of a sequence with minimal period T_{min} may have period $T_{min}, 2T_{min}, 3T_{min}, \ldots$. When we discuss the annihilator set, it is better to specify the period of an annihilator to avoid confusion. Thus the annihilator set is defined with respect to a specific T as follows.

$$AnnSet_T(\mathbf{s}) = \{\mathbf{a} | \mathbf{a} \in \Omega_T, \mathbf{s} \cdot \mathbf{a} = \mathbf{0}, \mathbf{a} \neq \mathbf{0}\}$$

where Ω_T is the set of all sequences of period T. Then the whole annihilator set, which is the set of all possible annihilators, is a union of $AnnSet_T(\mathbf{s})$:

$$AnnSet(\mathbf{s}) = \{\mathbf{a} | \mathbf{s} \cdot \mathbf{a} = \mathbf{0}, \mathbf{a} \neq \mathbf{0}\} = \bigcup_{T_{min}|T} AnnSet_T(\mathbf{s}).$$

We find the annihilator set with respect to T has the following two properties.

Property 1. Let $\mathbf{s}$ be a sequence with minimal period T_{min}. Let w be its weight with respect to T_{min}. For a positive integer T such that $T_{min}|T$, its annihilator set $AnnSet_T(\mathbf{s})$ is a principal ideal generated by $\mathbf{s} + \mathbf{1}$ in the ring Ω_T.

Property 2. The cardinality of the set is $|AnnSet_T(\mathbf{s})| = 2^{T - wT/T_{min}}$.

Proof. Under termwise addition and multiplication, for any $\mathbf{a_0}, \mathbf{a_1} \in AnnSet_T(\mathbf{s})$, we have $(\mathbf{a_0} + \mathbf{a_1}) \cdot \mathbf{s} = \mathbf{a_0} \cdot \mathbf{s} + \mathbf{a_1} \cdot \mathbf{s} = \mathbf{0}$ and for any $\mathbf{a} \in AnnSet_T(\mathbf{s}), \mathbf{z} \in \Omega_T$, we have $\mathbf{z} \cdot \mathbf{a} \cdot \mathbf{s} = \mathbf{a} \cdot \mathbf{z} \cdot \mathbf{s} = \mathbf{0}$. Thus $AnnSet_T(\mathbf{s})$ is an ideal in the ring Ω_T. Moreover for any $\mathbf{z} \in \Omega_T$, we have $\mathbf{z} \cdot (\mathbf{s} + \mathbf{1}) \cdot \mathbf{s} = \mathbf{0}$ and for any $\mathbf{a} \in AnnSet_T(\mathbf{s})$, we have $\mathbf{a} = \mathbf{a} + \mathbf{a} \cdot \mathbf{s} = \mathbf{a}(\mathbf{1} + \mathbf{s})$; thus the set $AnnSet_T(\mathbf{s})$ is a principal ideal generated by $\mathbf{s} + \mathbf{1}$.

The number of zeroes of the sequence $\mathbf{s}$ in time span T is $T - w \cdot T/T_{min}$. Thus there are $2^{T - wT/T_{min}}$ possibilities for a sequence of period T to be an annihilator of $\mathbf{s}$. The cardinality of the annihilator set is then $|AnnSet_T(\mathbf{s})| = 2^{T - wT/T_{min}}$. □

In the proof of Property 1, we show that an annihilator $\mathbf{a}$ of the sequence $\mathbf{s}$ gives the relation $\mathbf{a} = \mathbf{a}(\mathbf{s} + \mathbf{1})$. Let $l(\mathbf{a})$ be the spectral weight of $\mathbf{a}$. For the sub-assumption S1 of low spectral weight relation, it is required that $l(\mathbf{a}) + l(\mathbf{a}) < l(\mathbf{s} + \mathbf{1})$; for the sub-assumption S2 of low spectral annihilators, it is required that $l(\mathbf{a}) < l(\mathbf{s})$. Since $O(l(\mathbf{a})) = O(2l(\mathbf{a}))$ and $|l(\mathbf{s}) - l(\mathbf{s} + \mathbf{1})| = 1$, if a sufficiently low spectral weight annihilator exists, then a sufficiently low spectral weight relation also exists.

In turn, by Property 1, for a spectral weight relation $\mathbf{zs} = \mathbf{a}$ for some $\mathbf{a}$ and $\mathbf{z}$, $\mathbf{a}$ is found as an annihilator of $\mathbf{s} + \mathbf{1}$. Using the relation in sub-assumption S1 requires that $l(\mathbf{a}) + l(\mathbf{z}) < l(\mathbf{s})$ while using the annihilator in Sub-assumption S2 requires only $l(\mathbf{a}) < l(\mathbf{s})$. Thus if a low spectral weight relation exists, a low spectral weight annihilator must exist.

Therefore the existence of low spectral weight relation is equivalent to that of low spectral weight annihilator. When deciding if key streams of a stream cipher can fulfill the assumption of the spectra attack, it is sufficient to decide the existence of just one of them. For spectra attacks, Sub-assumption S1 can be reduced to Sub-assumption S2 without loss of efficiency.

By Property 2, the cardinality of the annihilator set with respect to T grows with the period T, but the ratio $|AnnSet_{l \cdot T_{min}}(\mathbf{s})|/|\Omega_{l \cdot T_{min}}| = 2^{-wT/T_{min}}$ shrinks. Thus we are more unlikely to find an annihilator if we look for it in the set of sequences with larger multiple of period T.

Proposition 2. *Let s be a sequence with minimal period T_{min}. The probability of any sequence of period $l \cdot T_{min}$ being an annihilator of s approaches to zero when the positive integer l approaches to infinity.*

6 Upper Bound of Spectral Immunity

Spectral immunity is of great importance in describing the difficulty of recovering the key stream by spectra attacks. The complexity of spectra attacks grows with the spectral weight of the annihilator of the key stream. Spectral immunity is defined as the lowest spectral weight of all the annihilators. As a result, it determines the least complexity that spectra attacks need to recover the key stream. Thus we use spectral immunity to measure the security level of a stream cipher against spectra attacks.

This section studies spectral immunity and mainly its upper bound. This general upper bound gives a general security level that a stream cipher, in defense to spectra attacks, at most could achieve. The upper bound is given in period of the key stream, one of the design parameters for a stream cipher. As a result, according to this upper bound, in order to defend spectra attacks, a stream cipher should have each of its key streams get a minimal period greater than 2^{128}.

The spectral immunity is first proposed in [8] and is generalized in [10]. These two definitions of spectral immunity are given here for reference and both have been adapted in order to be consistent with the symbols and definitions in this paper.

Definition 2. *For a periodic sequence* $\mathbf{s}$*, spectral immunity (SI) is the lowest spectral weight of all annihilators of* $\mathbf{s}$ *and all annihilators of* $\mathbf{s}+\mathbf{1}$*. Namely,* $SI(\mathbf{s}) = \min_{\mathbf{a} \in AnnSet(\mathbf{s}) \bigcup AnnSet(\mathbf{s}+\mathbf{1})} l(\mathbf{a})$.

Definition 3. *For a periodic sequence* $\mathbf{s}$*, let T be one of its period value. Then spectral immunity with respect to T (SI_T) is the lowest spectral weight of all annihilators of period T of the sequence* $\mathbf{s}$ *and all annihilators of period T of the sequence* $\mathbf{s}+\mathbf{1}$*. Namely,* $SI_T(\mathbf{s}) = \min_{\mathbf{a} \in AnnSet_T(\mathbf{s}) \bigcup AnnSet_T(\mathbf{s}+\mathbf{1})} l(\mathbf{a})$.

The term "spectral immunity" here refers to the least spectral weight of all annihilators. As there is no known result for the relationship between the spectral weight of a sequence and the period of it, it is better to call the general definition that includes annihilators of all possible periods to be "spectral immunity" and to call the other "spectral immunity with respect to T". Obviously, these two kinds of spectral immunity have such relationship that

$$SI(\mathbf{s}) = \min_{T_{min}|T} SI_T(\mathbf{s})$$

where T_{min} is the minimal period of the sequence $\mathbf{s}$. In the rest of this section, the term "spectral immunity" will always refer to Definition 2.

In order to assess the spectral immunity, we need a way to calculate the spectral weight of periodic sequence. In the proof of Proposition 1, we have shown that for any sequence $\mathbf{a}$ of period T, $\{A_0, A_1, \ldots, A_{T-1}\}$ is its spectra and its spectral weight is $v = \sum_{g \in G} n_g$ where the set G of integers is $G = \{g | A_g \neq 0, g \in \Gamma_2(T)\}$.

Since any spectrum A_g can be uniquely represented by $A_g = \sum_{j=0}^{n_g-1} e_{g,j}\alpha^{g,j}$ where coefficients the $\{e_{g,j}|0 \leq j \leq n_g - 1\}$ are in $\mathbb{F}_2$ and $\{\alpha^{gj}|0 \leq j \leq n_g - 1\}$ is a basis of $\mathbb{F}_2^{n_g}$ (we have shown that in the proof of Proposition 1), $A_g = 0$ if and only if $\{e_{g,j}|0 \leq j \leq n_g - 1\}$ are all 0.

Then the set G of integers is

$$\begin{aligned} G &= \{g | A_g \neq 0, g \in \Gamma_2(T)\} \\ &= \{g | \prod_{j=0}^{n_g-1} (1 + e_{g,j}) \neq 0, g \in \Gamma_2(T)\}. \end{aligned}$$

It follows that the spectral weight of the periodic sequence $\mathbf{a}$ is

$$\begin{aligned} v &= \sum_{g \in G} n_g \\ &= \sum_{g \in \Gamma_2(T)} n_g(1 + \prod_{j=0}^{n_g-1} (1 + e_{g,j})). \end{aligned}$$

We represent the periodic sequence $\mathbf{a}$ by all those coefficients $\bigcup_{g \in \Gamma_2(T)}\{e_{g,0}, e_{g,1}, \ldots, e_{g,n_g-1}\}$ involved in the calculation of spectral weight of $\mathbf{a}$. Let $u_{g,j,t} = \mathrm{Tr}_1^{n_g}(\alpha^{gj} \cdot \alpha^{gt})$. Substitute A_g by $\sum_{j=0}^{n_g-1} e_{g,j}\alpha^{g,j}$ and then the trace representation of $\mathbf{a}$ equals to

$$\begin{aligned} a_t &= \sum_{g \in \Gamma_2(T)} \mathrm{Tr}_1^{n_g}(A_g\alpha^{gt}) \\ &= \sum_{g \in \Gamma_2(T)} \mathrm{Tr}_1^{n_g}(\sum_{j=0}^{n_g-1} e_{g,j}\alpha^{g \cdot j}\alpha^{gt}), \\ &= \sum_{g \in \Gamma_2(T)} \sum_{j=0}^{n_g-1} e_{g,j}\, \mathrm{Tr}_1^{n_g}(\alpha^{g \cdot j}\alpha^{gt}), t = 0, 1, \ldots, T-1 \\ &= \sum_{g \in \Gamma_2(T)} \sum_{j=0}^{n_g-1} e_{g,j}u_{g,j,t}, t = 0, 1, \ldots, T-1 \end{aligned} \tag{2}$$

Let $\mathbf{u_{g,j}}$ be the sequence $u_{g,j,0}, u_{g,j,1}, u_{g,j,2}, \ldots$. Then $\mathbf{a}$ is a linear combination of sequences in the set $U = \bigcup_{g\in\Gamma_2(T)}\{\mathbf{u_{g,0}}, \mathbf{u_{g,1}}, \ldots, \mathbf{u_{g,n_g-1}}\}$ with coefficients $\bigcup_{g\in\Gamma_2(T)}\{e_{g,0}, e_{g,1}, \ldots, e_{g,n_g-1}\}$.

Note that the last equality of Equation (2) is actually a unique representation of $\mathbf{a}$. Let Ω_T be the linear space which contains all sequences of period T under termwise addition and scalar multiplication. Rank$(\Omega_T) = T = |U|$ where the second equality results from the definition of cyclotomic cosets modulo T. Since any sequence of period T can be expressed in sequences from the set U, U is a basis of Ω_T. Therefore the coefficients $\bigcup_{g\in\Gamma_2(T)}\{e_{g,0}, e_{g,1}, \ldots, e_{g,n_g-1}\}$ are uniquely determined by the periodic sequence $\mathbf{a}$ without the necessity to do the discrete Fourier transform and so is the spectral weight of $\mathbf{a}$.

Lemma 1. *Let $u_{g,j,t} = Tr_1^{n_g}(\alpha^{gj}\cdot\alpha^{gt})$ where g is a coset leader of a cyclotomic coset modulo T and n_g is the size of the coset led by the leader g, $0 \le j \le n_g - 1$. Let $\mathbf{u_{g,j}}$ be the sequence $u_{g,j,0}, u_{g,j,1}, u_{g,j,2}, \ldots$.*

Then any sequence $\mathbf{a}$ of period T can be expressed as a linear combination of sequences in the set $U = \bigcup_{g\in\Gamma_2(T)}\{\mathbf{u_{g,0}}, \mathbf{u_{g,1}}, \ldots, \mathbf{u_{g,n_g-1}}\}$:

$$\mathbf{a} = \sum_{g\in\Gamma_2(T)}\sum_{j=0}^{n_g-1} e_{g,j}\mathbf{u_{g,j}}$$

where the coefficients $\bigcup_{g\in\Gamma_2(T)}\{e_{g,0}, e_{g,1}, \ldots, e_{g,n_g-1}\} \in \mathbb{F}_2^T$. By those coefficients, the spectral weight of the sequence $\mathbf{a}$ is

$$v = \sum_{g\in\Gamma_2(T)} n_g(1 + \prod_{j=0}^{n_g-1}(1 + e_{g,j})).$$

Now that we are able to calculate the spectral weight of any sequence of period T, we are going to study the spectral immunity with respect to T first and then applies it to the more general spectral immunity.

Suppose the period T satisfies that $T|2^n - 1$ for some odd n. Let A^* and B^* be two subsets of U:

$$A^* = \bigcup_{\substack{h\in\Gamma_2(T),\\ 1\le wt_2(hR)\le\frac{n-1}{2}}}\{\mathbf{u_{h,0}}, \mathbf{u_{h,1}}, \ldots, \mathbf{u_{h,n_h-1}}\}$$

$$B^* = \bigcup_{\substack{h\in\Gamma_2(T),\\ \frac{n+1}{2}\le wt_2(hR)\le n-1}}\{\mathbf{u_{h,0}}, \mathbf{u_{h,1}}, \ldots, \mathbf{u_{h,n_h-1}}\}$$

we are going to show that for any sequence of period T, one of its annihilators is either a linear combination of sequences in $A^*\cup\{\mathbf{u_{0,0}}\}$ or that of sequences in $B^*\cup\{\mathbf{u_{0,0}}\}$. And thus the spectral immunity with respect to T is at most the spectral weight of this annihilator.

Before that, we are going to find some properties of the two sets A^* and B^* in order to calculate the spectral weight of this annihilator.

Property 3. $A^* \cup B^* = U^* = U \backslash \{\mathbf{u_{0,0}}\}$ and $|A^*| = |B^*| = (|U|-1)/2 = (T-1)/2$ where $R = (2^n-1)/T$ and $wt_2(\cdot)$ denotes the Hamming weight of an integer.

Proof. For an integer $g \in \Gamma_2(T)$and$g \neq 0$, $(2^n-1)/T \leq gR \leq (2^n-1)-(2^n-1)/T < 2^n-1$; the Hamming weight of gR satisfies that $1 \leq wt_2(gR) \leq n-1$. Thus $U^* = \bigcup_{0 \neq g \in \Gamma_2(T)} \{\mathbf{u_{g,0}}, \mathbf{u_{g,1}}, \ldots, \mathbf{u_{g,n_g-1}}\} = A^* \cup B^*$.

The number of elements in A^* is

$$|A^*| = \sum_{\substack{h \in \Gamma_2(T),\\ 1 \leq wt_2(hR) \leq \frac{n-1}{2}}} n_h.$$

Let Δ_A be the set of integers

$$\Delta_A = \{h | 1 \leq wt_2(hR) \leq \frac{n-1}{2} \text{ and } 1 \leq h \leq T-1\}.$$

For any positive integer $h \leq T-1$, $wt_2(hR) = wt_2(2hR \mod 2^n-1) = wt_2((2h \mod T)R)$. It follows that Δ_A is equivalent to the union of cyclotomic cosets of which leaders are of certain Hamming weight:

$$\Delta_A = \bigcup_{\substack{h \in \Gamma_2(T),\\ 1 \leq wt_2(hR) \leq \frac{n-1}{2}}} \{h, 2h, \ldots, 2^{n_h-1}h\}$$

where the product in Δ_A is taken modulo T. The number of elements in Δ_A is equal to that of elements in $|A^*|$:

$$|\Delta_A| = \sum_{\substack{h \in \Gamma_2(T),\\ 1 \leq wt_2(hR) \leq \frac{n-1}{2}}} n_h = |A^*|.$$

Similarly, let Δ_B be the set of integers

$$\Delta_B = \{h | \frac{n-1}{2} \leq wt_2(hR) \leq n-1 \text{ and } 1 \leq h \leq T-1\}$$

and then the number of elements in Δ_B is also equal to that of elements in $|B^*|$: $|\Delta_B| = |B^*|$.

There is a one-to-one correspondence between Δ_A and Δ_B. Let i be an integer and $i = T-h, h \in \Delta_A$. Then $i \in \Delta_B$ as $wt_2(iR) = wt_2((T-h)R) = n - wt_2(hR)$. Similarly for any integer $h \in \Delta_B$, $T-h \in \Delta_A$. Therefore, $|\Delta_A| = |\Delta_B|$.

Then $|A^*| = |B^*|$. And as $A^* \cup B^* = U^*$, $|A^*| = |B^*| = (|U|-1)/2 = (T-1)/2$. $\square$

Let A be the set $A^* \cup \{\mathbf{u_{0,0}}\}$ and let B be the set $B^* \cup \{\mathbf{u_{0,0}}\}$. For the set A, by Lemma 1, the linear combination of its members has spectral weight v_A at most $(T+1)/2$:

$$v_A \leq n_0 + \sum_{\substack{h \in \Gamma_2(T),\\ 1 \leq wt_2(hR) \leq \frac{n-1}{2}}} n_h \cdot 1 = 1 + |A^*| = (T+1)/2.$$

Similarly, for the set B, the linear combination of its members has spectral weight v_B also at most $(T+1)/2$:

$$v_B \leq n_0 + \sum_{\substack{h \in \Gamma_2(T), \\ \frac{n+1}{2} \leq wt_2(hR) \leq n-1}} n_h \cdot 1 = 1 + |B^*| = (T+1)/2.$$

Since we have found an upper-bound of the spectral weight of the linear combination of sequences in A or that in B, we are going to show the upper-bound of the spectral immunity with respect to T. Our result is summarized in the following theorem, of which the proof shows how to find an annihilator for any sequence to be a linear combination of sequences in either A or that of sequences in B.

Theorem 1. *For some odd integer n, let T be an integer such that $T|2^n - 1$. The spectral immunity with respect to T of a sequence $\mathbf{s}$ of period T is upper-bounded by $(T+1)/2$.*

Proof. Consider two sets A and $B \cdot \mathbf{s} = \{\mathbf{bs} | \mathbf{b} \in B\}$.

If $|B \cdot \mathbf{s}| < |B| = (T+1)/2$, then there exists two sequences $\mathbf{b_1}, \mathbf{b_2}$ in B such that $\mathbf{b_1 s} = \mathbf{b_2 s}$. $(\mathbf{b_1} + \mathbf{b_2})\mathbf{s} = \mathbf{0}$ and $\mathbf{b_1} + \mathbf{b_2}$ is therefore an annihilator of the sequence $\mathbf{s}$.

If $A \cap B \cdot \mathbf{s} \neq \emptyset$, there exists two sequences $\mathbf{a_1} \in A$ and $\mathbf{b_1} \in B$ such that $\mathbf{a_1} = \mathbf{b_1 s}$. Since $\mathbf{a_1 s} = \mathbf{b_1 s} \cdot \mathbf{s} = \mathbf{b_1 s}$, $\mathbf{a_1}$ is an annihilator of the sequence $\mathbf{s} + \mathbf{1}$.

If both conditions do not hold, i.e., $|B \cdot \mathbf{s}| = (T+1)/2$ and $A \cap B \cdot \mathbf{s} = \emptyset$, then $A \cup B \cdot \mathbf{s}$ contains $T+1$ different elements. Since the rank of Ω_T is T, there must exist a sum of $N \leq T$ sequences in $A \cup B \cdot \mathbf{s}$, which is equal to $\mathbf{0}$. At least one of those N sequences is in $B \cdot \mathbf{s}$; otherwise the linear dependency exists among sequences of A, a contradiction. Then suppose

$$\begin{aligned} (\mathbf{a_1} + \mathbf{a_2} + \cdots + \mathbf{a_p}) + (\mathbf{b_1 s} + \mathbf{b_2 s} + \cdots + \mathbf{b_q s}) = \mathbf{0}, 1 \leq p, q \leq (T+1)/2 \\ \text{or } (\mathbf{b_1 s} + \mathbf{b_2 s} + \cdots + \mathbf{b_q s}) = \mathbf{0}, 1 \leq p, q \leq (T+1)/2. \end{aligned} \quad (3)$$

Let $\mathbf{a} = \sum_{i=0}^{p} \mathbf{a_i}$ and let $\mathbf{b} = \sum_{i=0}^{q} \mathbf{b_i}$. The equation is reduced to $\mathbf{a} + \mathbf{bs} = \mathbf{0}$ or $\mathbf{bs} = \mathbf{0}$. Then either the sequence $\mathbf{a}$ is an annihilator of the sequence $\mathbf{s} + \mathbf{1}$ or the sequence $\mathbf{b}$ is an annihilator of the sequence $\mathbf{s}$.

Now that $\mathbf{s}$ must have an annihilator which is a linear combination of sequences in A or that of sequences in B, its spectral immunity is upper-bounded by the spectral weight of that annihilator. Since that spectral weight is at most $(T+1)/2$, the spectral immunity with respect to T of the sequence $\mathbf{s}$ is upper-bounded by $(T+1)/2$. □

Corollary 1. *For some even integer n, let T be a positive integer satisfying $T|2^n - 1$. Then the spectral immunity with respect to T of a sequence $\mathbf{s}$ of period T is upper-bounded by $\sum_{\substack{h \in \Gamma_2(T), \\ 0 \leq wt_2(hR) \leq \frac{n}{2}}} n_h$.*

In particular, if no integer $h, 1 \leq h \leq T-1$ satisfies $wt_2(hR \mod (2^n - 1)) = n/2$ then the spectral immunity is upper-bounded by $T/2$.

Corollary 2. *For some odd integer n, if the minimal period T_{min} of a sequence* $\mathbf{s}$ *satisfies $T_{min}|2^n - 1$, then the spectral immunity of* $\mathbf{s}$ *is upper-bounded by $(T_{min}+1)/2$.*

Proof. The spectral immunity $SI(\mathbf{s}) = \min_{T_{min}|T} SI_T(\mathbf{s}) \leq SI_{T_{min}}(\mathbf{s}) \leq (T_{min}+1)/2$. □

The results above show that to recover any periodic sequence, fast discrete fourier spectra attacks need data bits of a number no more than half of the period of the sequence. Those key streams with small period or small minimal period are vulnerable to fast discrete fourier spectra attacks. Therefore a sufficiently large lower bound of the minimal periods of key streams is important in the future design of a stream cipher in order to resist fast discrete fourier spectra attacks.

The proof of Theorem 1 also shows that if the periodic sequence $\mathbf{s}$ satisfies $|B \cdot \mathbf{s}| < |B| = (T+1)/2$ then it must have an annihilator of spectral weight no greater than $2n$ and that the periodic sequence $\mathbf{s}$ satisfies $A \cap B \cdot \mathbf{s} \neq \emptyset$, then it must have an annihilator of spectral weight no greater than n.

7 Conclusion

In this paper we find that low spectral weight annihilators are essential for fast discrete Fourier spectra attacks as they does not only fulfill the assumption of those attacks but also profile the least data complexity of those attacks against stream ciphers. We give a formal definition of annihilator and get a necessary and sufficient condition to decide the (non)existence of annihilator for a periodic sequence. We study the properties of annihilators and notice that the existence of low spectral weight annihilator is equivalent to the existence of low spectral weight relation, the general assumption of fast discrete Fourier spectra attacks. Finally we give an upper bound of spectral immunity for any periodic sequence and a general method to find annihilators for any periodic sequence. This general method can give low spectral weight annihilators when the periodic sequence satisfies some condition.

Two questions on annihilators of fast discrete Fourier spectra attacks are left open here. One is how to decide the (non)existence of low spectral weight annihilator efficiently for a sequence. It appears to be difficult to have an algorithm to fully decide the (non)existence, which both show the flexibility of fast discrete Fourier attacks for attackers and the difficulty to defend those attacks for designers. The other is the probability of any sequence to have a low spectral weight annihilator. It is essential for the resistance of a stream cipher to fast discrete Fourier spectra attacks in general but it seems to be a much harder problem than the first one.

References

1. Al-Hinai, S.Z., Dawson, E., Henricksen, M., Simpson, L.: On the Security of the LILI Family of Stream Ciphers Against Algebraic Attacks. In: Pieprzyk, J., Ghodosi, H., Dawson, E. (eds.) ACISP 2007. LNCS, vol. 4586, pp. 11–28. Springer, Heidelberg (2007)

2. Billet, O., Gilbert, H.: Resistance of SNOW 2.0 Against Algebraic Attacks. In: Menezes, A. (ed.) CT-RSA 2005. LNCS, vol. 3376, pp. 19–28. Springer, Heidelberg (2005)
3. Cho, J.Y., Pieprzyk, J.: Algebraic Attacks on SOBER-t32 and SOBER-t16 without Stuttering. In: Roy, B., Meier, W. (eds.) FSE 2004. LNCS, vol. 3017, pp. 49–64. Springer, Heidelberg (2004)
4. Courtois, N.T.: Higher Order Correlation Attacks, XL Algorithm and Cryptanalysis of Toyocrypt. In: Lee, P.J., Lim, C.H. (eds.) ICISC 2002. LNCS, vol. 2587, pp. 182–199. Springer, Heidelberg (2003)
5. Courtois, N.T.: Fast Algebraic Attacks on Stream Ciphers with Linear Feedback. In: Boneh, D. (ed.) CRYPTO 2003. LNCS, vol. 2729, pp. 176–194. Springer, Heidelberg (2003)
6. Courtois, N., Meier, W.: Algebraic Attacks on Stream Ciphers with Linear Feedback. In: Biham, E. (ed.) EUROCRYPT 2003. LNCS, vol. 2656, pp. 345–359. Springer, Heidelberg (2003)
7. Du, Y., Pei, D.: Count of Annihilators of Boolean Functions with Given Algebraic Immunity. In: IEEE International Conference on Wireless Communications, Networking and Information Security (WCNIS), Beijing, China, pp. 640–643 (2010)
8. Gong, G., Ronjom, S., Helleseth, T., Hu, H.: Fast Discrete Fourier Spectra Attacks on Stream Ciphers. IEEE Trans. Inform. Theory 57(8), 5555–5565 (2011)
9. Golomb, S.W., Gong, G.: Signal Design for Good Correlation: For Wireless Communication, Cryptography and Radar. Cambridge University Press, Cambridge (2005)
10. Helleseth, T., Rønjom, S.: Simplifying Algebraic Attacks with Univariate Analysis. In: Information Theory and Applications Workshop (ITA), La Jolla, pp. 1–7 (2011)
11. Lidl, R., Niederreiter, H.: Finite Fields, Encyclopedia of Mathematics and its Applications, 2nd edn., vol. 20. Cambridge University Press, Cambridge (1997)
12. Meier, W., Pasalic, E., Carlet, C.: Algebraic Attacks and Decomposition of Boolean Functions. In: Cachin, C., Camenisch, J.L. (eds.) EUROCRYPT 2004. LNCS, vol. 3027, pp. 474–491. Springer, Heidelberg (2004)

Meet-in-the-Middle Attack on Reduced Versions of the Camellia Block Cipher*

Jiqiang Lu[1,**], Yongzhuang Wei[2,3], Enes Pasalic[4], and Pierre-Alain Fouque[5]

[1] Institute for Infocomm Research, Agency for Science, Technology and Research, 1 Fusionopolis Way, Singapore 138632
lvjiqiang@hotmail.com, jlu@i2r.a-star.edu.sg
[2] Guilin University of Electronic Technology, Guilin City, Guangxi Province 541004, China
[3] State Key Lab of Information Security, Institute of Software, Chinese Academy of Sciences, Beijing 100190, China
walker_wei@msn.com
[4] University of Primorska FAMNIT, Koper, Slovenia
enes.pasalic6@gmail.com
[5] Département d'Informatique, École Normale Supérieure,
[6] 45 Rue d'Ulm, Paris 75005, France
Pierre-Alain.Fouque@ens.fr

Abstract. The Camellia block cipher has a 128-bit block length and a user key of 128, 192 or 256 bits long, which employs a total of 18 rounds for a 128-bit key and 24 rounds for a 192 or 256-bit key. It is a Japanese CRYPTREC-recommended e-government cipher, a European NESSIE selected cipher, and an ISO international standard. In this paper, we describe a few 5 and 6-round properties of Camellia and finally use them to give (higher-order) meet-in-the-middle attacks on 10-round Camellia with the $\mathrm{FL}/\mathrm{FL}^{-1}$ functions under 128 key bits, 11-round Camellia with the $\mathrm{FL}/\mathrm{FL}^{-1}$ and whitening functions under 192 key bits and 12-round Camellia with the $\mathrm{FL}/\mathrm{FL}^{-1}$ and whitening functions under 256 key bits.

Keywords: Block cipher, Camellia, Meet-in-the-middle attack.

1 Introduction

Camellia [1] is a 128-bit block cipher with a user key length of 128, 192 or 256 bits, which employs a total of 18 rounds if a 128-bit key is used and a total of 24 rounds if a 192/256-bit key is used. It has a Feistel structure with key-dependent logical functions $\mathrm{FL}/\mathrm{FL}^{-1}$ inserted after every six rounds, plus four

* The work was supported by the French ANR project SAPHIR II (No. ANR-08-VERS-014), the Natural Science Foundation of China (No. 61100185), Guangxi Natural Science Foundation (No. 2011GXNSFB018071), and the Foundation of Guangxi Key Lab of Wireless Wideband Communication and Signal Processing (No. 11101).

** The author was with École Normale Supérieure (France) when an earlier version of this work, comprising the MitM results without whitening functions, was completed.

G. Hanaoka and T. Yamauchi (Eds.): IWSEC 2012, LNCS 7631, pp. 197–215, 2012.

additional whitening operations at both ends. Camellia became a CRYPTREC e-government recommended cipher [8] in 2002, a NESSIE selected block cipher [25] in 2003, and was adopted as an ISO international standard [16] in 2005. In this work, we consider the version of Camellia that has the FL/FL^{-1} functions, and for simplicity, we denote by Camellia-128/192/256 the three versions of Camellia that use 128, 192 and 256 key bits, respectively.

The security of Camellia has been analysed against a variety of cryptanalytic techniques, including differential cryptanalysis [5], truncated differential cryptanalysis [17], higher-order differential cryptanalysis [17, 20], linear cryptanalysis [24], integral cryptanalysis [9, 15, 19], boomerang attack [27], rectangle attack [4], collision attack [26] and impossible differential cryptanalysis [3, 18]; and many cryptanalytic results on Camellia have been published, of which impossible differential cryptanalysis is the most efficient technique (in terms of the numbers of attacked rounds), that broke 11-round Camellia-128, 12-round Camellia-192 and 14-round Camellia-256 [2, 21], presented most recently at FSE 2012 and ISPEC 2012.[1]

The meet-in-the-middle (MitM) attack was introduced in 1977 by Diffie and Hellman [11]. It usually treats a block cipher $\mathbf{E} : \{0,1\}^n \times \{0,1\}^k \to \{0,1\}^n$ as a cascade of two sub-ciphers $\mathbf{E} = \mathbf{E}^a \circ \mathbf{E}^b$. Given a guess for the subkeys used in $\mathbf{E}^a$ and $\mathbf{E}^b$, if a plaintext produces just after $\mathbf{E}^a$ the same value as the corresponding ciphertext produces just before $\mathbf{E}^b$, then this guess for the subkeys is likely to be correct; otherwise, this guess must be incorrect. Thus, we can find the correct subkey, given a sufficient number of matching plaintext-ciphertext pairs in a known-plaintext attack scenario. In a chosen-plaintext attack scenario, things may get better, and as in [10], by choosing a set of plaintexts with a particular property we may be able to express the concerned value-in-the-middle as a function of plaintext and a smaller number of unknown constants than the number of unknown constants (of the same length) from the subkey involved.

In 2011 Lu et al. [23] proposed an extension of the MitM attack, known as the higher-order MitM (HO-MitM) attack, which is based on using multiple plaintexts to cancel some key-dependent component(s) or parameter(s) when constructing a basic unit of "value-in-the-middle". The HO-MitM attack technique can lead to some better cryptanalytic results than the MitM attack technique in certain circumstances. In particular, Lu et al. found some 5 and 6-round HO-MitM properties of Camellia that were used to break 10-round Camellia-128, 11-round Camellia-192 and 12-round Camellia-256, but the corresponding 5 and 6-round MitM properties can enable us to break only 12-round Camellia-256.

In this paper, we analyse the security of Camellia (with the FL/FL^{-1} functions) against the MitM attack in detail, following the work in [23]. In all those 5 and 6-round (higher-order) MitM properties of Camellia owing to Lu et al. [23], the basic unit of value-in-the-middle is one byte long. Nevertheless, we observe

[1] When the earlier version of our work was completed, the best previously published results on Camellia with FL/FL^{-1} functions were square attack on 9-round Camellia-128 [12], impossible differential attack on 10-round Camellia-192 [7], and higher-order differential and impossible differential attacks on 11-round Camellia-256 [7, 13].

Table 1. Main cryptanalytic results on Camellia with FL/FL^{-1} functions

Cipher	Attack Type	Rounds	Data	Memory	Time	Source
Camellia-128	Square	9	2^{48}CP	2^{53}Bytes	2^{122}Enc.	[12]
	Impossible differential	10	2^{118}CP	2^{93}Bytes	2^{118}Enc.	[22]
		11	$2^{120.5}$CP	$2^{115.5}$Bytes	$2^{123.8}$Enc.	[2]§
		11†	2^{122}CP	2^{102}Bytes	2^{122}Enc.	[21]§
	HO-MitM (256 inputs)	10	2^{93}CP	2^{109}Bytes	$2^{118.6}$Enc.	[23]
	(2 inputs)	10	2^{56}CP	2^{90}Bytes	$2^{121.5}$Enc.	Sect. 4.2
	MitM	10	2^{56}CP	2^{105}Bytes	$2^{121.5}$Enc.	Sect. 3.2
Camellia-192	Impossible differential	10	2^{121}CP	$2^{155.2}$Bytes	2^{144}Enc.	[7]
		10†	2^{121}CP	$2^{155.2}$Bytes	$2^{175.3}$Enc.	[7]
		11	2^{118}CP	2^{141}Bytes	$2^{163.1}$Enc.	[22]
		12	$2^{120.6}$CP	$2^{171.6}$Bytes	$2^{171.4}$Enc.	[2]§
		12†	2^{123}CP	2^{160}Bytes	$2^{187.2}$Enc.	[21]§
	HO-MitM (256 inputs)	11	2^{94}CP	2^{174}Bytes	$2^{180.2}$Enc.	[23]
	(2 inputs)	11†	2^{56}CP	2^{165}Bytes	$2^{173.4}$Enc.	Sect. 4.3
	MitM	11	2^{80}CP	2^{105}Bytes	$2^{189.4}$Enc.	Sect. 3.3
		11†	2^{56}CP	2^{185}Bytes	$2^{185.2}$Enc.	Sect. 3.4
Camellia-256	Higher-order differential	11‡	2^{93}CP	2^{98}Bytes	$2^{255.6}$Enc.	[13, 22]
	Impossible differential	11†	2^{121}CP	2^{166}Bytes	$2^{206.8}$Enc.	[7]
		13†	2^{123}CP	2^{208}Bytes	$2^{251.1}$Enc.	[21]§
		14	$2^{121.2}$CP	$2^{180.2}$Bytes	$2^{238.3}$Enc.	[2]§
		14	2^{120}CC	2^{125}Bytes	$2^{250.5}$Enc.	[21]§
	HO-MitM (256 inputs)	12	2^{94}CP	2^{174}Bytes	$2^{237.3}$Enc.	[23]
	(2 inputs)	12†	2^{19}CP	2^{221}Bytes	$2^{223.2}$Enc.	[6]§,Sect. 4
	(2 inputs)	12†	2^{56}CP	2^{165}Bytes	$2^{237.9}$Enc.	Sect. 4.4
	MitM	12	2^{56}CP	2^{185}Bytes	$2^{219.9}$Enc.	Sect. 3.5
		12†	2^{56}CP	2^{185}Bytes	$2^{239.9}$Enc.	Sect. 3.6

§: Newly emerging results; †: Include whitening operations; ‡: Can include whitening operations by making use of an equivalent structure of Camellia.

that if we consider only a smaller number of bits of the concerned byte, instead of the whole 8 bits, a few 5 and 6-round MitM properties with a smaller number of unknown 1-bit constant parameters can be obtained. This is owing to the fact that an output bit of the FL^{-1} function only relies on a small fraction of the bits of the subkey used in the FL^{-1} function (as well as a few input bits to FL^{-1}), thus reducing the number of unknown 1-bit constant parameters when we consider a fraction of the bits of the concerned byte. As a consequence, the 5 and 6-round MitM properties can be used to conduct MitM attacks on 10-round Camellia-128 with only FL/FL^{-1} functions, 11-round Camellia-192 with FL/FL^{-1} and whitening functions and 12-round Camellia-256 with FL/FL^{-1} and whitening functions. At last, we brief 5 and 6-round HO-MitM properties obtained from the 5 and 6-round MitM properties by taking XOR under two plaintexts to cancel several 1-bit constant parameters, which can be used to conduct HO-MitM attacks on the same numbers of rounds as the MitM attacks. Table 1 summarises previous, our and the newly emerging main cryptanalytic results on Camellia, where CP and CC refer respectively to the numbers of chosen plaintexts and chosen ciphertexts, and Enc. refers to the required number of encryption operations of the relevant reduced version of Camellia.

The remainder of the paper is organised as follows. In the next section, we describe the notation and the Camellia block cipher. We present our MitM results on Camellia in Section 3, and give our HO-MitM results on Camellia in Section 4. Concluding remarks are given in Section 5.

2 Preliminaries

In this section we give the notation used throughout this paper, and then briefly describe the Camellia block cipher.

2.1 Notation

The bits of a value are numbered from left to right, starting with 1. We use the following notation throughout this paper.

$\oplus$	bitwise logical exclusive OR (XOR) of two bit strings of the same length
$\cap$	bitwise logical AND of two bit strings of the same length
$\cup$	bitwise logical OR of two bit strings of the same length
$\lll$	left rotation of a bit string
$\|\|$	bit string concatenation
$\circ$	functional composition. When composing functions X and Y, X $\circ$ Y denotes the function obtained by first applying X and then Y
$\overline{X}$	bitwise logical complement of a bit string X
$X[i_1,\cdots,i_j]$	the j-bit string of bits $(i_1,\cdots,i_j)$ of a bit string X

2.2 The Camellia Block Cipher

Camellia [1] has a Feistel structure, a 128-bit block length, and a user key length of 128, 192 or 256 bits. It uses the following five functions:

- **S** : $\{0,1\}^{64} \rightarrow \{0,1\}^{64}$ is a non-linear substitution constructed by applying eight 8×8-bit S-boxes $S_1, S_2, S_3, S_4, S_5, S_6, S_7$ and S_8 in parallel to the input.
- **P** : $GF(2^8)^8 \rightarrow GF(2^8)^8$ is a linear permutation which is equivalent to pre-multiplication by a 8×8 byte matrix P; the matrix P and its reverse P^{-1} are as follows.

$$P = \begin{pmatrix} 1&0&1&1&0&1&1&1 \\ 1&1&0&1&1&0&1&1 \\ 1&1&1&0&1&1&0&1 \\ 0&1&1&1&1&1&1&0 \\ 1&1&0&0&0&1&1&1 \\ 0&1&1&0&1&0&1&1 \\ 0&0&1&1&1&1&0&1 \\ 1&0&0&1&1&1&1&0 \end{pmatrix}, \quad P^{-1} = \begin{pmatrix} 0&1&1&1&0&1&1&1 \\ 1&0&1&1&1&0&1&1 \\ 1&1&0&1&1&1&0&1 \\ 1&1&1&0&1&1&1&0 \\ 1&1&0&0&1&0&1&1 \\ 0&1&1&0&1&1&0&1 \\ 0&0&1&1&1&1&1&0 \\ 1&0&0&1&0&1&1&1 \end{pmatrix}.$$

- $\mathbf{F}: \{0,1\}^{64} \times \{0,1\}^{64} \rightarrow \{0,1\}^{64}$ is a Feistel function. If X and Y are 64-bit blocks, $\mathbf{F}(X,Y) = \mathbf{P}(\mathbf{S}(X \oplus Y))$.
- $\mathbf{FL}/\mathbf{FL}^{-1}: \{0,1\}^{64} \times \{0,1\}^{64} \rightarrow \{0,1\}^{64}$ are key-dependent linear functions. If $X = (X_L||X_R)$ and $Y = (Y_L||Y_R)$ are 64-bit blocks, then $\mathbf{FL}(X,Y) = ((((X_L \cap Y_L) \lll 1 \oplus X_R) \cup Y_R) \oplus X_L)||((X_L \cap Y_L) \lll 1 \oplus X_R)$, and $\mathbf{FL}^{-1}(X,Y) = (X_L \oplus (X_R \cup Y_R))||(((X_L \oplus (X_R \cup Y_R)) \cap Y_L) \lll 1 \oplus X_R)$.

Camellia uses a total of four 64-bit whitening subkeys KW_j, $2\lfloor\frac{N_r-6}{6}\rfloor$ 64-bit subkeys KI_l for the $\mathbf{FL}$ and $\mathbf{FL}^{-1}$ functions, and N_r 64-bit round subkeys K_i, $(1 \leqslant j \leqslant 4, 1 \leqslant l \leqslant 2\lfloor\frac{N_r-6}{6}\rfloor, 1 \leqslant i \leqslant N_r)$, all derived from a N_k-bit key K, where N_r is 18 for Camellia-128, and 24 for Camellia-192/256, N_k is 128 for Camellia-128, 192 for Camellia-192, and 256 for Camellia-256. The key schedule is as follows. First, generate two 128-bit strings K_L and K_R from K in the following way: For Camellia-128, K_L is the 128-bit key K, and K_R is zero; for Camellia-192, K_L is the left 128 bits of K, and K_R is the concatenation of the right 64 bits of K and the complement of the right 64 bits of K; and for Camellia-256, K_L is the left 128 bits of K, and K_R is the right 128 bits of K. Second, depending on the key size, generate one or two 128-bit strings K_A and K_B from (K_L, K_R) by a non-linear transformation (see [1] for its detail). Finally, the subkeys are as follows.[2]

- For Camellia-128: $K_2 = (K_A \lll 0)[65 \sim 128], K_3 = (K_L \lll 15)[1 \sim 64], K_9 = (K_A \lll 45)[1 \sim 64], K_{10} = (K_L \lll 60)[65 \sim 128], K_{11} = (K_A \lll 60)[1 \sim 64], \cdots$.
- For Camellia-192/256: $K_7 = (K_B \lll 30)[1 \sim 64], K_8 = (K_B \lll 30)[65 \sim 128], K_{13} = (K_R \lll 60)[1 \sim 64], K_{14} = (K_R \lll 60)[65 \sim 128], K_{15} = (K_B \lll 60)[1 \sim 64], K_{16} = (K_B \lll 60)[65 \sim 128], K_{17} = (K_L \lll 77)[1 \sim 64], K_{18} = (K_L \lll 77)[65 \sim 128], K_{21} = (K_A \lll 94)[1 \sim 64], K_{22} = (K_A \lll 94)[65 \sim 128], K_{23} = (K_L \lll 111)[1 \sim 64], \cdots$.

Below is the encryption procedure Camellia, where P is a 128-bit plaintext, represented as 16 bytes, and L_0, R_0, L_i, R_i, $\widehat{L}_i$ and $\widehat{R}_i$ are 64-bit variables.

1. $L_0||R_0 = P \oplus (KW_1||KW_2)$
2. For $i = 1$ to N_r:
   if $i = 6$ or 12 (or 18 for Camellia-192/256),
   $\widehat{L}_i = \mathbf{F}(L_{i-1}, K_i) \oplus R_{i-1}$, $\widehat{R}_i = L_{i-1}$;
   $L_i = \mathbf{FL}(\widehat{L}_i, KI_{\frac{i}{3}-1})$, $R_i = \mathbf{FL}^{-1}(\widehat{R}_i, KI_{\frac{i}{3}})$;
   else
   $L_i = \mathbf{F}(L_{i-1}, K_i) \oplus R_{i-1}$, $R_i = L_{i-1}$;
3. Ciphertext $C = (R_{N_r} \oplus KW_3)||(L_{N_r} \oplus KW_4)$.

We refer to the ith iteration of Step 2 in the above description as Round i, and write $K_{i,j}$ for the j-th byte of K_i, $(1 \leqslant j \leqslant 8)$.

[2] Here we give only the subkeys concerned in this paper, $(K_A \lll 0)[65 \sim 128]$ represents bits $(65, 66, \cdots, 128)$ of $(K_A \lll 0)$, and so on.

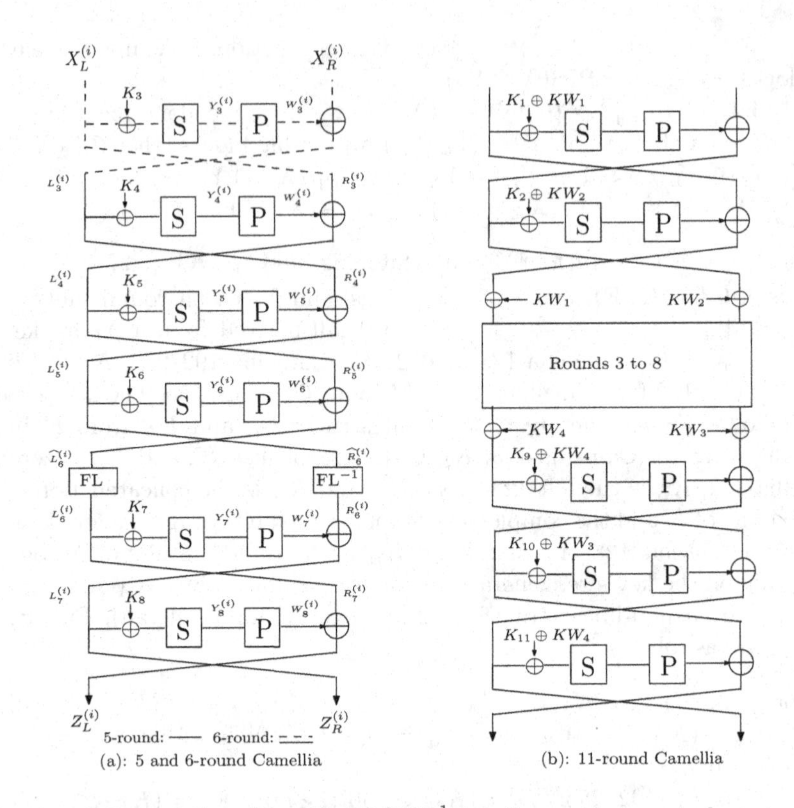

Fig. 1. 5 and 6-round Camellia with $\mathbf{FL}/\mathbf{FL}^{-1}$ functions and an equivalent structure of 11-round Camellia with whitening operations

3 MitM Attacks on 10-Round Camellia-128, 11-Round Camellia-192 and 12-Round Camellia-256

In this section we first give the 5 and 6-round MitM properties and then present our MitM attacks on Camellia with $\mathbf{FL}/\mathbf{FL}^{-1}$ functions.

3.1 MitM Properties for 5 and 6-Round Camellia

We assume the 5-round Camellia is from Rounds 4 to 8, and the 6-round Camellia is from Rounds 3 to 8; see Fig. 1-(a). The MitM properties are as follows, and their proof is given in the Appendix.

Proposition 1. *Suppose a set of 256 sixteen-byte values* $X^{(i)} = (X_L^{(i)} || X_R^{(i)}) = (m_1, m_2, m_3, m_4, m_5, m_6, m_7, m_8, x^{(i)}, m_9, m_{10}, m_{11}, m_{12}, m_{13}, m_{14}, m_{15})$ *with* $x^{(i)}$ *taking all the possible values in* $\{0,1\}^8$ *and the other 15 bytes* $m_1, m_2, \cdots, m_{15}$ *fixed to arbitrary values,* $(i = 1, \cdots, 256)$. *Then:*

1. *If* $Z^{(i)} = (Z_L^{(i)}||Z_R^{(i)})$ *is the result of encrypting* $X^{(i)}$ *using Rounds 4 to 8 with the* $\mathbf{FL}/\mathbf{FL}^{-1}$ *functions between Rounds 6 and 7, then* $\mathbf{P}^{-1}(Z_R^{(i)})[49 \sim (49+\omega)]$ *can be expressed with a function of* $x^{(i)}$ *and* $100 + 15 \times \omega$ *constant 1-bit parameters* $c_1, c_2, \cdots, c_{100+15\times\omega}$*, written* $\Theta_{c_1,c_2,\cdots,c_{100+15\times\omega}}(x^{(i)})$*, where* $0 \leqslant \omega \leqslant 6$.
2. *If* $Z^{(i)} = (Z_L^{(i)}||Z_R^{(i)})$ *is the result of encrypting* $X^{(i)}$ *using Rounds 3 to 8 with the* $\mathbf{FL}/\mathbf{FL}^{-1}$ *functions between Rounds 6 and 7, then* $\mathbf{P}^{-1}(Z_R^{(i)})[41 \sim (41+\omega)]$ *can be expressed with a function of* $x^{(i)}$ *and* $164 + 15 \times \omega$ *constant 1-bit parameters* $c'_1, c'_2, \cdots, c'_{164+15\times\omega}$*, written* $\Upsilon_{c'_1,c'_2,\cdots,c'_{164+15\times\omega}}(x^{(i)})$*, where* $0 \leqslant \omega \leqslant 6$.

3.2 Attacking 10-Round Camellia-128 without Whitening Functions

A simple analysis on the key schedule of Camellia-128 reveals the following property.

Property 1. *For Camellia-128, given a value of* $(K_{2,1}, K_{2,2}, K_{2,3}, K_{2,5}, K_{2,8}, K_{3,1})$ *there are only* 60 *unknown bits of* $(K_{9,7}, K_{10,3}, K_{10,4}, K_{10,5}, K_{10,6}, K_{10,8}, K_{11})$.

The 5-round MitM property given in Proposition 1-1 allows us to break 10-round Camellia-128 with $\mathbf{FL}/\mathbf{FL}^{-1}$ functions, but without the whitening functions. Below is the procedure for attacking Rounds 2 to 11, where the 5-round MitM property with $\omega = 0$ is used from Rounds 4 to 8, and the approach used to choose plaintexts with δ was introduced in [22].

1. For each of 2^{100} possible values of the 100 one-bit parameters $c_1, c_2, \cdots, c_{100}$, precompute $\Theta_{c_1,c_2,\cdots,c_{100}}(z)$ sequentially for $z = 0, 1, \cdots, 255$. Store the 2^{100} 256-bit sequences in a hash table $\mathcal{L}_\Theta$.
2. Randomly choose six 8-bit constants $\gamma_1, \gamma_2, \cdots, \gamma_6$, and define a secret parameter δ to be $\delta = \mathrm{S}_4(\gamma_1 \oplus K_{2,4}) \oplus \mathrm{S}_6(\gamma_2 \oplus K_{2,6}) \oplus \mathrm{S}_7(\gamma_3 \oplus K_{2,7}) \oplus \gamma_4 \oplus \gamma_5 \oplus \gamma_6$.
3. Guess a value for $(K_{2,1}, K_{2,2}, K_{2,3}, K_{2,5}, K_{2,8}, K_{3,1}, \delta)$, and we denote the guessed value by $(K^*_{2,1}, K^*_{2,2}, K^*_{2,3}, K^*_{2,5}, K^*_{2,8}, K^*_{3,1}, \delta^*)$. Then for $x = 0, 1, \cdots, 255$, choose plaintext $P^{(x)} = (P_L^{(x)}, P_R^{(x)})$ in the following way, where $\alpha_1, \alpha_2, \cdots, \alpha_5, \beta_1, \beta_2, \cdots, \beta_7$ are randomly chosen 8-bit constants:

$$P_L^{(x)} = \begin{pmatrix} \mathrm{S}_1(x \oplus K^*_{3,1}) \oplus \alpha_1 \\ \mathrm{S}_1(x \oplus K^*_{3,1}) \oplus \alpha_2 \\ \mathrm{S}_1(x \oplus K^*_{3,1}) \oplus \alpha_3 \\ \gamma_1 \\ \mathrm{S}_1(x \oplus K^*_{3,1}) \oplus \alpha_4 \\ \gamma_2 \\ \gamma_3 \\ \mathrm{S}_1(x \oplus K^*_{3,1}) \oplus \alpha_5 \end{pmatrix}^{\mathrm{T}},$$

$$P_R^{(x)} = \mathbf{P}\begin{pmatrix} \mathrm{S}_1(\mathrm{S}_1(x \oplus K^*_{3,1}) \oplus \alpha_1 \oplus K^*_{2,1}) \\ \mathrm{S}_2(\mathrm{S}_1(x \oplus K^*_{3,1}) \oplus \alpha_2 \oplus K^*_{2,2}) \\ \mathrm{S}_3(\mathrm{S}_1(x \oplus K^*_{3,1}) \oplus \alpha_3 \oplus K^*_{2,3}) \\ \gamma_4 \\ \mathrm{S}_5(\mathrm{S}_1(x \oplus K^*_{3,1}) \oplus \alpha_4 \oplus K^*_{2,5}) \\ \gamma_5 \\ \gamma_6 \\ \mathrm{S}_8(\mathrm{S}_1(x \oplus K^*_{3,1}) \oplus \alpha_5 \oplus K^*_{2,8}) \end{pmatrix}^{\mathrm{T}} \oplus \begin{pmatrix} x \oplus \delta^* \\ \beta_1 \\ \beta_2 \\ \beta_3 \\ \beta_4 \\ \beta_5 \\ \beta_6 \\ \beta_7 \end{pmatrix}^{\mathrm{T}}.$$

In a chosen-plaintext attack scenario, obtain the ciphertexts for the plaintexts; we denote by $C^{(x)}$ the ciphertext for plaintext $P^{(x)}$.

4. Guess a value for $(K_{9,7}, K_{10,3}, K_{10,4}, K_{10,5}, K_{10,6}, K_{10,8}, K_{11})$, and we denote the guessed value by $(K^*_{9,7}, K^*_{10,3}, K^*_{10,4}, K^*_{10,5}, K^*_{10,6}, K^*_{10,8}, K^*_{11})$. Then, partially decrypt every ciphertext $C^{(x)}$ with $(K^*_{10,3}, K^*_{10,4}, K^*_{10,5}, K^*_{10,6}, K^*_{10,8}, K^*_{11})$ to get the corresponding value for bytes $(1, 2, \cdots, 8, 15)$ just before Round 10, and we denote it by $(L_9^{(x)}, R_{9,7}^{(x)})$; compute $T^{(x)} = \mathbf{P}^{-1}(L_9^{(x)})[49] \oplus \mathrm{S}_7(R_{9,7}^{(x)} \oplus K^*_{9,7})[49]$. Next, check whether the sequence $(T^{(0)}, T^{(1)}, \cdots, T^{(255)})$ matches a sequence in $\mathcal{L}_\Theta$; if yes, record the guessed value $(K^*_{2,1}, K^*_{2,2}, K^*_{2,3}, K^*_{2,5}, K^*_{2,8}, K^*_{3,1}, K^*_{9,7}, K^*_{10,3}, K^*_{10,4}, K^*_{10,5}, K^*_{10,6}, K^*_{10,8}, K^*_{11})$ and execute Step 5; otherwise, repeat Step 1 with another subkey guess (if all the subkey possibilities are tested in Step 4, repeat Step 3 with another subkey guess).
5. For every recorded value for $(K_{10,3}, K_{10,4}, K_{10,5}, K_{10,6}, K_{10,8})$, exhaustively search the remaining 11 key bytes.

The attack requires 2^{56} chosen plaintexts. The one-off precomputation requires a memory of $2^{100} \times 256 \times \frac{1}{8} = 2^{105}$ bytes, and has a time complexity of $2^{100} \times 256 \times 2 \times \frac{1}{10} \approx 2^{109.7}$ 10-round Camellia-128 encryptions under the rough estimate that a computation of $\Theta_{c_1, c_2, \cdots, c_{100}}(z)$ equals 2 one-round Camellia-128 encryptions in terms of time. If the guessed value $(K^*_{2,1}, K^*_{2,2}, K^*_{2,3}, K^*_{2,5}, K^*_{2,8}, K^*_{3,1}, \delta^*)$ is correct, the input to Round 4 must have the form $(m_1, m_2, m_3, m_4, m_5, m_6, m_7, m_8, x, m_9, m_{10}, m_{11}, m_{12}, m_{13}, m_{14}, m_{15})$, where $m_1, \cdots, m_{15}$ are indeterminate constants.

Step 3 has a time complexity of about $2^{56} \times 256 \times \frac{1+5}{8 \times 10} \approx 2^{60.3}$ 10-round Camellia-128 encryptions. Folllowing Property 1, we learn that the time complexity of Step 4 is approximately $2^{56+60} \times 256 \times \frac{8+5+1}{8 \times 10} \approx 2^{121.5}$ 10-round Camellia-128 encryptions. In Step 4, if the guessed value $(K^*_{2,1}, K^*_{2,2}, K^*_{2,3}, K^*_{2,5}, K^*_{2,8}, K^*_{3,1}, \delta^*, K^*_{9,7}, K^*_{10,3}, K^*_{10,4}, K^*_{10,5}, K^*_{10,6}, K^*_{10,8}, K^*_{11})$ is correct, the sequence $(T^{(0)}, T^{(1)}, \cdots, T^{(255)})$ must match a sequence in $\mathcal{L}_\Theta$; if the guessed value $(K^*_{2,1}, K^*_{2,2}, K^*_{2,3}, K^*_{2,5}, K^*_{2,8}, K^*_{3,1}, \delta^*, K^*_{9,7}, K^*_{10,3}, K^*_{10,4}, K^*_{10,5}, K^*_{10,6}, K^*_{10,8}, K^*_{11})$ is wrong, the probability that the sequence $(T^{(0)}, T^{(1)}, \cdots, T^{(255)})$ matches a sequence in $\mathcal{L}_\Theta$ is $1 - \binom{2^{100}}{0}(2^{-256})^0(1-2^{-256})^{2^{100}} \approx 2^{-256} \times 2^{100} = 2^{-156}$, (assuming the event has a binomial distribution). Consequently, it is expected that at most $2^{56+60} \times 2^{-156} = 2^{-40}$ values for $(K_{2,1}, K_{2,2}, K_{2,3}, K_{2,5}, K_{2,8}, K_{3,1}, K_{9,7}, K_{10,3}, K_{10,4}, K_{10,5}, K_{10,6}, K_{10,8}, K_{11})$ are recorded in Step 4. Since a total of 40 bits of K_L can be known from the recorded $(K_{10,3}, K_{10,4}, K_{10,5}, K_{10,6}, K_{10,8})$, Step 5 takes at most 2^{88} 10-round Camellia-128 encryptions to find the correct 128-bit user key.

Therefore, the attack has a memory complexity of 2^{105} bytes and a total time complexity of approximately $2^{121.5}$ 10-round Camellia-128 encryptions.

Note that we can also attack Rounds 8 to 17 (without whitening functions) by applying the 5-round MitM property with $\omega = 0$ from Rounds 10 to 14. This attack has the same data and memory complexity as the above 10-round Camellia-128 attack, but has a total time complexity of approximately $2^{56+65} \times 256 \times \frac{8+5+1}{8 \times 10} \approx 2^{126.5}$ 10-round Camellia-128 encryptions.

3.3 Attacking 11-Round Camellia-192 without Whitening Functions

Both the 5 and 6-round MitM properties given in Proposition 1 can be used to attack 11-round Camellia-192 with $\mathbf{FL}/\mathbf{FL}^{-1}$ functions, excluding the whitening functions. We first brief an attack on Rounds 13 to 23 using the 5-round MitM property with $\omega = 0$, where we guess $(K_{13}, K_{14}, K_{15,1}, K_{21,7}, K_{22,3}, K_{22,4}, K_{22,5}, K_{22,6}, K_{22,8}, K_{23})$. Note that the following property holds for Camellia-192.

Property 2. *For Camellia-192, there is no overlapping bit between* $(K_{13}, K_{14}, K_{15,1})$ *and* $(K_{21,7}, K_{22,3}, K_{22,4}, K_{22,5}, K_{22,6}, K_{22,8}, K_{23})$.

The attack is very similar to the above 10-round Camellia-128 attack, except that we use a different approach to choose plaintexts: Denote by $(K_{13}^*, K_{14}^*, K_{15,1}^*)$ a guess for $(K_{13}, K_{14}, K_{15,1})$, and then for $x = 0, 1, \cdots, 255$, choose plaintext $P^{(x)} = (P_L^{(x)}, P_R^{(x)})$ as below, where $\alpha_1, \alpha_2, \cdots, \alpha_8, \beta_1, \beta_2, \cdots, \beta_7$ are randomly chosen 8-bit constants.

$$P_L^{(x)} = \mathbf{P}\begin{pmatrix} S_1(S_1(x \oplus K_{15,1}^*) \oplus \alpha_1 \oplus K_{14,1}^*) \\ S_2(S_1(x \oplus K_{15,1}^*) \oplus \alpha_2 \oplus K_{14,2}^*) \\ S_3(S_1(x \oplus K_{15,1}^*) \oplus \alpha_3 \oplus K_{14,3}^*) \\ S_4(\alpha_4 \oplus K_{14,4}^*) \\ S_5(S_1(x \oplus K_{15,1}^*) \oplus \alpha_5 \oplus K_{14,5}^*) \\ S_6(\alpha_6 \oplus K_{14,6}^*) \\ S_7(\alpha_7 \oplus K_{14,7}^*) \\ S_8(S_1(x \oplus K_{15,1}^*) \oplus \alpha_8 \oplus K_{14,8}^*) \end{pmatrix}^{\mathrm{T}} \oplus \begin{pmatrix} x \\ \beta_1 \\ \beta_2 \\ \beta_3 \\ \beta_4 \\ \beta_5 \\ \beta_6 \\ \beta_7 \end{pmatrix}^{\mathrm{T}},$$

$$P_R^{(x)} = \mathbf{F}(P_L^{(x)}, K_{13}^*) \oplus \begin{pmatrix} S_1(x \oplus K_{15,1}^*) \oplus \alpha_1 \\ S_1(x \oplus K_{15,1}^*) \oplus \alpha_2 \\ S_1(x \oplus K_{15,1}^*) \oplus \alpha_3 \\ \alpha_4 \\ S_1(x \oplus K_{15,1}^*) \oplus \alpha_5 \\ \alpha_6 \\ \alpha_7 \\ S_1(x \oplus K_{15,1}^*) \oplus \alpha_8 \end{pmatrix}^{\mathrm{T}}.$$

There are $2^{64+8} = 2^{72}$ possible values for $(K_{13}, K_{14}, K_{15,1})$. Similarly, the attack requires $256 \times 2^{72} = 2^{80}$ chosen plaintexts and a memory of $2^{100} \times 256 \times \frac{1}{8} = 2^{105}$ bytes, and has a total time complexity of approximately $2^{100} \times 256 \times 2 \times \frac{1}{11} + 2^{72+112} \times 256 \times \frac{8+5+1}{8 \times 11} \approx 2^{189.4}$ 11-round Camellia-192 encryptions.

We can use the 6-round MitM property to break Rounds 13 to 23. We choose $\omega = 0$. The attack is similar to the 10-round Camellia-128 attack described in Section 3.2, except the following two points: (1) There are 164 one-bit parameters $c_1', c_2', \cdots, c_{164}'$ in the off-line precomputation phase; and (2) We append three rounds (i.e., Rounds 21 to 23) after the 6-round MitM property. There are only 2^{40} possible values for $(K_{13,1}, K_{13,2}, K_{13,3}, K_{13,5}, K_{13,8}, K_{14,1})$, and thus the attack requires $256 \times 2^{40+8} = 2^{56}$ chosen plaintexts. After a similar analysis, we get that the off-line precomputation requires a memory of $2^{164} \times 256 \times \frac{1}{8} = 2^{169}$ bytes and has a time complexity of $2^{164} \times 256 \times 3 \times \frac{1}{11} \approx 2^{170.2}$ 11-round Camellia-192 encryptions under the rough estimate that a computation of $\Upsilon_{c_1', c_2', \cdots, c_{164}'}(\cdot)$ equals 3 one-round Camellia-192 encryptions in terms of time. The time complexity in the key-recovery phase is approximately $2^{48+112} \times 256 \times \frac{8+5+1}{8 \times 11} \approx 2^{165.4}$ 11-round Camellia-192 encryptions. We can obtain a data–memory–time tradeoff [14] version from this 11-round Camellia-192 attack, which has a data complexity of

$2^{59.4}$ chosen plaintexts, a memory complexity of $2^{167.6}$ bytes and a total time complexity of $2^{169.8}$ 11-round Camellia-192 encryptions.

3.4 Attacking 11-Round Camellia-192 with Whitening Functions

The 6-round MitM property can also be used to mount an MitM attack on 11-round Camellia-192 with $\mathbf{FL}/\mathbf{FL}^{-1}$ and whitening functions, by taking advantage of an equivalent structure of 11-round Camellia as depicted in Fig. 1-(b). Here we attack the first 11 rounds of Camellia-192, and choose $\omega = 1$.

Define equivalent round subkeys $\widehat{K}_1 = K_1 \oplus KW_1, \widehat{K}_2 = K_2 \oplus KW_2, \widehat{K}_9 = K_9 \oplus KW_4, \widehat{K}_{10} = K_{10} \oplus KW_3, \widehat{K}_{11} = K_{11} \oplus KW_4$. Below is the attack procedure.

1. For each of 2^{179} possible values of the 179 one-bit parameters $c'_1, c'_2, \cdots, c'_{179}$, precompute $\Upsilon_{c'_1,c'_2,\cdots,c'_{179}}(z)$ sequentially for $z = 0, 1, \cdots, 255$. Store the 2^{179} 512-bit sequences in a hash table $\mathcal{L}_\Upsilon$.
2. Randomly choose six 8-bit constants $\gamma_1, \gamma_2, \cdots, \gamma_6$, and define a secret parameter $\delta = KW_2[1 \sim 8] \oplus \mathrm{S}_4(\gamma_1 \oplus \widehat{K}_{1,4}) \oplus \mathrm{S}_6(\gamma_2 \oplus \widehat{K}_{1,6}) \oplus \mathrm{S}_7(\gamma_3 \oplus \widehat{K}_{1,7}) \oplus \gamma_4 \oplus \gamma_5 \oplus \gamma_6$.
3. Guess a value for $(\widehat{K}_{1,1}, \widehat{K}_{1,2}, \widehat{K}_{1,3}, \widehat{K}_{1,5}, \widehat{K}_{1,8}, K_{2,1}, \delta)$, and we denote the guessed value by $(\widehat{K}^*_{1,1}, \widehat{K}^*_{1,2}, \widehat{K}^*_{1,3}, \widehat{K}^*_{1,5}, \widehat{K}^*_{1,8}, K^*_{2,1}, \delta^*)$. Then for $x = 0, 1, \cdots,$ 255, choose plaintext $P^{(x)} = (P_L^{(x)}, P_R^{(x)})$ in the following way, where $\alpha_1, \alpha_2, \cdots, \alpha_5, \beta_1, \beta_2, \cdots, \beta_7$ are randomly chosen 8-bit constants:

$$P_L^{(x)} = \begin{pmatrix} \mathrm{S}_1(x \oplus K^*_{2,1}) \oplus \alpha_1 \\ \mathrm{S}_1(x \oplus K^*_{2,1}) \oplus \alpha_2 \\ \mathrm{S}_1(x \oplus K^*_{2,1}) \oplus \alpha_3 \\ \gamma_1 \\ \mathrm{S}_1(x \oplus K^*_{2,1}) \oplus \alpha_4 \\ \gamma_2 \\ \gamma_3 \\ \mathrm{S}_1(x \oplus K^*_{2,1}) \oplus \alpha_5 \end{pmatrix}^{\mathrm{T}},$$

$$P_R^{(x)} = \mathbf{P} \begin{pmatrix} \mathrm{S}_1(\mathrm{S}_1(x \oplus K^*_{2,1}) \oplus \alpha_1 \oplus \widehat{K}_{1,1}) \\ \mathrm{S}_2(\mathrm{S}_1(x \oplus K^*_{2,1}) \oplus \alpha_2 \oplus \widehat{K}_{1,2}) \\ \mathrm{S}_3(\mathrm{S}_1(x \oplus K^*_{2,1}) \oplus \alpha_3 \oplus \widehat{K}_{1,3}) \\ \gamma_4 \\ \mathrm{S}_5(\mathrm{S}_1(x \oplus K^*_{2,1}) \oplus \alpha_4 \oplus \widehat{K}_{1,5}) \\ \gamma_5 \\ \gamma_6 \\ \mathrm{S}_8(\mathrm{S}_1(x \oplus K^*_{2,1}) \oplus \alpha_5 \oplus \widehat{K}_{1,8}) \end{pmatrix}^{\mathrm{T}} \oplus \begin{pmatrix} x \oplus \delta^* \\ \beta_1 \\ \beta_2 \\ \beta_3 \\ \beta_4 \\ \beta_5 \\ \beta_6 \\ \beta_7 \end{pmatrix}^{\mathrm{T}}.$$

In a chosen-plaintext attack scenario, obtain the ciphertexts for the plaintexts; we denote by $C^{(x)}$ the ciphertext for plaintext $P^{(x)}$.
4. Guess a value for $(\mathbf{P}^{-1}(KW_3)[41 \sim 42], \widehat{K}_{9,6}, \widehat{K}_{10,2}, \widehat{K}_{10,3}, \widehat{K}_{10,5}, \widehat{K}_{10,7}, \widehat{K}_{10,8}, \widehat{K}_{11})$, and we denote the guessed value by $(\mathbf{P}^{-1}(KW_3)^*[41 \sim 42], \widehat{K}^*_{9,6}, \widehat{K}^*_{10,2}, \widehat{K}^*_{10,3}, \widehat{K}^*_{10,5}, \widehat{K}^*_{10,7}, \widehat{K}^*_{10,8}, \widehat{K}^*_{11})$. Then partially decrypt every ciphertext $C^{(x)}$ with $(\widehat{K}^*_{10,2}, \widehat{K}^*_{10,3}, \widehat{K}^*_{10,5}, \widehat{K}^*_{10,7}, \widehat{K}^*_{10,8}, \widehat{K}^*_{11})$ to get the corresponding value for bytes $(1, 2, \cdots, 8, 14)$ immediately before Round 10; and we denote it by $(L_9^{(i,x)}, R_{9,6}^{(i,x)})$. Next, compute

$$T^{(x)} = \mathbf{P}^{-1}(KW_3)^*[41 \sim 42] \oplus \mathbf{P}^{-1}(L_9^{(x)})[41 \sim 42] \oplus \mathrm{S}_6(R_{9,6}^{(x)} \oplus K^*_{9,6})[41 \sim 42].$$

Finally, check whether the sequence $(T^{(0)}, T^{(1)}, \cdots, T^{(255)})$ matches a sequence in $\mathcal{L}_\Upsilon$; if yes, record the guessed value $(\widehat{K}^*_{1,1}, \widehat{K}^*_{1,2}, \widehat{K}^*_{1,3}, \widehat{K}^*_{1,5}, \widehat{K}^*_{1,8}, K^*_{2,1}, \widehat{K}^*_{9,6}, \widehat{K}^*_{10,2}, \widehat{K}^*_{10,3}, \widehat{K}^*_{10,5}, \widehat{K}^*_{10,7}, \widehat{K}^*_{10,8}, \widehat{K}^*_{11})$ and execute Step 5; otherwise, repeat Step 4 with another subkey guess (if all the subkey possibilities are tested in Step 4, repeat Step 3 with another subkey guess).

5. For every recorded subkey guess, determine the correct user key.

The attack requires 2^{56} chosen plaintexts. The one-off precomputation requires a memory of $2^{179} \times 256 \times \frac{2}{8} = 2^{185}$ bytes, and has a time complexity of $2^{179} \times 256 \times 3 \times \frac{1}{11} \approx 2^{185.2}$ 11-round Camellia-192 encryptions. If the guessed value $(\widehat{K}^*_{1,1}, \widehat{K}^*_{1,2}, \widehat{K}^*_{1,3}, \widehat{K}^*_{1,5}, \widehat{K}^*_{1,8}, K^*_{2,1}, \delta^*)$ is correct, the input to Round 3 must have the form $(m_1, m_2, m_3, m_4, m_5, m_6, m_7, m_8, x, m_9, m_{10}, m_{11}, m_{12}, m_{13}, m_{14}, m_{15})$, where $m_1, m_2, \cdots, m_{15}$ are indeterminate constants.

Step 3 has a time complexity of about $2^{56} \times 256 \times \frac{1+5}{8\times 11} \approx 2^{60.2}$ 11-round Camellia-192 encryptions. Step 4 has a time complexity of approximately $2^{56+114} \times 256 \times \frac{8+5+1}{8\times 11} \approx 2^{175.4}$ 11-round Camellia-192 encryptions. In Step 4, for the correct guess of $(\mathbf{P}^{-1}(KW_3)[41 \sim 42], \delta, \widehat{K}_{1,1}, \widehat{K}_{1,2}, \widehat{K}_{1,3}, \widehat{K}_{1,5}, \widehat{K}_{1,8}, K_{2,1}, \widehat{K}_{9,6}, \widehat{K}_{10,2}, \widehat{K}_{10,3}, \widehat{K}_{10,5}, \widehat{K}_{10,7}, \widehat{K}_{10,8}, \widehat{K}_{11})$, the sequence $(T^{(0)}, T^{(1)}, \cdots, T^{(255)})$ must match a sequence in $\mathcal{L}_\Upsilon$; for a wrong guess of $(\mathbf{P}^{-1}(KW_3)[41 \sim 42], \delta, \widehat{K}_{1,1}, \widehat{K}_{1,2}, \widehat{K}_{1,3}, \widehat{K}_{1,5}, \widehat{K}_{1,8}, K_{2,1}, \widehat{K}_{9,6}, \widehat{K}_{10,2}, \widehat{K}_{10,3}, \widehat{K}_{10,5}, \widehat{K}_{10,7}, \widehat{K}_{10,8}, \widehat{K}_{11})$, the probability that the sequence $(T^{(0)}, T^{(1)}, \cdots, T^{(255)})$ matches a sequence in $\mathcal{L}_\Upsilon$ is approximately $1 - \binom{2^{179}}{0}(2^{-512})^0(1-2^{-512})^{2^{179}} \approx 2^{-512} \times 2^{179} = 2^{-333}$, (assuming the event has a binomial distribution). Consequently, it is expected that at most $2^{56+114} \times 2^{-333} = 2^{-163}$ values for $(\mathbf{P}^{-1}(KW_3)[41 \sim 42], \delta, \widehat{K}_{1,1}, \widehat{K}_{1,2}, \widehat{K}_{1,3}, \widehat{K}_{1,5}, \widehat{K}_{1,8}, K_{2,1}, \widehat{K}_{9,6}, \widehat{K}_{10,2}, \widehat{K}_{10,3}, \widehat{K}_{10,5}, \widehat{K}_{10,7}, \widehat{K}_{10,8}, \widehat{K}_{11})$ are recorded in Step 4, that is very likely to be the correct subkey guess. Since 8 bits of K_B can be known from $K_{2,1}$, we can find out the correct user key with a time complexity of at most $2^{120} \times \frac{6}{11} \approx 2^{119.2}$ 11-round Camellia-192 encryptions by using Property 4 from [22] (as well as the obtained relationship about the subkeys). Therefore, the attack has a memory complexity of 2^{185} bytes and a total time complexity of approximately $2^{185.2}$ 11-round Camellia-192 encryptions.

We can similarly attack two other series of 12-round Camellia-256 with $\mathbf{FL}/\mathbf{FL}^{-1}$ and whitening functions, i.e., Rounds 7 to 17 and Rounds 13 to 23.

3.5 Attacking 12-Round Camellia-256 without Whitening Functions

We can use the 6-round MitM property given in Proposition 1-2 to mount an MitM attack on 12-round Camellia-256 with $\mathbf{FL}/\mathbf{FL}^{-1}$ functions, excluding the whitening functions. We attack Rounds 7 to 18, and choose $\omega = 1$, where we guess $(K_{7,1}, K_{7,2}, K_{7,3}, K_{7,5}, K_{7,8}, K_{8,1}, K_{15,6}, K_{16,2}, K_{16,3}, K_{16,5}, K_{16,7}, K_{16,8}, K_{17}, K_{18})$, plus a secret 8-bit parameter δ with a similar meaning as the one from the above 10-round Camellia-128 attack. We have the following property for Camellia-256.

Property 3. *For Camellia-256, given a value for* $(K_{7,1}, K_{7,2}, K_{7,3}, K_{7,5}, K_{7,8}, K_{8,1})$ *there are only 158 unknown bits for* $(K_{15,6}, K_{16,2}, K_{16,3}, K_{16,5}, K_{16,7}, K_{16,8}, K_{17}, K_{18})$.

Similarly, the attack requires 2^{56} chosen plaintexts and a memory of $2^{179} \times 256 \times \frac{2}{8} = 2^{185}$ bytes, and has a total time complexity of $2^{179} \times 256 \times 3 \times \frac{1}{12} + 2^{56+158} \times 256 \times \frac{8+8+5+1}{8 \times 12} \approx 2^{219.9}$ 12-round Camellia-256 encryptions.

It is noteworthy that we can also break two other series of 12-round Camellia-256 with $\mathbf{FL}/\mathbf{FL}^{-1}$ functions, namely Rounds 1 to 12 and Rounds 13 to 24. Similarly, the attack has the same data and memory complexity as the above 12-round Camellia-256 attack, but has a total time complexity of approximately $2^{56+176} \times 256 \times \frac{8+8+5+1}{8 \times 12} \approx 2^{237.9}$ 12-round Camellia-256 encryptions.

3.6 Attacking 12-Round Camellia-256 with Whitening Functions

The 6-round MitM property can enable us to conduct an MitM attack on 12-round Camellia-256 with $\mathbf{FL}/\mathbf{FL}^{-1}$ and whitening functions, by making use of an equivalent structure of 12-round Camellia similar to the 11-round structure depicted in Fig. 1-(b). Here we attack Rounds 1 to 12, and choose $\omega = 1$. The attack is basically the version of the 11-round Camellia-192 attack given in Section 3.4 when one more round is appended at the end. As a result, the attack requires 2^{56} chosen plaintexts and a memory of 2^{185} bytes, and has a total time complexity of at most $2^{56+178} \times 256 \times \frac{8+8+5+1}{8 \times 12} \approx 2^{239.9}$ 12-round Camellia-256 encryptions.

4 HO-MitM Attacks on 10-Round Camellia-128, 11-Round Camellia-192 and 12-Round Camellia-256

It can be easily seen from the proof of the 5 and 6-round MitM properties that a few 1-bit constants can be cancelled if we take XOR under two different inputs; such a resulting attack is termed a HO-MitM attack by definition in [23] (As mentioned in [23], this type of HO-MitM attacks appeared under the name of MitM attacks before). In this section we briefly describe certain of these HO-MitM attacks based on 5 and 6-round HO-MitM properties obtained by taking XOR under two different inputs in the above 5 and 6-round MitM properties.

4.1 HO-MitM Properties for 5 and 6-Round Camellia

Because $A \oplus A = 0$, $(A \cap C) \oplus (B \cap C) = (A \oplus B) \cap C$ and $(A \cup C) \oplus (B \cup C) = (A \oplus B) \oplus (A \oplus B) \cap C$, where A, B, C are blocks of the same length, from the proof in the Appendix we learn that: (1) If we take XOR between two inputs from the 5-round MitM property with $\omega = 0$, then fifteen 1-bit constant parameters can be cancelled, namely $KI_2[42, 49, 50], b_1[2], b_2[2], b_3[1, 2], b_4[1], b_5[1, 2], b_6[1, 2], b_7[1, 2], b_8[1]$; and (2) If we take XOR between two inputs from the 6-round MitM property with $\omega = 1$, then twenty 1-bit constant parameters can be cancelled, namely $\widehat{e}_1[2, 3], \widehat{e}_2[1, 2, 3], \widehat{e}_3[1, 2], \widehat{e}_4[2, 3], \widehat{e}_5[1, 2, 3], \widehat{e}_6[1, 2, 3], \widehat{e}_7[1, 2], \widehat{e}_8[1, 2, 3]$. More formally, we have the following 5 and 6-round HO-MitM properties.

Proposition 2. *Suppose $X^{(i)}$ is defined as in Proposition 1. Let $i_1, i_2 \in \{1, 2, \cdots, 256\}$ and $i_1 \neq i_2$, then:*

1. *If* $Z^{(i)} = (Z_L^{(i)} || Z_R^{(i)})$ *is the result of encrypting* $X^{(i)}$ *using Rounds 4 to 8 with the* $\mathbf{FL}/\mathbf{FL}^{-1}$ *functions between Rounds 6 and 7, then* $\mathbf{P}^{-1}(Z_R^{(i_1)} \oplus Z_R^{(i_2)})[49]$ *can be expressed with a function of* $x^{(i_1)}, x^{(i_2)}$ *and 85 constant 1-bit parameters.*
2. *If* $Z^{(i)} = (Z_L^{(i)} || Z_R^{(i)})$ *is the result of encrypting* $X^{(i)}$ *using Rounds 3 to 8 with the* $\mathbf{FL}/\mathbf{FL}^{-1}$ *functions between Rounds 6 and 7, then* $\mathbf{P}^{-1}(Z_R^{(i_1)} \oplus Z_R^{(i_2)})[41 \sim 42]$ *can be expressed with a function of* $x^{(i_1)}, x^{(i_2)}$ *and 159 constant 1-bit parameters.*

4.2 Attacking 10-Round Camellia-128 without Whitening Functions

We can use Proposition 2-1 to make a HO-MitM attack corresponding to the MitM attack on 10-round Camellia-128 given in Section 3.2, here we fix i_1 to a value and let i_2 take all the other 255 values. The HO-MitM attack requires 2^{56} chosen plaintexts and a memory of $2^{85} \times 255 \times \frac{1}{8} \approx 2^{90}$ bytes, and has a time complexity of approximately $2^{85} \times 256 \times 2 \times \frac{1}{10} + 2^{56+60} \times 256 \times \frac{8+5+1}{8 \times 10} \approx 2^{121.5}$ 10-round Camellia-128 encryptions.

4.3 Attacking 11-Round Camellia-192 with Whitening Functions

Based on Proposition 2-2, the HO-MitM attack on the first 11 rounds of Camellia-192 with $\mathbf{FL}/\mathbf{FL}^{-1}$ and whitening functions, corresponding to the MitM attack on 11-round Camellia-192 given in Section 3.4, requires 2^{56} chosen plaintexts and a memory of $2^{159} \times 255 \times \frac{2}{8} \approx 2^{165}$ bytes, and has a time complexity of approximately $2^{159} \times 256 \times 3 \times \frac{1}{11} + 2^{56+112} \times 256 \times \frac{8+5+1}{8 \times 11} \approx 2^{173.4}$ 11-round Camellia-192 encryptions. Note that we do not need to guess $\mathbf{P}^{-1}(KW_3)[41 \sim 42]$, since it is cancelled after an XOR operation.

4.4 Attacking 12-Round Camellia-256 with Whitening Functions

Similar to the MitM attack on 12-round Camellia-256 given in Section 3.6, Proposition 2-2 can also be used to conduct a HO-MitM attack on the first 12 rounds of Camellia-256 with $\mathbf{FL}/\mathbf{FL}^{-1}$ and whitening functions, which requires 2^{56} chosen plaintexts and a memory of $2^{159} \times 255 \times \frac{2}{8} \approx 2^{165}$ bytes, and has a time complexity of approximately $2^{159} \times 256 \times 3 \times \frac{1}{12} + 2^{56+176} \times 256 \times \frac{8+8+5+1}{8 \times 12} \approx 2^{237.9}$ 12-round Camellia-256 encryptions.

We notice that recently Chen and Li [6] published an MitM attack on 12-round Camellia-256 with $\mathbf{FL}/\mathbf{FL}^{-1}$ and whitening functions, which is actually a HO-MitM attack by definition in [23], building on a 7-round property with 224 constant 1-bit parameters. When constructing the 7-round property, Chen and Li cancelled four 1-bit constant parameters by taking XOR under two different inputs. Likewise, we observe that eight other 1-bit constant parameters were cancelled actually, too. Thus, the 7-round property involves 221 constant 1-bit parameters, and the resulting attack requires 2^{19} chosen plaintexts and a memory of 2^{221} bytes and has a time complexity of $2^{223.2}$ 12-round Camellia-256 encryptions.

5 Concluding Remarks

In this paper, we have analysed the security of Camellia against the MitM attack in detail, following the work in [23]. We have presented 5 and 6-round MitM properties of Camellia, that can be used to conduct MitM attacks on 10-round Camellia-128 with the FL/FL^{-1} functions, 11-round Camellia-192 with the FL/FL^{-1} and whitening functions and 12-round Camellia-256 with the FL/FL^{-1} and whitening functions. We have also described 5 and 6-round HO-MitM properties of Camellia, obtained from the 5 and 6-round MitM properties by taking XOR under two inputs to cancel some constant parameters, which can be used to break the same numbers of rounds as the MitM attacks.

Our results show that as far as Camellia is concerned, the semi-advanced MitM attack technique is more efficient than or at least as efficient as the advanced cryptanalytic techniques studied, except impossible differential cryptanalysis; in this latter case the MitM attacks are one or two rounds inferior to the best newly emerging impossible differential cryptanalysis results from [2,21].

We attribute these MitM attacks to the fact that the FL^{-1} function does not have a good avalanche effect (i.e., an output bit relies on a large number of the bits of the input and the subkey used). If the FL^{-1} function were modified to have a good avalanche effect, then those MitM properties would involve a large number of unknown 1-bit constant parameters, and the resulting MitM attacks would be ineffective for the resulting cipher, but nevertheless it does not necessarily resist the HO-MitM attack technique, for those HO-MitM attacks described in [23] work as long as that integral property of Camellia holds (canceling the FL^{-1} function). Actually, if the FL/FL^{-1} functions had had a good avalanche effect, the Camellia cipher could also have withstood the best currently known cryptanalytic results that are the newly emerging impossible differential cryptanalysis results from [2,21]. In this sense, the FL/FL^{-1} functions do play an important role in the security of Camellia.

Acknowledgments. The authors thank the anonymous referees for their comments on this paper.

References

1. Aoki, K., Ichikawa, T., Kanda, M., Matsui, M., Moriai, S., Nakajima, J., Tokita, T.: *Camellia*: A 128-Bit Block Cipher Suitable for Multiple Platforms - Design and Analysis. In: Stinson, D.R., Tavares, S. (eds.) SAC 2000. LNCS, vol. 2012, pp. 39–56. Springer, Heidelberg (2001)
2. Bai, D., Li, L.: New Impossible Differential Attacks on Camellia. In: Ryan, M.D., Smyth, B., Wang, G. (eds.) ISPEC 2012. LNCS, vol. 7232, pp. 80–96. Springer, Heidelberg (2012)
3. Biham, E., Biryukov, A., Shamir, A.: Cryptanalysis of Skipjack reduced to 31 rounds using impossible differentials. In: Stern, J. (ed.) EUROCRYPT 1999. LNCS, vol. 1592, pp. 12–23. Springer, Heidelberg (1999)

4. Biham, E., Dunkelman, O., Keller, N.: The Rectangle Attack - Rectangling the Serpent. In: Pfitzmann, B. (ed.) EUROCRYPT 2001. LNCS, vol. 2045, pp. 340–357. Springer, Heidelberg (2001)
5. Biham, E., Shamir, A.: Differential cryptanalysis of DES-like cryptosystems. Journal of Cryptology 4(1), 3–72 (1991)
6. Chen, J., Li, L.: Low Data Complexity Attack on Reduced Camellia-256. In: Susilo, W., Mu, Y., Seberry, J. (eds.) ACISP 2012. LNCS, vol. 7372, pp. 101–114. Springer, Heidelberg (2012)
7. Chen, J., Jia, K., Yu, H., Wang, X.: New Impossible Differential Attacks of Reduced-Round Camellia-192 and Camellia-256. In: Parampalli, U., Hawkes, P. (eds.) ACISP 2011. LNCS, vol. 6812, pp. 16–33. Springer, Heidelberg (2011)
8. CRYPTREC — Cryptography Research and Evaluatin Committees, report 2002 (2003)
9. Daemen, J., Knudsen, L.R., Rijmen, V.: The Block Cipher SQUARE. In: Biham, E. (ed.) FSE 1997. LNCS, vol. 1267, pp. 149–165. Springer, Heidelberg (1997)
10. Demirci, H., Selçuk, A.A.: A Meet-in-the-Middle Attack on 8-Round AES. In: Nyberg, K. (ed.) FSE 2008. LNCS, vol. 5086, pp. 116–126. Springer, Heidelberg (2008)
11. Diffie, W., Hellman, M.: Exhaustive cryptanalysis of the NBS data encryption standard. Computer 10(6), 74–84 (1977)
12. Lei, D., Chao, L., Feng, K.: New Observation on Camellia. In: Preneel, B., Tavares, S. (eds.) SAC 2005. LNCS, vol. 3897, pp. 51–64. Springer, Heidelberg (2006)
13. Hatano, Y., Sekine, H., Kaneko, T.: Higher Order Differential Attack of Camellia(II). In: Nyberg, K., Heys, H.M. (eds.) SAC 2002. LNCS, vol. 2595, pp. 39–56. Springer, Heidelberg (2003)
14. Hellman, M.E.: A cryptanalytic time–memory trade-off. IEEE Transcations on Information Theory 26(4), 401–406 (1980)
15. Hu, Y., Zhang, Y., Xiao, G.: Integral cryptanalysis of SAFER+. Electronics Letters 35(17), 1458–1459 (1999)
16. International Standardization of Organization (ISO), International Standard – ISO/IEC 18033-3, Information technology – Security techniques – Encryption algorithms – Part 3: Block ciphers (2005)
17. Knudsen, L.R.: Truncated and Higher Order Differentials. In: Preneel, B. (ed.) FSE 1994. LNCS, vol. 1008, pp. 196–211. Springer, Heidelberg (1995)
18. Knudsen, L.R.: DEAL — a 128-bit block cipher. Technical report, Department of Informatics, University of Bergen, Norway (1998)
19. Knudsen, L.R., Wagner, D.: Integral Cryptanalysis. In: Daemen, J., Rijmen, V. (eds.) FSE 2002. LNCS, vol. 2365, pp. 112–127. Springer, Heidelberg (2002)
20. Lai, X.: Higher order derivatives and differential cryptanalysis. In: Communications and Cryptography, pp. 227–233. Academic Publishers (1994)
21. Liu, Y., Li, L., Gu, D., Wang, X., Liu, Z., Chen, J., Li, W.: New Observations on Impossible Differential Cryptanalysis of Reduced-Round Camellia. In: Canteaut, A. (ed.) FSE 2012. LNCS, vol. 7549, pp. 90–109. Springer, Heidelberg (2012)
22. Lu, J., Wei, Y., Kim, J., Fouque, P.-A.: Cryptanalysis of reduced versions of the Camellia block cipher. In: Miri, A., Vaudenay, S. (eds.) Pre-proceedings of SAC 2011 (2011), `http://sac2011.ryerson.ca/SAC2011/LWKF.pdf`; An editorially revised version is to appear in IET Information Security
23. Lu, J., Wei, Y., Kim, J., Pasalic, E.: The higher-order meet-in-the-middle attack and its application to the Camellia block cipher. Presented in part at the First Asian Workshop on Symmetric Key Cryptography, ASK 2011 (2000), `https://sites.google.com/site/jiqiang/HO-MitM.pdf`

24. Matsui, M.: Linear Cryptanalysis Method for DES Cipher. In: Helleseth, T. (ed.) EUROCRYPT 1993. LNCS, vol. 765, pp. 386–397. Springer, Heidelberg (1994)
25. NESSIE — New European Schemes for Signatures, Integrity, and Encryption, final report of European project IST-1999-12324 (2004)
26. Wenling, W., Dengguo, F., Hua, C.: Collision Attack and Pseudorandomness of Reduced-Round Camellia. In: Handschuh, H., Hasan, M.A. (eds.) SAC 2004. LNCS, vol. 3357, pp. 252–266. Springer, Heidelberg (2004)
27. Wagner, D.: The Boomerang Attack. In: Knudsen, L.R. (ed.) FSE 1999. LNCS, vol. 1636, pp. 156–170. Springer, Heidelberg (1999)

Appendix: Proof of Proposition 1

First, we have the following property for the $\mathbf{FL}/\mathbf{FL}^{-1}$ functions.

Property 4 (from [23]). *Let $x_1, x_2, \cdots, x_8, y_1, y_2, \cdots, y_8$ be 8-bit blocks and KI be a 64-bit subkey.*

1. *If $(y_1||y_2||\cdots||y_8) = \mathbf{FL}(x_1||x_2||\cdots||x_8, KI)$, then*

$$\begin{aligned}
y_1 &= ((((x_1[2\sim8]||x_2[1]) \cap KI[2\sim9]) \oplus x_5) \cup KI[33\sim40]) \oplus x_1,\\
y_2 &= ((((x_2[2\sim8]||x_3[1]) \cap KI[10\sim17]) \oplus x_6) \cup KI[41\sim48]) \oplus x_2,\\
y_3 &= ((((x_3[2\sim8]||x_4[1]) \cap KI[18\sim25]) \oplus x_7) \cup KI[49\sim56]) \oplus x_3,\\
y_4 &= ((((x_4[2\sim8]||x_1[1]) \cap KI[26\sim32,1]) \oplus x_8) \cup KI[57\sim64]) \oplus x_4,\\
y_5 &= ((x_1[2\sim8]||x_2[1]) \cap KI[2\sim9]) \oplus x_5,\\
y_6 &= ((x_2[2\sim8]||x_3[1]) \cap KI[10\sim17]) \oplus x_6,\\
y_7 &= ((x_3[2\sim8]||x_4[1]) \cap KI[18\sim25]) \oplus x_7,\\
y_8 &= ((x_4[2\sim8]||x_1[1]) \cap KI[26\sim32,1]) \oplus x_8.
\end{aligned}$$

2. *If $(y_1||y_2||\cdots||y_8) = \mathbf{FL}^{-1}(x_1||x_2||\cdots||x_8, KI)$, then*

$$\begin{aligned}
y_1 &= (x_5 \cup KI[33\sim40]) \oplus x_1,\\
y_2 &= (x_6 \cup KI[41\sim48]) \oplus x_2,\\
y_3 &= (x_7 \cup KI[49\sim56]) \oplus x_3,\\
y_4 &= (x_8 \cup KI[57\sim64]) \oplus x_4,\\
y_5 &= ((((x_5[2\sim8]||x_6[1]) \cup KI[34\sim41]) \oplus (x_1[2\sim8]||x_2[1])) \cap\\
&\quad KI[2\sim9]) \oplus x_5,\\
y_6 &= ((((x_6[2\sim8]||x_7[1]) \cup KI[42\sim49]) \oplus (x_2[2\sim8]||x_3[1])) \cap\\
&\quad KI[10\sim17]) \oplus x_6,\\
y_7 &= ((((x_7[2\sim8]||x_8[1]) \cup KI[50\sim57]) \oplus (x_3[2\sim8]||x_4[1])) \cap\\
&\quad KI[18\sim25]) \oplus x_7,\\
y_8 &= ((((x_8[2\sim8]||x_5[1]) \cup KI[58\sim64,33]) \oplus (x_4[2\sim8]||x_1[1])) \cap\\
&\quad KI[26\sim32,1]) \oplus x_8.
\end{aligned}$$

When encrypting $X^{(i)}$, we denote by $Y_t^{(i)}$ the value immediately after the **S** operation of Round t, and by $W_t^{(i)}$ the value immediately after the **P** operation of Round t, $(3 \leqslant t \leqslant 8)$.

We have Eq. (1) for Rounds 4 to 8 and have Eq. (2) for Rounds 3 to 8.

$$\mathbf{P}^{-1}(Z_R^{(i)}) = \mathbf{P}^{-1}(\mathbf{FL}^{-1}(X_L^{(i)} \oplus W_5^{(i)}, KI_2)) \oplus Y_7^{(i)}. \tag{1}$$

$$\mathbf{P}^{-1}(Z_R^{(i)}) = \mathbf{P}^{-1}(\mathbf{FL}^{-1}(X_R^{(i)} \oplus W_3^{(i)} \oplus W_5^{(i)}, KI_2)) \oplus Y_7^{(i)}. \tag{2}$$

We first prove Proposition 1-1, and focus on encrypting $X^{(i)}$ through Rounds 4 to 8 below. The output of Round 4 is as follows, where $a_1, a_2, \cdots, a_8$ are 8-bit constants completely determined by $m_1, m_2, \cdots, m_{15}$ and K_4.

$$L_4^{(i)} = (x^{(i)} \oplus a_1, a_2, a_3, a_4, a_5, a_6, a_7, a_8),\ R_4^{(i)} = (m_1, m_2, m_3, m_4, m_5, m_6, m_7, m_8).$$

The output of Round 5 is as follows, where $b, b_1, \cdots, b_8$ are 8-bit constants completely determined by $m_1, m_2, \cdots, m_8$, $a_1, a_2, \cdots, a_8$ and K_5:

$$L_5^{(i)} = (L_{5,1}^{(i)}, L_{5,2}^{(i)}, L_{5,3}^{(i)}, L_{5,4}^{(i)}, L_{5,5}^{(i)}, L_{5,6}^{(i)}, L_{5,7}^{(i)}, L_{5,8}^{(i)}),\ R_5^{(i)} = (x^{(i)} \oplus a_1, a_2, a_3, \cdots, a_8),$$

with

$$L_{5,1}^{(i)} = \mathrm{S}_1(x^{(i)} \oplus b) \oplus b_1,\ L_{5,2}^{(i)} = \mathrm{S}_1(x^{(i)} \oplus b) \oplus b_2,\ L_{5,3}^{(i)} = \mathrm{S}_1(x^{(i)} \oplus b) \oplus b_3,$$
$$L_{5,4}^{(i)} = b_4,\quad L_{5,5}^{(i)} = \mathrm{S}_1(x^{(i)} \oplus b) \oplus b_5,\ L_{5,6}^{(i)} = b_6,$$
$$L_{5,7}^{(i)} = b_7,\quad L_{5,8}^{(i)} = \mathrm{S}_1(x^{(i)} \oplus b) \oplus b_8.$$

The output immediately before the FL/FL^{-1} functions is as follows, where $d_1 = b_1 \oplus K_{6,1}, d_2 = b_2 \oplus K_{6,2}, d_3 = b_3 \oplus K_{6,3}, d_4 = b_5 \oplus K_{6,5}, d_5 = b_8 \oplus K_{6,8}$; and $e_1, e_2, \cdots, e_8$ are 8-bit constants completely determined by $a_1, a_2, \cdots, a_8$ and $b_1, b_2, \cdots, b_8$:

$$\widehat{L}_6^{(i)} = (\widehat{L}_{6,1}^{(i)}, \widehat{L}_{6,2}^{(i)}, \widehat{L}_{6,3}^{(i)}, \widehat{L}_{6,4}^{(i)}, \widehat{L}_{6,5}^{(i)}, \widehat{L}_{6,6}^{(i)}, \widehat{L}_{6,7}^{(i)}, \widehat{L}_{6,8}^{(i)}),\ \widehat{R}_6^{(i)} = (L_{5,1}^{(i)}, L_{5,2}^{(i)}, \cdots, L_{5,8}^{(i)}),$$

with

$$\begin{aligned}
\widehat{L}_{6,1}^{(i)} &= \mathrm{S}_1(\mathrm{S}_1(x^{(i)} \oplus b) \oplus d_1) \oplus \mathrm{S}_3(\mathrm{S}_1(x^{(i)} \oplus b) \oplus d_3) \oplus \mathrm{S}_8(\mathrm{S}_1(x^{(i)} \oplus b) \oplus d_5) \oplus \\
&\quad x^{(i)} \oplus e_1, \\
\widehat{L}_{6,2}^{(i)} &= \mathrm{S}_1(\mathrm{S}_1(x^{(i)} \oplus b) \oplus d_1) \oplus \mathrm{S}_2(\mathrm{S}_1(x^{(i)} \oplus b) \oplus d_2) \oplus \mathrm{S}_5(\mathrm{S}_1(x^{(i)} \oplus b) \oplus d_4) \oplus \\
&\quad \mathrm{S}_8(\mathrm{S}_1(x^{(i)} \oplus b) \oplus d_5) \oplus e_2, \\
\widehat{L}_{6,3}^{(i)} &= \mathrm{S}_1(\mathrm{S}_1(x^{(i)} \oplus b) \oplus d_1) \oplus \mathrm{S}_2(\mathrm{S}_1(x^{(i)} \oplus b) \oplus d_2) \oplus \mathrm{S}_3(\mathrm{S}_1(x^{(i)} \oplus b) \oplus d_3) \oplus \\
&\quad \mathrm{S}_5(\mathrm{S}_1(x^{(i)} \oplus b) \oplus d_4) \oplus \mathrm{S}_8(\mathrm{S}_1(x^{(i)} \oplus b) \oplus d_5) \oplus e_3, \\
\widehat{L}_{6,4}^{(i)} &= \mathrm{S}_2(\mathrm{S}_1(x^{(i)} \oplus b) \oplus d_2) \oplus \mathrm{S}_3(\mathrm{S}_1(x^{(i)} \oplus b) \oplus d_3) \oplus \mathrm{S}_5(\mathrm{S}_1(x^{(i)} \oplus b) \oplus d_4) \oplus e_4, \\
\widehat{L}_{6,5}^{(i)} &= \mathrm{S}_1(\mathrm{S}_1(x^{(i)} \oplus b) \oplus d_1) \oplus \mathrm{S}_2(\mathrm{S}_1(x^{(i)} \oplus b) \oplus d_2) \oplus \mathrm{S}_8(\mathrm{S}_1(x^{(i)} \oplus b) \oplus d_5) \oplus e_5, \\
\widehat{L}_{6,6}^{(i)} &= \mathrm{S}_2(\mathrm{S}_1(x^{(i)} \oplus b) \oplus d_2) \oplus \mathrm{S}_3(\mathrm{S}_1(x^{(i)} \oplus b) \oplus d_3) \oplus \mathrm{S}_5(\mathrm{S}_1(x^{(i)} \oplus b) \oplus d_4) \oplus \\
&\quad \mathrm{S}_8(\mathrm{S}_1(x^{(i)} \oplus b) \oplus d_5) \oplus e_6, \\
\widehat{L}_{6,7}^{(i)} &= \mathrm{S}_3(\mathrm{S}_1(x^{(i)} \oplus b) \oplus d_3) \oplus \mathrm{S}_5(\mathrm{S}_1(x^{(i)} \oplus b) \oplus d_4) \oplus \mathrm{S}_8(\mathrm{S}_1(x^{(i)} \oplus b) \oplus d_5) \oplus e_7, \\
\widehat{L}_{6,8}^{(i)} &= \mathrm{S}_1(\mathrm{S}_1(x^{(i)} \oplus b) \oplus d_1) \oplus \mathrm{S}_5(\mathrm{S}_1(x^{(i)} \oplus b) \oplus d_4) \oplus e_8.
\end{aligned}$$

By Property 4-1, we know that $\mathbf{FL}(\widehat{L}_6^{(i)}, KI_1)[49 \sim 56]$ is determined only by $\widehat{L}_{6,3}^{(i)}, \widehat{L}_{6,4}^{(i)}, \widehat{L}_{6,7}^{(i)}, KI_1[18 \sim 25]$. Thus, $Y_7^{(i)}[49 \sim (49+\omega)] = \mathrm{S}_7(\mathbf{FL}(\widehat{L}_6^{(i)}, KI_1)[49 \sim 56] \oplus K_{7,7})[49 \sim (49+\omega)]$ is determined only by $(x^{(i)}, b, d_1, d_2, \cdots, d_5, e_3, e_4, l_1, KI_1[26 \sim 32, 1])$, where $l_1 = e_7 \oplus K_{7,7}$.

Since $X_L^{(i)} \oplus W_5^{(i)} = \widehat{R}_6^{(i)}$, by Property 4-2 we know that $\mathbf{P}^{-1}(\mathbf{FL}^{-1}(X_L^{(i)} \oplus W_5^{(i)}, KI_2))[49 \sim (49+\omega)] = \mathbf{P}^{-1}(\mathbf{FL}^{-1}(\widehat{R}_6^{(i)}, KI_2))[49 \sim (49+\omega)]$ is determined only by $(x^{(i)}, b, b_1[2 \sim (2+\omega)], b_2[2 \sim (2+\omega)], b_3[1 \sim (2+\omega)], b_4[1 \sim (1+\omega)], b_5[1 \sim (2+\omega)], b_6[1 \sim (2+\omega)], b_7[1 \sim (2+\omega)], b_8[1 \sim (1+\omega)], KI_2[2 \sim (2+\omega), 10 \sim (10+\omega), 18 \sim (18+\omega), 34 \sim (34+\omega), 42 \sim (42+\omega), 49 \sim (50+\omega), 57 \sim (57+\omega)])$.

So $\mathbf{P}^{-1}(\mathbf{FL}^{-1}(X_L^{(i)} \oplus W_5^{(i)}, KI_2))[49 \sim (49+\omega)] \oplus Y_7^{(i)}[49 \sim (49+\omega)]$ is determined by $x^{(i)}$ and $b, d_1, d_2, \cdots, d_5,\ e_3, e_4, l_1, b_1[2 \sim (2+\omega)], b_2[2 \sim (2+\omega)], b_3[1 \sim (2+\omega)], b_4[1 \sim (1+\omega)], b_5[1 \sim (2+\omega)], b_6[1 \sim (2+\omega)], b_7[1 \sim (2+\omega)], b_8[1 \sim (1+\omega)], KI_1[26 \sim 32, 1], KI_2[2 \sim (2+\omega), 10 \sim (10+\omega), 18 \sim (18+\omega), 34 \sim (34+\omega), 42 \sim (42+\omega), 49 \sim (50+\omega), 57 \sim (57+\omega)])$, a total of $100 + 15 \times \omega$ constant 1-bit parameters. Proposition 1-1 follows from Eq. (1).

We next prove Proposition 1-2. The output $(L_3^{(i)}, R_3^{(i)})$ of Round 3 is as follows, where $\widehat{a}_1, \widehat{a}_2, \cdots, \widehat{a}_8$ are 8-bit constants completely determined by $m_1, m_2, \cdots, m_{15}$ and K_3.

$$L_3^{(i)} = (x^{(i)} \oplus \widehat{a}_1, \widehat{a}_2, \widehat{a}_3, \widehat{a}_4, \widehat{a}_5, \widehat{a}_6, \widehat{a}_7, \widehat{a}_8), R_3^{(i)} = (m_1, m_2, m_3, m_4, m_5, m_6, m_7, m_8).$$

The output $(L_4^{(i)}, R_4^{(i)})$ of Round 4 is as follows, where $\widehat{b}, \widehat{b}_1, \cdots, \widehat{b}_8$ are 8-bit constants completely determined by $m_1, m_2, \cdots, m_8,\ \widehat{a}_1, \widehat{a}_2, \cdots, \widehat{a}_8$ and K_4:

$$L_4^{(i)} = (L_{4,1}^{(i)}, L_{4,2}^{(i)}, L_{4,3}^{(i)}, L_{4,4}^{(i)}, L_{4,5}^{(i)}, L_{4,6}^{(i)}, L_{4,7}^{(i)}, L_{4,8}^{(i)}), R_4^{(i)} = (x^{(i)} \oplus \widehat{a}_1, \widehat{a}_2, \widehat{a}_3, \cdots, \widehat{a}_8),$$

with

$$\begin{array}{lll} L_{4,1}^{(i)} = \mathrm{S}_1(x^{(i)} \oplus \widehat{b}) \oplus \widehat{b}_1, & L_{4,2}^{(i)} = \mathrm{S}_1(x^{(i)} \oplus \widehat{b}) \oplus \widehat{b}_2, & L_{4,3}^{(i)} = \mathrm{S}_1(x^{(i)} \oplus \widehat{b}) \oplus \widehat{b}_3, \\ L_{4,4}^{(i)} = \widehat{b}_4, & L_{4,5}^{(i)} = \mathrm{S}_1(x^{(i)} \oplus \widehat{b}) \oplus \widehat{b}_5, & L_{4,6}^{(i)} = \widehat{b}_6, \\ L_{4,7}^{(i)} = \widehat{b}_7, & L_{4,8}^{(i)} = \mathrm{S}_1(x^{(i)} \oplus \widehat{b}) \oplus \widehat{b}_8. & \end{array}$$

The output $(L_5^{(i)}, R_5^{(i)})$ of Round 5 is as follows, where $\widehat{d}_1, \widehat{d}_2, \cdots, \widehat{d}_5$ are 8-bit constants completely determined by $\widehat{b}_1, \widehat{b}_2, \cdots, \widehat{b}_8$ and K_5; and $\widehat{e}_1, \widehat{e}_2, \cdots, \widehat{e}_8$ are 8-bit constants completely determined by $\widehat{a}_1, \widehat{a}_2, \cdots, \widehat{a}_8,\ \widehat{b}_1, \widehat{b}_2, \cdots, \widehat{b}_8$ and K_5:

$$L_5^{(i)} = (L_{5,1}^{(i)}, L_{5,2}^{(i)}, L_{5,3}^{(i)}, L_{5,4}^{(i)}, L_{5,5}^{(i)}, L_{5,6}^{(i)}, L_{5,7}^{(i)}, L_{5,8}^{(i)}), R_5^{(i)} = (L_{4,1}^{(i)}, L_{4,2}^{(i)}, \cdots, L_{4,8}^{(i)}),$$

with

$$\begin{aligned} L_{5,1}^{(i)} = {} & \mathrm{S}_1(\mathrm{S}_1(x^{(i)} \oplus \widehat{b}) \oplus \widehat{d}_1) \oplus \mathrm{S}_3(\mathrm{S}_1(x^{(i)} \oplus \widehat{b}) \oplus \widehat{d}_3) \oplus \mathrm{S}_8(\mathrm{S}_1(x^{(i)} \oplus \widehat{b}) \oplus \widehat{d}_5) \oplus \\ & x^{(i)} \oplus \widehat{e}_1, \\ L_{5,2}^{(i)} = {} & \mathrm{S}_1(\mathrm{S}_1(x^{(i)} \oplus \widehat{b}) \oplus \widehat{d}_1) \oplus \mathrm{S}_2(\mathrm{S}_1(x^{(i)} \oplus \widehat{b}) \oplus \widehat{d}_2) \oplus \mathrm{S}_5(\mathrm{S}_1(x^{(i)} \oplus \widehat{b}) \oplus \widehat{d}_4) \oplus \\ & \mathrm{S}_8(\mathrm{S}_1(x^{(i)} \oplus \widehat{b}) \oplus \widehat{d}_5) \oplus \widehat{e}_2, \\ L_{5,3}^{(i)} = {} & \mathrm{S}_1(\mathrm{S}_1(x^{(i)} \oplus \widehat{b}) \oplus \widehat{d}_1) \oplus \mathrm{S}_2(\mathrm{S}_1(x^{(i)} \oplus \widehat{b}) \oplus \widehat{d}_2) \oplus \mathrm{S}_3(\mathrm{S}_1(x^{(i)} \oplus \widehat{b}) \oplus \widehat{d}_3) \oplus \end{aligned}$$

$$\begin{aligned}
& S_5(S_1(x^{(i)} \oplus \widehat{b}) \oplus \widehat{d}_4) \oplus S_8(S_1(x^{(i)} \oplus \widehat{b}) \oplus \widehat{d}_5) \oplus \widehat{e}_3, \\
L_{5,4}^{(i)} =\ & S_2(S_1(x^{(i)} \oplus \widehat{b}) \oplus S_3(S_1(x^{(i)} \oplus \widehat{b}) \oplus \widehat{d}_3) \oplus S_5(S_1(x^{(i)} \oplus \widehat{b}) \oplus \widehat{d}_4) \oplus \widehat{e}_4, \\
L_{5,5}^{(i)} =\ & S_1(S_1(x^{(i)} \oplus \widehat{b}) \oplus \widehat{d}_1) \oplus S_2(S_1(x^{(i)} \oplus \widehat{b}) \oplus \widehat{d}_2) \oplus S_8(S_1(x^{(i)} \oplus \widehat{b}) \oplus \widehat{d}_5) \oplus \widehat{e}_5, \\
L_{5,6}^{(i)} =\ & S_2(S_1(x^{(i)} \oplus \widehat{b}) \oplus \widehat{d}_2) \oplus S_3(S_1(x^{(i)} \oplus \widehat{b}) \oplus \widehat{d}_3) \oplus S_5(S_1(x^{(i)} \oplus \widehat{b}) \oplus \widehat{d}_4) \oplus \\
& S_8(S_1(x^{(i)} \oplus \widehat{b}) \oplus \widehat{d}_5) \oplus \widehat{e}_6, \\
L_{5,7}^{(i)} =\ & S_3(S_1(x^{(i)} \oplus \widehat{b}) \oplus \widehat{d}_3) \oplus S_5(S_1(x^{(i)} \oplus \widehat{b}) \oplus \widehat{d}_4) \oplus S_8(S_1(x^{(i)} \oplus \widehat{b}) \oplus \widehat{d}_5) \oplus \widehat{e}_7, \\
L_{5,8}^{(i)} =\ & S_1(S_1(x^{(i)} \oplus \widehat{b}) \oplus \widehat{d}_1) \oplus S_5(S_1(x^{(i)} \oplus \widehat{b}) \oplus \widehat{d}_4) \oplus \widehat{e}_8.
\end{aligned}$$

By Property 4-1, we know that $\mathbf{FL}(\widehat{L}_6^{(i)}, KI_1)[41 \sim 48]$ is determined only by $\widehat{L}_{6,2}^{(i)}, \widehat{L}_{6,3}^{(i)}, \widehat{L}_{6,6}^{(i)}$, $KI_1[10 \sim 17]$, where

$$\begin{aligned}
\widehat{L}_{6,2}^{(i)} =\ & S_1(L_{5,1}^{(i)} \oplus K_{6,1}) \oplus S_2(L_{5,2}^{(i)} \oplus K_{6,2}) \oplus S_4(L_{5,4}^{(i)} \oplus K_{6,4}) \oplus S_5(L_{5,5}^{(i)} \oplus K_{6,5}) \oplus \\
& S_7(L_{5,7}^{(i)} \oplus K_{6,7}) \oplus S_8(L_{5,8}^{(i)} \oplus K_{6,8}) \oplus S_1(x^{(i)} \oplus \widehat{b}) \oplus \widehat{b}_2, \\
\widehat{L}_{6,3}^{(i)} =\ & S_1(L_{5,1}^{(i)} \oplus K_{6,1}) \oplus S_2(L_{5,2}^{(i)} \oplus K_{6,2}) \oplus S_3(L_{5,3}^{(i)} \oplus K_{6,3}) \oplus S_5(L_{5,5}^{(i)} \oplus K_{6,5}) \oplus \\
& S_6(L_{5,6}^{(i)} \oplus K_{6,6}) \oplus S_8(L_{5,8}^{(i)} \oplus K_{6,8}) \oplus S_1(x^{(i)} \oplus \widehat{b}) \oplus \widehat{b}_3, \\
\widehat{L}_{6,6}^{(i)} =\ & S_2(L_{5,2}^{(i)} \oplus K_{6,2}) \oplus S_3(L_{5,3}^{(i)} \oplus K_{6,3}) \oplus S_5(L_{5,5}^{(i)} \oplus K_{6,5}) \oplus S_7(L_{5,7}^{(i)} \oplus K_{6,7}) \oplus \\
& S_8(L_{5,8}^{(i)} \oplus K_{6,8}) \oplus \widehat{b}_6.
\end{aligned}$$

Letting $\widehat{n}_l = \widehat{e}_l \oplus K_{6,l}$ and $\widehat{o}_1 = \widehat{b}_6 \oplus K_{7,6}$, $(l = 1, 2, \cdots, 8)$, then we can learn that $Y_7^{(i,j)}[41 \sim (41+\omega)]$ is determined only by $(x^{(i)}, \widehat{b}, \widehat{b}_2, \widehat{b}_3, \widehat{o}_1, \widehat{d}_1, \widehat{d}_2, \cdots, \widehat{d}_5, \widehat{n}_1, \widehat{n}_2, \cdots, \widehat{n}_8, KI_1[10 \sim 17])$.

Since $\mathbf{FL}^{-1}(X_R^{(i)} \oplus W_3^{(i)} \oplus W_5^{(i)}, KI_2) = R_6^{(i)}$, then $\mathbf{P}^{-1}(\mathbf{FL}^{-1}(X_R^{(i)} \oplus W_3^{(i)} \oplus W_5^{(i)}, KI_2))[41 \sim (41+\omega)] = \mathbf{P}^{-1}(\mathbf{FL}^{-1}(\widehat{R}_6^{(i)}, KI_2))[41 \sim (41+\omega)]$ is determined only by $(x^{(i)}, \widehat{b}, \widehat{d}_1, \widehat{d}_2, \cdots, \widehat{d}_5, \widehat{e}_1[2 \sim (2+\omega)], \widehat{e}_2[1 \sim (2+\omega)], \widehat{e}_3[1 \sim (1+\omega)], \widehat{e}_4[2 \sim (2+\omega)], \widehat{e}_5[1 \sim (2+\omega)], \widehat{e}_6[1 \sim (2+\omega)], \widehat{e}_7[1 \sim (1+\omega)], \widehat{e}_8[1 \sim (2+\omega)], KI_2[2 \sim (2+\omega), 10 \sim (10+\omega), 26 \sim (26+\omega), 34 \sim (34+\omega), 41 \sim (42+\omega), 49 \sim (49+\omega), 58 \sim (58+\omega)])$.

Hence, $\mathbf{P}^{-1}(\mathbf{FL}(X_R^{(i)} \oplus W_4^{(i)} \oplus W_6^{(i)}, KI_1))[41 \sim (41+\omega)] \oplus Y_7^{(i)}[41 \sim (41+\omega)]$ is determined by $x^{(i)}$ and $\widehat{b}, \widehat{b}_2, \widehat{b}_3, \widehat{o}_1, \widehat{d}_1, \widehat{d}_2, \cdots, \widehat{d}_5, \widehat{e}_1[2 \sim (2+\omega)], \widehat{e}_2[1 \sim (2+\omega)], \widehat{e}_3[1 \sim (1+\omega)], \widehat{e}_4[2 \sim (2+\omega)], \widehat{e}_5[1 \sim (2+\omega)], \widehat{e}_6[1 \sim (2+\omega)], \widehat{e}_7[1 \sim (1+\omega)], \widehat{e}_8[1 \sim (2+\omega)], \widehat{n}_1, \widehat{n}_2, \cdots, \widehat{n}_8, KI_1[10 \sim 17], KI_2[2 \sim (2+\omega), 10 \sim (10+\omega), 26 \sim (26+\omega), 34 \sim (34+\omega), 41 \sim (42+\omega), 49 \sim (49+\omega), 58 \sim (58+\omega)])$, a total of $164 + 15 \times \omega$ constant 1-bit parameters. The result follows from Eq. (2). □

Efficient Concurrent Oblivious Transfer in Super-Polynomial-Simulation Security

Susumu Kiyoshima[1], Yoshifumi Manabe[1,2], and Tatsuaki Okamoto[1,3]

[1] Graduate School of Informatics, Kyoto University, Japan
kiyoshima@ai.soc.i.kyoto-u.ac.jp
[2] NTT Communication Science Laboratories, Japan
manabe.yoshifumi@lab.ntt.co.jp
[3] NTT Secure Platform Laboratories, Japan
okamoto.tatsuaki@lab.ntt.co.jp

Abstract. In this paper, we show a concurrent oblivious transfer protocol in super-polynomial-simulation (SPS) security. Our protocol does not require any setup and does not assume any independence among the inputs. In addition, our protocol is efficient since it does not use any inefficient primitives such as general zero-knowledge proofs for all NP statements. This is the first concurrent oblivious transfer protocol that achieves both of these properties simultaneously. The security of our protocol is based on the decisional Diffie-Hellman (DDH) assumption.

1 Introduction

1.1 Background

Oblivious Transfer. *Oblivious transfer protocols* [31] have been extensively studied in cryptography due to their usefulness in protocol constructions. Oblivious transfer protocols[1] enable the *receiver* to receive one of two values from the *sender*. The sender cannot know which value the receiver received, whereas the receiver can learn only one value and cannot learn anything about the other value. Numerous protocols have been constructed using oblivious transfer protocols. In fact, oblivious transfer is *complete* for secure computation, i.e., we can compute any function securely given an oblivious transfer protocol [17,18].

Oblivious transfer protocols against malicious adversaries can be obtained by transforming oblivious transfer protocols against semi-honest adversaries to protocols against malicious adversaries using the protocol compiler of Goldreich et al. [13]. However, the protocols obtained in this way are highly inefficient since they use general zero-knowledge proofs for NP statements. As a result, the task of constructing efficient oblivious transfer protocols, which are indispensable for practical purposes, has attracted much attention. Efficient "fully-simulatable"[2]

---

[1] In this paper, we consider 1-out-of-2 oblivious transfer protocols.

[2] If we consider the half-simulation definition [24], there exist many highly-efficient protocols, e.g., [1,23].

G. Hanaoka and T. Yamauchi (Eds.): IWSEC 2012, LNCS 7631, pp. 216–232, 2012.

oblivious transfer protocols are shown in [15, 19, 20]. In addition, there exist black-box transformations, which do not use general zero-knowledge proofs, from semi-honest oblivious transfer to malicious oblivious transfer [8, 16, 17, 22, 27].

Concurrent Security. All of the above protocols achieve only *stand-alone security*, i.e., they are secure only when a single instance of the protocols is executed at a time. More realistic and desirable security is *concurrent security*, which guarantees that the protocol remains secure even when several instances of the protocol are executed at the same time in an arbitrary schedule.

Unfortunately, in the standard model (with adaptively-chosen inputs and no trusted setup), we cannot construct concurrent oblivious transfer protocols with black-box simulation [21]. As a result, existent concurrent oblivious transfer protocols have been constructed in other models. For example, as noted in [21], the concurrent oblivious transfer of [11] circumvents the impossibility result by considering a model where all the inputs in all the executions are independent of each other. *Universally composable* (UC) oblivious transfer protocols [9, 14, 28] achieve UC security, which implies concurrent security, in models with setups such as *common reference strings* (CRS). Although these models are reasonable in some situations, it is desirable to achieve concurrent security in the standard model.

Super-Polynomial-Simulation Security. *Super-polynomial-simulation* (SPS) security [25, 29] enables us to achieve concurrent security in the standard model. SPS security is a relaxed notion of security in the *simulation paradigm*. Before explaining further about SPS security, we introduce the simulation paradigm.

In the simulation paradigm, we define the *real world* and the *ideal world*. In the real world, the parties carry out a task (or *functionality*) by executing a protocol. In the ideal world, the parties carry out the task by interacting with an incorruptible trusted third party called the *ideal functionality*. Then, the protocol is said to be secure if for any adversary who can perform some attacks in the real world there exists an adversary (or *simulator*) who can perform essentially the same attacks in the ideal world. In the case of oblivious transfer, we define the ideal functionality as follows. The ideal functionality $\mathcal{F}$ receives m_0 and m_1 from the sender and $\sigma \in \{0, 1\}$ from the receiver. Then, $\mathcal{F}$ sends m_σ to the receiver. Clearly, $\mathcal{F}$ carries out the task in a perfectly-secure fashion. Then, the security in the simulation paradigm means that, if some attacks can be performed on the protocol by the adversary, essentially the same attacks can be performed even on $\mathcal{F}$ by the simulator.

In standard security definitions of the simulation paradigm, we restrict the running time of the simulator to polynomial time. In SPS security, we relax this security definition by allowing the simulator to run in super-polynomial time. Thus, SPS security guarantees that, if the adversary can perform some attacks in the real world, the simulator can perform essentially the same attacks *in super-polynomial time*. Although SPS security is a relaxed notion of security, it guarantees sufficient security *if the ideal functionality is information-theoretically secure*, i.e., if the ideal functionality is secure against computationally-unbounded adversaries. Clearly, the above oblivious transfer ideal functionality is information-theoretically secure.

SPS security was introduced to construct constant-round concurrent zero-knowledge proofs [25, 26]. SPS security was also used in the UC framework, and it was shown that there exist protocols that compute any functionality in the standard model [2,6,12,30]. Hence, using these protocols, we can construct concurrent oblivious transfer protocols in the standard model. However, the protocols obtained in this way are inefficient, since they use general zero-knowledge proofs for all NP statements. Therefore, for practical purposes, we believe that more work is needed on efficient concurrent oblivious transfer protocols in the standard model.

1.2 Our Result

In this paper, we present a concurrently-secure oblivious transfer protocol secure under SPS security. Our protocol does not require any setup and does not assume any independence among the inputs. In addition, our protocol is efficient since it does not use any inefficient primitives such as general zero-knowledge proofs for all NP statements. To the best of our knowledge, our protocol is the first concurrent oblivious transfer protocol that achieves both of these properties simultaneously. The security of our protocol is based on the decisional Diffie-Hellman (DDH) assumption.

Our Technique. Here, we give a brief overview of our protocol.

We construct our protocol and prove its security in the *UC-SPS framework* [12,30]. The UC-SPS framework is the same as the UC framework [3] except that in the UC-SPS framework we allow the simulator to run in super-polynomial time.

Our protocol is based on the UC oblivious transfer of [28], which is secure in the CRS model. In the protocol of [28], the CRS is $(g_0, h_0, g_1, h_1) \in \mathbb{G}^4$ for cyclic group $\mathbb{G}$. The protocol of [28] has the following properties.

- If (g_0, h_0, g_1, h_1) is a non-DDH tuple, the sender can break the receiver's security with trapdoor $(\log_{g_0} h_0, \log_{g_1} h_1)$.
- If (g_0, h_0, g_1, h_1) is a DDH tuple, the receiver can break the sender's security with trapdoor $\log_{g_0} g_1$.

In [28], the simulator is constructed using these two properties.

In our protocol, the receiver chooses group $\mathbb{G}$ and its elements $g_0, h_0, g_1 \in \mathbb{G}$. Then, the sender and the receiver execute a coin-toss protocol and generate a random element $h_1 \in \mathbb{G}$. Finally, the sender and receiver execute the oblivious transfer protocol of [28] using (g_0, h_0, g_1, h_1) as the CRS. We note that, because of the security of the coin-toss protocol, (g_0, h_0, g_1, h_1) is a non-DDH tuple with overwhelming probability. Then, our protocol has the following properties.

– Super-polynomial-time senders can break the receiver's security by computing trapdoor $(\log_{g_0} h_0, \log_{g_1} h_1)$ in super-polynomial time.
– Super-polynomial-time receivers can let (g_0, h_0, g_1, h_1) be a DDH tuple by cheating in the coin-toss protocol in super-polynomial time. Then, the receivers can break the sender's security with trapdoor $\log_{g_0} g_1$.

Then, we construct a simulator using these two properties.

Although the idea of our protocol is relatively simple, the security proof is not so simple. The reason is that the simulator runs in super-polynomial time whereas we assume only an assumption for polynomial-time adversaries. Therefore, we cannot use simple reduction to prove the indistinguishability between the real-world execution (which runs in polynomial time) and the ideal-world execution (which runs in super-polynomial time). To overcome this problem, we use the technique of [12]. The idea is that we define a hybrid execution in which we use rewinding instead of the super-polynomial power. Then, since the running time of the hybrid execution is polynomial time, we can use the DDH assumption to prove the indistinguishability between the real execution and the hybrid execution. In contrast, since we can show the indistinguishability between the hybrid execution and the ideal execution without any computational assumption, the super-polynomial-time simulator does not cause any problem.

2 Preliminaries

2.1 Notations

Let $\mathbb{N}$ denote the set of all positive integers. For any $q \in \mathbb{N}$, let $\mathbb{Z}_q$ denote the set $\{0, \ldots, q-1\}$. For any set X, let $x \stackrel{\mathsf{U}}{\leftarrow} X$ denote that x is an element of X chosen uniformly at random. For any random variable X, let $x \stackrel{\mathsf{R}}{\leftarrow} X$ denote that x is a value chosen at random according to the probability distribution of X. For any randomized algorithm Algo, let $\mathsf{Algo}(x)$ denote a random variable for the output of Algo on input x with a uniformly-chosen random tape. For any random variable X, let $\mathsf{Algo}(X)$ denote a random variable for the output of Algo on input $x \stackrel{\mathsf{R}}{\leftarrow} X$ with a uniformly-chosen random tape.

Let λ denote a security parameter. Let $\epsilon(\lambda)$ denote an arbitrary negligible function in λ. That is, for any constant $c > 0$, there exists $N \in \mathbb{N}$ such that for any $n > N$ we have $\epsilon(n) < 1/n^c$. For any probability ensembles $\mathcal{X} = \{X_k\}_{k \in \mathbb{N}}$ and $\mathcal{Y} = \{Y_k\}_{k \in \mathbb{N}}$, let $\mathcal{X} \stackrel{\mathsf{c}}{\approx} \mathcal{Y}$ denote that $\mathcal{X}$ and $\mathcal{Y}$ are computationally indistinguishable. That is, we have $\mathcal{X} \stackrel{\mathsf{c}}{\approx} \mathcal{Y}$ if and only if for any probabilistic polynomial-time distinguisher $\mathcal{D}$ we have

$$|\Pr[\mathcal{D}(X_\lambda) = 1] - \Pr[\mathcal{D}(Y_\lambda) = 1]| < \epsilon(\lambda)$$

for a sufficiently large λ.

2.2 The Assumption

In this paper, we use the DDH assumption. Let GenG be a probabilistic polynomial-time algorithm that, on input 1^λ, outputs a description of cyclic group $\mathbb{G}$, its order q, and generator $g \in \mathbb{G}$. Then, the DDH assumption on GenG is defined as follows.

Definition 1 (DDH assumption). *We say that the DDH assumption holds on* GenG *if for any probabilistic polynomial-time algorithm $\mathcal{A}$ we have*

$$\left| \Pr\left[\mathcal{A}(\mathbb{G}, q, g, g^x, g^y, g^{xy}) = 1 \,\middle|\, \begin{array}{l} (\mathbb{G}, q, g) \xleftarrow{\mathsf{R}} \mathsf{GenG}(1^\lambda); \\ x, y \xleftarrow{\mathsf{U}} \mathbb{Z}_q; \end{array} \right] - \Pr\left[\mathcal{A}(\mathbb{G}, q, g, g^x, g^y, g^z) = 1 \,\middle|\, \begin{array}{l} (\mathbb{G}, q, g) \xleftarrow{\mathsf{R}} \mathsf{GenG}(1^\lambda); \\ x, y, z \xleftarrow{\mathsf{U}} \mathbb{Z}_q; \end{array} \right] \right| < \epsilon(\lambda).$$

2.3 UC Framework

In this section, we briefly review the UC framework. For full details, see [3].

The model for protocol execution consists of *environment* $\mathcal{Z}$, *adversary* $\mathcal{A}$, and the parties running protocol π. In the execution of the protocol, the environment $\mathcal{Z}$ is first invoked on external input z. Environment $\mathcal{Z}$ adaptively gives inputs to the parties and receives outputs from them. In addition, $\mathcal{Z}$ communicates freely with $\mathcal{A}$ throughout the execution of the protocol. On inputs from $\mathcal{Z}$, the parties execute π by sending messages to each other. Adversary $\mathcal{A}$ sees all communications between the parties and controls the schedule of the communications. In addition, adversary $\mathcal{A}$ can *corrupt* some parties. After corruption, $\mathcal{A}$ receives all internal information of the corrupted parties. Moreover, from now on, $\mathcal{A}$ can fully control the corrupted parties. In this paper, we assume that there exist authenticated communication channels[3]. Thus, the adversary cannot change the contents of messages sent by the honest parties. In addition, in this paper we consider only static adversaries. In other words, we assume that the adversary corrupts parties only at the beginning of the protocol execution. The protocol execution ends when $\mathcal{Z}$ outputs a bit. Let $\mathsf{Exec}_{\pi,\mathcal{A},\mathcal{Z}}(\lambda, z)$ denote a random variable for the output of $\mathcal{Z}$ on security parameter $\lambda \in \mathbb{N}$ and input $z \in \{0,1\}^*$ with a uniformly-chosen random tape. Let $\mathsf{Exec}_{\pi,\mathcal{A},\mathcal{Z}}$ denote the ensemble $\{\mathsf{Exec}_{\pi,\mathcal{A},\mathcal{Z}}(\lambda, z)\}_{\lambda\in\mathbb{N}, z\in\{0,1\}^*}$.

The security of protocol π is defined using the *ideal protocol.* In the execution of the ideal protocol, all parties simply hand their inputs to *ideal functionality* $\mathcal{F}$. Ideal functionality $\mathcal{F}$ carries out the desired task securely and gives outputs to the parties. The parties simply forward these outputs to $\mathcal{Z}$. Let *dummy parties* denote the parties in the ideal protocol. Adversary $\mathcal{S}$ in the execution of the ideal protocol is often called the *simulator.* Let $\pi(\mathcal{F})$ denote the ideal protocol for functionality $\mathcal{F}$. Let $\mathsf{Ideal}_{\mathcal{F},\mathcal{S},\mathcal{Z}}$ denote the ensemble $\mathsf{Exec}_{\pi(\mathcal{F}),\mathcal{S},\mathcal{Z}}$.

---

[3] This is not essential since authentication can be realized by a protocol, given a standard authentication infrastructure [4].

Then, the security of π is defined by comparing the execution of π (referred to as the *real world*) and the execution of $\pi(\mathcal{F})$ (referred to as the *ideal world*).

Definition 2 (UC-realization). *Let π be a protocol and $\mathcal{F}$ be an ideal functionality. We say that π **UC-realizes** $\mathcal{F}$ if for any adversary $\mathcal{A}$ there exists a simulator $\mathcal{S}$ such that for any environment $\mathcal{Z}$ we have*

$$\mathsf{Exec}_{\pi,\mathcal{A},\mathcal{Z}} \overset{c}{\approx} \mathsf{Ideal}_{\mathcal{F},\mathcal{S},\mathcal{Z}} \ .$$

2.4 UC-SPS Framework

The UC-SPS framework [2,12,30] is the same as the UC framework except that we allow the simulator to run in super-polynomial time. The running time of the other machines is implicitly assumed to be polynomial time.

The UC realization is generalized naturally to the UC-SPS framework as follows.

Definition 3 (UC-SPS-realization). *Let π be a protocol and $\mathcal{F}$ be an ideal functionality. We say that π **UC-SPS-realizes** $\mathcal{F}$ if for any adversary $\mathcal{A}$ there exists a super-polynomial-time simulator $\mathcal{S}$ such that for any environment $\mathcal{Z}$ we have*

$$\mathsf{Exec}_{\pi,\mathcal{A},\mathcal{Z}} \overset{c}{\approx} \mathsf{Ideal}_{\mathcal{F},\mathcal{S},\mathcal{Z}} \ .$$

In general, the UC theorem [3] does not hold in the UC-SPS framework. Thus, stand-alone security in the UC-SPS framework does not imply concurrent security.

3 Concurrent Oblivious Transfer Protocol

In this section, we construct a concurrently-secure oblivious transfer protocol in the UC-SPS framework and prove its security.

As noted in Section 2.4, we cannot use the UC theorem in the UC-SPS framework to prove concurrent security. We therefore prove concurrent security by defining the concurrent oblivious transfer ideal functionality $\mathcal{F}_{\mathsf{cOT}}$ and proving that our protocol UC-SPS-realizes $\mathcal{F}_{\mathsf{cOT}}$. The concurrent oblivious transfer ideal functionality $\mathcal{F}_{\mathsf{cOT}}$, which is based on the (stand-alone) oblivious transfer ideal functionality [5], is shown in Fig. 1 [4]. Functionality $\mathcal{F}_{\mathsf{cOT}}$ captures concurrent security since, with a single run of $\mathcal{F}_{\mathsf{cOT}}$, the sender can send several values to the receiver. Thus, by constructing a protocol that UC-SPS-realizes $\mathcal{F}_{\mathsf{cOT}}$, we obtain a concurrent oblivious transfer protocol. Here, *ssid* in $\mathcal{F}_{\mathsf{cOT}}$ is the *subsession ID*, which is used to distinguish among the different subsessions that take place within a single run of $\mathcal{F}_{\mathsf{cOT}}$. We note that $\mathcal{F}_{\mathsf{cOT}}$ is different from the

[4] We say that functionality $\mathcal{F}$ generates *delayed output* v to party P if $\mathcal{F}$ first sends to $\mathcal{S}$ a note that it is ready to generate an output to P and, after $\mathcal{S}$ replies to the note, $\mathcal{F}$ sends v to P. If the output is *private*, then v is not mentioned in this note [3].

Functionality $\mathcal{F}_{\mathsf{cOT}}$

$\mathcal{F}_{\mathsf{cOT}}$ proceeds as follows, running with sender P_S, receiver P_R, and simulator $\mathcal{S}$.

- Upon receiving input $(\mathsf{Send}, sid, ssid, m_0, m_1)$ from P_S, if message $(\mathsf{Send}, sid, ssid, \ldots)$ was previously received, then do nothing. Else if $sid = (P_S, P_R, sid')$ for some sid' and P_R, then record $(ssid, m_0, m_1)$ and send $(\mathsf{Input}_S, sid, ssid)$ to $\mathcal{S}$. Furthermore, if a value $(ssid, \sigma)$ is recorded, then generate private delayed output $(\mathsf{Output}, sid, ssid, m_\sigma)$ to P_R.
- Upon receiving input $(\mathsf{Receive}, sid, ssid, \sigma)$ for $\sigma \in \{0,1\}$ from P_R, if message $(\mathsf{Receive}, sid, ssid, \ldots)$ was previously received, then do nothing. Else if $sid = (P_S, P_R, sid')$ for some sid' and P_S, then record $(ssid, \sigma)$ and send $(\mathsf{Input}_R, sid, ssid)$ to $\mathcal{S}$. Furthermore, if a value $(ssid, m_0, m_1)$ is recorded, then generate private delayed output $(\mathsf{Output}, sid, ssid, m_\sigma)$ to P_R.

Fig. 1. The concurrent oblivious transfer functionality $\mathcal{F}_{\mathsf{cOT}}$

multi-session oblivious transfer functionality $\hat{\mathcal{F}}_{\mathsf{OT}}$ [7,28], with which any party can concurrently send messages to other parties. Unlike $\hat{\mathcal{F}}_{\mathsf{OT}}$, here only a fixed party P_S can interact with $\mathcal{F}_{\mathsf{cOT}}$ as a sender[5], and as a result $\mathcal{F}_{\mathsf{cOT}}$ does not capture any kind of non-malleability.

3.1 Protocols

First, we show a challenge-response based extractable commitment scheme $\langle C, R\rangle$, and then we show our concurrent oblivious transfer protocol Π_{OT}, which uses $\langle C, R\rangle$ as a primitive.

Extractable Commitment Scheme $\langle C, R\rangle$. Let Com be a non-interactive perfectly-binding commitment scheme[6]. Then the extractable commitment scheme $\langle C, R\rangle$, which is used in literature such as [12,27,29], is defined as follows.

Commit Phase. Sender C commits to element a of group $\mathbb{G}$ for receiver R as follows.

(1) $C \Rightarrow R$: For each $i \in \{1, 2, \ldots, k = \omega(\log\lambda)\}$, C chooses $\alpha_i \overset{\mathsf{U}}{\leftarrow} \mathbb{G}$ and computes $A_i^{(0)} \overset{\mathsf{R}}{\leftarrow} \mathsf{Com}(\alpha_i)$ and $A_i^{(1)} \overset{\mathsf{R}}{\leftarrow} \mathsf{Com}(a\alpha_i^{-1})$. Then C sends these $\{(A_i^{(0)}, A_i^{(1)})\}_{i=1}^k$ to R.

(2) $R \Rightarrow C$: Receiver R chooses $r_1, \ldots, r_k \overset{\mathsf{U}}{\leftarrow} \{0,1\}$ and sends them to C.

(3) $C \Rightarrow R$: Sender C opens all of $\{A_i^{(r_i)}\}_{i=1}^k$ to R.

---

[5] In this paper, we define $\mathcal{F}_{\mathsf{cOT}}$ in such a way that only a single sender and a single receiver can interact with $\mathcal{F}_{\mathsf{cOT}}$. Our protocol remains secure even when we modify $\mathcal{F}_{\mathsf{cOT}}$ so that (a) a single sender and multiple receivers can interact with $\mathcal{F}_{\mathsf{cOT}}$ or (b) multiple senders and a single receiver can interact with $\mathcal{F}_{\mathsf{cOT}}$.

[6] We can construct an efficient non-interactive perfectly-binding commitment scheme under the DDH assumption using ElGamal encryption.

Open Phase. Sender C sends a, and opens all of $\{(A_i^{(0)}, A_i^{(1)})\}_{i=1}^k$ to R.

It is known that $\langle C, R\rangle$ is a perfectly-binding commitment scheme [27].

Concurrent Oblivious Transfer Protocol Π_{OT}. Our concurrent oblivious transfer protocol Π_{OT} is described below. Here, we use algorithm GenG described in Section 2.2.

When the input of the sender is $(\mathsf{Send}, sid, ssid, m_0, m_1)$ and the input of the receiver is $(\mathsf{Receive}, sid, ssid, \sigma)$, sender P_S and receiver P_R do the following. (For simplicity, we assume $m_0, m_1 \in \{0,1\}$. It is easy to modify our protocol so that the sender can send any $m_0, m_1 \in \{0,1\}^{O(\log \lambda)}$. In addition, if there is an efficiently-decodable encoding scheme from $\{0,1\}^\lambda$ to $\mathbb{G}$ for any $\mathbb{G} \xleftarrow{\mathsf{R}} \mathsf{GenG}(1^\lambda)$, the sender can send any $m_0, m_1 \in \{0,1\}^\lambda$.)

(1) $P_R \Rightarrow P_S$: Receiver P_R computes $(\mathbb{G}, q, g_0) \xleftarrow{\mathsf{R}} \mathsf{GenG}(1^\lambda)$. Next, P_R chooses $x, y \xleftarrow{\mathsf{U}} \mathbb{Z}_q$ and sets $h_0 := g_0^x$, $g_1 := g_0^y$. Then P_R sends $(sid, ssid, \mathbb{G}, q, g_0, h_0, g_1)$ to P_S.

(2) $P_S \Leftrightarrow P_R$: Sender P_S chooses $a \xleftarrow{\mathsf{U}} \mathbb{G}$. Then P_S commits to a for P_R using $\langle C, R\rangle$. In other words, P_S and P_R do the following.
  (2.1) $P_S \Rightarrow P_R$: For each $i \in \{1, 2, \ldots, k = \omega(\log \lambda)\}$, P_S chooses $\alpha_i \xleftarrow{\mathsf{U}} \mathbb{G}$ and computes $A_i^{(0)} \xleftarrow{\mathsf{R}} \mathsf{Com}(\alpha_i)$ and $A_i^{(1)} \xleftarrow{\mathsf{R}} \mathsf{Com}(a\alpha_i^{-1})$. Then P_S sends $(sid, ssid, (A_1^{(0)}, A_1^{(1)}), \ldots, (A_k^{(0)}, A_k^{(1)}))$ to P_R.
  (2.2) $P_R \Rightarrow P_S$: Receiver P_R chooses $r_1, \ldots, r_k \xleftarrow{\mathsf{U}} \{0,1\}$ and sends $(sid, ssid, r_1, \ldots, r_k)$ to P_S.
  (2.3) $P_S \Rightarrow P_R$: Sender P_S opens all of $\{A_i^{(r_i)}\}_{i=1}^k$ to P_R. If P_S fails to open one of these commitments, P_R aborts the protocol.

(3) $P_R \Rightarrow P_S$: Receiver P_R chooses $b \xleftarrow{\mathsf{U}} \mathbb{G}$ and sends $(sid, ssid, b)$ to P_S.

(4) $P_S \Rightarrow P_R$: Sender P_S opens the commitment of $\langle C, R\rangle$ in step (2). If P_S fails to open the commitment, P_R aborts the protocol.

(5) P_S and P_R set $h_1 := ab$.

(6) $P_R \Rightarrow P_S$: Receiver P_R chooses $r \xleftarrow{\mathsf{U}} \mathbb{Z}_q$ and sets $g := g_\sigma^r, h := h_\sigma^r$. Then P_R sends $(sid, ssid, g, h)$ to P_S.

(7) $P_S \Rightarrow P_R$: For each $i \in \{0,1\}$, sender P_S chooses $s_i, t_i \xleftarrow{\mathsf{U}} \mathbb{Z}_q$, sets $(u_i, v_i) := (g_i^{s_i} h_i^{t_i}, g^{s_i} h^{t_i})$, and sets $c_i := (u_i, v_i g_0^{m_i})$. Then, P_S sends $(sid, ssid, c_0, c_1)$ to P_R.

(8) Receiver P_R parses c_σ as $(c_{\sigma,0}, c_{\sigma,1})$. Next, P_R sets $\tilde{m}_\sigma := 1$ if $c_{\sigma,1}/c_{\sigma,0}^r = g_0$ and sets $\tilde{m}_\sigma := 0$ otherwise. Then, P_R outputs $(\mathsf{Output}, sid, ssid, \tilde{m}_\sigma)$.

3.2 Security Proof

In this section, we prove the following theorem.

Theorem 4. *Assume that the DDH assumption holds. Then, protocol Π_{OT} UC-SPS-realizes $\mathcal{F}_{\mathrm{cOT}}$.*

Proof. We need to show that for any adversary $\mathcal{A}$ there exists a super-polynomial-time simulator $\mathcal{S}$ such that for any environment $\mathcal{Z}$ we have

$$\mathsf{Exec}_{\Pi_{\mathsf{OT}},\mathcal{A},\mathcal{Z}} \stackrel{\mathsf{c}}{\approx} \mathsf{Ideal}_{\mathcal{F}_{\mathsf{cOT}},\mathcal{S},\mathcal{Z}} \ . \quad (1)$$

In the real world, the sender sends several values concurrently using Π_{OT}. The schedule of the message delivery is determined by adversary $\mathcal{A}$. In the ideal world, the sender sends several values using $\mathcal{F}_{\mathsf{cOT}}$. A single run of $\mathcal{F}_{\mathsf{cOT}}$ consists of several subsessions, where a single value is sent in each subsession.

First, we show the description of simulator $\mathcal{S}$ for adversary $\mathcal{A}$. Simulator $\mathcal{S}$ internally invokes $\mathcal{A}$ and forwards every message from $\mathcal{Z}$ to the internal $\mathcal{A}$. For each message that internal $\mathcal{A}$ outputs to $\mathcal{Z}$, simulator $\mathcal{S}$ simply forwards it to external $\mathcal{Z}$. Furthermore, $\mathcal{S}$ internally simulates a real world with $\mathcal{A}$ as follows.

Case 1. Corrupted P_S and Honest P_R. Since internal $\mathcal{A}$ behaves as the sender on behalf of corrupted P_S, simulator $\mathcal{S}$ needs to interact with $\mathcal{A}$ as the receiver. In addition, $\mathcal{S}$ needs to extract both of the sender's values and send them to $\mathcal{F}_{\mathsf{cOT}}$. Toward this, $\mathcal{S}$ does the following for each subsession.

- Simulator $\mathcal{S}$ starts the subsession in the same way as honest P_R does. That is, $\mathcal{S}$ computes $(\mathbb{G}, q, g_0) \stackrel{\mathsf{R}}{\leftarrow} \mathsf{GenG}(1^\lambda)$, chooses $x, y \stackrel{\mathsf{U}}{\leftarrow} \mathbb{Z}_q$, sets $h_0 := g_0^x$, $g_1 := g_0^y$, and sends $(\mathbb{G}, q, g_0, h_0, g_1)$ to internal $\mathcal{A}$.
- Upon receiving $\{(A_i^{(0)}, A_i^{(1)})\}_{i=1}^k$ from $\mathcal{A}$, simulator $\mathcal{S}$ chooses $r'_1, \ldots, r'_k \stackrel{\mathsf{U}}{\leftarrow} \{0,1\}$ and extracts the committed values of $\{A_i^{(r'_i)}\}_{i=1}^k$ by breaking the hiding property of Com in super-polynomial time. Then, $\mathcal{S}$ chooses $r_1, \ldots, r_k \stackrel{\mathsf{U}}{\leftarrow} \{0,1\}$ and sends them to $\mathcal{A}$ in the same way as honest P_R does.
- If $\mathcal{A}$ opens the commitments of Com correctly in response to challenge $r_1, \ldots, r_k$, simulator $\mathcal{S}$ extracts committed value a of $\langle C, R\rangle$ by combining these opened values with the above extracted values[7]. Then $\mathcal{S}$ sends $b := a^{-1} g_0^{xy}$ to $\mathcal{A}$. Here, if $\mathcal{S}$ finds out that commitment $\{(A_i^{(0)}, A_i^{(1)})\}_{i=1}^k$ of $\langle C, R\rangle$ is invalid when $\mathcal{S}$ tries to extract a, simulator $\mathcal{S}$ sends $b \stackrel{\mathsf{U}}{\leftarrow} \mathbb{G}$ instead.
- When $\mathcal{A}$ opens the commitment of $\langle C, R\rangle$, simulator $\mathcal{S}$ verifies its validity in the same way as honest P_R does.
- Simulator $\mathcal{S}$ chooses $r \stackrel{\mathsf{U}}{\leftarrow} \mathbb{Z}_q$, sets $(g, h) := (g_1^r, h_1^r)$, and sends (g, h) to $\mathcal{A}$.
- Upon receiving $(c_0, c_1) = ((c_{0,0}, c_{0,1}), (c_{1,0}, c_{1,1}))$ from $\mathcal{A}$, simulator $\mathcal{S}$ sets $\tilde{m}_i := 1$ for each $i \in \{0,1\}$ if $c_{i,1}/c_{i,0}^{ry^{1-i}} = g_0$ and sets $\tilde{m}_i := 0$ otherwise. Then, simulator $\mathcal{S}$ sends $(\mathsf{Send}, sid, ssid, \tilde{m}_0, \tilde{m}_1)$ to $\mathcal{F}_{\mathsf{cOT}}$.

In summary, $\mathcal{S}$ extracts committed value a of $\langle C, R\rangle$ in super-polynomial time and sets $b := a^{-1} g_0^{xy}$. This will let $h_1 := ab = g_0^{xy}$. Then, we have

$$(g, h) = (g_1^r, h_1^r) = (g_0^{ry}, h_0^{ry}) \ .$$

Simulator $\mathcal{S}$ sets $\tilde{m}_i := 1$ for each $i \in \{0,1\}$ if $c_{i,1}/c_{i,0}^{ry^{1-i}} = g_0$ and sets $\tilde{m}_i := 0$ otherwise. Then, $\mathcal{S}$ sends $(\tilde{m}_0, \tilde{m}_1)$ to $\mathcal{F}_{\mathsf{cOT}}$.

[7] Since the probability that $(r_1, \ldots, r_k) = (r'_1, \ldots, r'_k)$ holds is negligible, we simply assume $(r_1, \ldots, r_k) \neq (r'_1, \ldots, r'_k)$ in what follows.

Case 2. Honest P_S and Corrupted P_R. Since internal $\mathcal{A}$ behaves as the receiver on behalf of corrupted P_R, simulator $\mathcal{S}$ needs to communicate with $\mathcal{A}$ as the sender knowing only one of the two values that honest P_S sent to $\mathcal{F}_{\mathsf{cOT}}$. Toward this, $\mathcal{S}$ does the following for each subsession.

Simulator $\mathcal{S}$ interacts with $\mathcal{A}$ as the honest sender from step (1) to step (6). Upon receiving (g, h) from $\mathcal{A}$, simulator $\mathcal{S}$ checks whether or not (g_0, h_0, g, h) is a DDH tuple in super-polynomial time. Next, $\mathcal{S}$ sets $\tilde{\sigma} := 0$ if (g_0, h_0, g, h) is a DDH tuple and sets $\tilde{\sigma} := 1$ otherwise. Then, $\mathcal{S}$ sends $(\mathsf{Receive}, sid, ssid, \tilde{\sigma})$ to $\mathcal{F}_{\mathsf{cOT}}$. Upon receiving $(\mathsf{Output}, sid, ssid, m)$ from $\mathcal{F}_{\mathsf{cOT}}$, simulator $\mathcal{S}$ carries out step (7) by letting $m_{\tilde{\sigma}} := m$ and $m_{1-\tilde{\sigma}} \xleftarrow{\mathsf{U}} \{0,1\}$.

Case 3. Honest P_S and Honest P_R. Simulator $\mathcal{S}$ interacts with $\mathcal{A}$ both as the sender and as the receiver. As the sender, $\mathcal{S}$ behaves honestly with input $(m_0 = 0, m_1 = 0)$. As the receiver, $\mathcal{S}$ behaves honestly with input $\sigma = 0$.

Next, we show that, if the above simulator $\mathcal{S}$ is used, we have (1) for each case.

Analysis of Case 1. We need to show that for any probabilistic polynomial-time distinguisher $\mathcal{D}$ and any polynomial p, we have

$$|\Pr[\mathcal{D}(\mathsf{Exec}_{\Pi_{\mathsf{OT}},\mathcal{A},\mathcal{Z}}(\lambda)) = 1] - \Pr[\mathcal{D}(\mathsf{Ideal}_{\mathcal{F}_{\mathsf{cOT}},\mathcal{S},\mathcal{Z}}(\lambda)) = 1]| < \frac{1}{p(\lambda)} \tag{2}$$

for a sufficiently large λ.

Let ℓ be an upper bound of the number of subsessions and let $\delta(\lambda) := 3\ell \cdot p(\lambda)$. We define the indices of the subsessions based on the order in which the messages of step (2.2) appear in the interaction between P_S and P_R. That is, the message of step (2.2) of subsession 1 appears before the message of step (2.2) of subsession 2, and the message of step (2.2) of subsession 2 appears before the message of step (2.2) of subsession 3, and so on.

To prove that we have (2), we use a hybrid argument by defining machines $B_0, \ldots, B_{2\ell+1}$. First, we describe the idea behind our argument. In the ideal world, simulator $\mathcal{S}$ extracts committed value a of $\langle C, R \rangle$ in step (2) of each subsession. Let us call this committed value a the *trapdoor secret* of each subsession. Now, machine B_0 internally executes the real-world protocol and machine $B_{2\ell+1}$ internally executes the ideal-world protocol. In the sequence of hybrid machines, we change B_0 into $B_{2\ell+1}$ step by step by increasing the number of subsessions of which the trapdoor secrets are extracted. That is, we will define $B_{2(i-1)}$ $(i = 1, \ldots, \ell)$ so that the trapdoor secrets of subsession j $(j = 1, \ldots, i-1)$ are extracted and used as in the ideal world. Then, we will define B_{2i-1} by modifying $B_{2(i-1)}$ in such a way that the trapdoor secret of subsession i is also extracted (but not used). Each hybrid machine records these extracted trapdoor secrets in a list, a-List. We note that the hybrid machines, except $B_{2\ell+1}$, are designed so that they do not use their super-polynomial power to extract the trapdoor secrets[8]. Instead, they use polynomial-time rewinding and extract the trapdoor secrets using the extractability of $\langle C, R\rangle$[9].

Now, let us define hybrid machines $B_0, \ldots, B_{2\ell+1}$. First, we introduce some notations. The hybrid machines, except $B_{2\ell+1}$, internally execute the real-world protocol repeatedly with different randomness. That is, they internally invoke machines such as $\mathcal{Z}$ and $\mathcal{A}$, execute the protocol, rewind all the machines, execute the protocol again, rewind all the machines again, and so on. We let a *thread* denote a single execution of the protocol. A thread begins when internal $\mathcal{Z}$ is invoked, and the thread ends when internal $\mathcal{Z}$ outputs a bit. Each hybrid machine outputs whatever internal $\mathcal{Z}$ outputs in the last thread. Let us call this last thread the *main thread* of each hybrid machine.

Machine B_0. As its main thread, machine B_0 internally executes the real-world protocol by internally invoking $\mathcal{Z}$, $\mathcal{A}$, P_S, and P_R. Machine B_0 simply outputs whatever the internal $\mathcal{Z}$ outputs.

Machine B_{2i-1} $(i = 1, \ldots, \ell)$. First, B_{2i-1} runs in the same way as $B_{2(i-1)}$, but B_{2i-1} does not output (and does not halt) even after the main thread of $B_{2(i-1)}$ ends. At the time, the trapdoor secret of subsession j $(j = 1, \ldots, i-1)$ on the main thread of $B_{2(i-1)}$ is extracted and recorded in the a-List. After the main thread of $B_{2(i-1)}$, machine B_{2i-1} rewinds this main thread[10] and executes it δ times with the same randomness except in step (2.2) of subsession i. Let us call these δ threads the *look-ahead threads.* In each look-ahead thread, challenge $r_1, \ldots, r_k$ in step (2.2) of subsession i is chosen fleshly.

In the case that $\mathcal{A}$ opens the commitments of Com correctly in step (2.3) of subsession i in the main thread of $B_{2(i-1)}$ and in at least one of the δ look-ahead threads, B_{2i-1} extracts trapdoor secret a of subsession i by combining the opened values of these two threads. Then, B_{2i-1} adds a pair (i, a) to the a-List. If B_{2i-1} finds out that the commitment of $\langle C, R\rangle$ is invalid when it tries to extract a, then B_{2i-1} adds $(i, \perp)$ to the a-List instead.

In the case that $\mathcal{A}$ does not open the commitments of Com correctly in step (2.3) of subsession i in all δ look-ahead threads but opens them correctly in the main thread of $B_{2(i-1)}$, machine B_{2i-1} outputs $\perp$ and halts. Let us call this event $\mathsf{RewindAbort}_i$.

After all look-ahead threads end, if $\mathsf{RewindAbort}_i$ does not occur, B_{2i-1} executes the main thread of $B_{2(i-1)}$ once again with exactly the same randomness. This thread is the main thread of B_{2i-1}. The output of B_{2i-1} is whatever internal $\mathcal{Z}$ outputs in this thread.

Remark 5. We note that B_{2i-1} can execute each look-ahead thread without any problem such as recursive rewinding. To see this, observe that each look-ahead

[8] If hybrid machines are super-polynomial-time machines, it is difficult to show the indistinguishability between the outputs of hybrid machines based on assumptions for polynomial-time adversaries.

[9] The technique of replacing the super-polynomial power with the polynomial-time rewinding is used in [6,12].

[10] That is, B_{2i-1} rewinds all the machines such as $\mathcal{Z}$ and $\mathcal{A}$.

thread proceeds in exactly the same way as the main thread of $B_{2(i-1)}$ until step (2.2) of subsession i. In particular, the message of step (2.1) in subsession j ($j = 1, \ldots, i-1$) in each look-ahead thread is the same as the message in the main thread of $B_{2(i-1)}$. This means that the trapdoor secret of subsession j ($j = 1, \ldots, i-1$) in each look-ahead thread is the same as the trapdoor secret in the main thread of $B_{2(i-1)}$. Thus, the values that are extracted and recorded in the a-List before the rewinding are valid even after the rewinding. Therefore, since B_{2i-1} can also use them in the look-ahead threads, there is no recursive rewinding.

Machine B_{2i} *(*$i = 1, \ldots, \ell$*).* B_{2i} runs in the same way as B_{2i-1} except that, in step (3) of subsession i in the main thread, internal P_R sets $b := a^{-1} g_0^{xy}$ if (i, a) is recorded in the a-List for $a \neq \perp$. In the case of $a = \perp$, internal P_R sets $b \overset{\mathsf{U}}{\leftarrow} \mathbb{G}$ as in B_{2i-1}.

Machine $B_{2\ell+1}$*.* $B_{2\ell+1}$ internally executes the ideal-world protocol by internally invoking $\mathcal{Z}$, $\mathcal{S}$, the dummy party P_S and P_R. Machine $B_{2\ell+1}$ outputs whatever the internal $\mathcal{Z}$ outputs.

Next, we show the indistinguishability among the outputs of hybrid machines. Below, we let $\mathsf{Exec}_i(\lambda)$ denote the random variable for the output of machine B_i.

$B_{2(i-1)}$ *and* B_{2i-1} *(*$i = 1, \ldots, \ell$*).* If $\mathsf{RewindAbort}_i$ does not occur in B_{2i-1}, the output of $B_{2(i-1)}$ and the output of B_{2i-1} are identical since their main threads are the same. $\mathsf{RewindAbort}_i$ occurs in B_{2i-1} if $\mathcal{A}$ does not open the commitments in step (2.3) on subsession i in all δ look-ahead threads but opens them correctly in the main thread. Since $\mathcal{A}$ opens these commitments correctly in each look-ahead thread with the same probability as in the main thread, we can show that $\mathsf{RewindAbort}_i$ occurs in B_{2i-1} with probability at most $1/\delta$. Thus, for any probabilistic polynomial-time distinguisher $\mathcal{D}$, we have

$$\left|\Pr\left[\mathcal{D}(\mathsf{Exec}_{2(i-1)}(\lambda)) = 1\right] - \Pr\left[\mathcal{D}(\mathsf{Exec}_{2i-1}(\lambda)) = 1\right]\right| \leq \frac{1}{\delta(\lambda)} \ . \qquad (3)$$

B_{2i-1} *and* B_{2i} *(*$i = 1, \ldots, \ell$*).* B_{2i} is the same as B_{2i-1} except that B_{2i} sets $b := a^{-1} g_0^{xy}$ instead of $b \overset{\mathsf{U}}{\leftarrow} \mathbb{G}$ in step (3) of subsession i on the main thread. Thus, from the DDH assumption, for any probabilistic polynomial-time distinguisher $\mathcal{D}$, we have

$$\left|\Pr\left[\mathcal{D}(\mathsf{Exec}_{2i-1}(\lambda)) = 1\right] - \Pr\left[\mathcal{D}(\mathsf{Exec}_{2i}(\lambda)) = 1\right]\right| < \epsilon(\lambda) \ . \qquad (4)$$

$B_{2\ell}$ *and* $B_{2\ell+1}$*.* In $B_{2\ell}$, all the trapdoor secrets are extracted and used as in $B_{2\ell+1}$. Machine $B_{2\ell}$ uses rewinding to extract the trapdoor secrets, whereas machine $B_{2\ell+1}$ uses its super-polynomial power. In order to show the indistinguishability, it suffices to show that the honest receiver's outputs and the computed trapdoor secrets in $B_{2\ell}$ are the same as the ones in $B_{2\ell+1}$.

First, we show the indistinguishability between $B_{2\ell}$ and $B_{2\ell+1}$ under the condition that $\mathsf{RewindAbort}_i$ does not occur in $B_{2\ell}$ for all i. In this case, in each subsession, trapdoor secret a that $B_{2\ell}$ records in the a-List and trapdoor secret a that $\mathcal{S}$ computes in $B_{2\ell+1}$ are identically distributed. To see this, observe that in both machines we can think that the trapdoor secret a is computed by combining two responses of $\langle C, R\rangle$ for two different challenges. In addition, since we have

$$(g, h) = (g_1^r, h_1^r) = (g_0^{ry}, h_0^{ry}) \qquad \text{(since we have } h_1 = g_0^{xy}\text{)}$$

in $B_{2\ell+1}$, the receiver outputs the same value in $B_{2\ell}$ and $B_{2\ell+1}$. Therefore, we conclude that the view of $\mathcal{Z}$ in the main threads of $B_{2\ell}$ and the view of $\mathcal{Z}$ in $B_{2\ell+1}$ are identical if $\mathsf{RewindAbort}_i$ does not occur in $B_{2\ell}$ for all i.

Next, we compute the probability that $\mathsf{RewindAbort}_i$ occurs in $B_{2\ell}$ for some i. From (3) and (4), we have

$$|\Pr[\mathcal{D}(\mathsf{Exec}_0(\lambda)) = 1] - \Pr[\mathcal{D}(\mathsf{Exec}_{2\ell}(\lambda)) = 1]| \leq \frac{\ell}{\delta(\lambda)} + \epsilon(\lambda) \;, \tag{5}$$

for any probabilistic polynomial-time distinguisher $\mathcal{D}$. Since $\mathsf{RewindAbort}_i$ does not occur in B_0 for all i, we conclude that $\mathsf{RewindAbort}_i$ occurs in $B_{2\ell}$ for some i with probability at most $\ell/\delta(\lambda) + \epsilon(\lambda)$.

Combining the above, we conclude that for any probabilistic polynomial-time distinguisher $\mathcal{D}$ we have

$$|\Pr[\mathcal{D}(\mathsf{Exec}_{2\ell}(\lambda)) = 1] - \Pr[\mathcal{D}(\mathsf{Exec}_{2\ell+1}(\lambda)) = 1]| \leq \frac{\ell}{\delta(\lambda)} + \epsilon(\lambda) \;. \tag{6}$$

Finishing the Analysis of Case 1. From (5) and (6), for any probabilistic polynomial-time distinguisher $\mathcal{D}$, we have

$$|\Pr[\mathcal{D}(\mathsf{Exec}_0(\lambda)) = 1] - \Pr[\mathcal{D}(\mathsf{Exec}_{2\ell+1}(\lambda)) = 1]| \leq \frac{2\ell}{\delta(\lambda)} + \epsilon(\lambda) \;.$$

By substituting $\mathsf{Exec}_0(\lambda) = \mathsf{Exec}_{\Pi_{\mathrm{OT}},\mathcal{A},\mathcal{Z}}(\lambda)$, $\mathsf{Exec}_{2\ell+1}(\lambda) = \mathsf{Ideal}_{\mathcal{F}_{\mathrm{cOT}},\mathcal{S},\mathcal{Z}}(\lambda)$ and $\delta(\lambda) = 3\ell \cdot p(\lambda)$, we have (2).

Analysis of Case 2. In the real world, honest sender P_S interacts with $\mathcal{A}$ (via the corrupted receiver) using m_0 and m_1, which P_S received as an input from $\mathcal{Z}$. In the ideal world, simulator $\mathcal{S}$ interacts with internal $\mathcal{A}$ honestly using m_0 and m_1, where $m_{\tilde{\sigma}}$ is received from $\mathcal{F}_{\mathrm{cOT}}$ and $m_{1-\tilde{\sigma}}$ is chosen uniformly at random. Thus, in the view of $\mathcal{Z}$, the only possible difference between the real world and the ideal world is the value of $c_{1-\tilde{\sigma}} = (u_{1-\tilde{\sigma}}, v_{1-\tilde{\sigma}} g_0^{m_{1-\tilde{\sigma}}})$. In what follows, we let $\mu := 1 - \tilde{\sigma}$.

First, we show the indistinguishability under the condition that (g_0, h_0, g_1, h_1) is a non-DDH tuple in each subsession both in the real world and in the ideal world. In this case, at least one of (g_0, h_0, g, h) and (g_1, h_1, g, h) is also a non-DDH tuple in each subsession. From the definition of $\tilde{\sigma}$, this means that (g_μ, h_μ, g, h) is a non-DDH tuple. That is, there exist $\alpha, \beta, \gamma \in \mathbb{Z}_q$ such that $(h_\mu, g, h) =$

$(g_\mu^\alpha, g_\mu^\beta, g_\mu^\gamma)$ and $\alpha\beta \neq \gamma$. Using this, we can show that v_μ is uniformly random for $\mathcal{Z}$. To see this, observe that we have

$$u_\mu = g_\mu^{s_\mu} h_\mu^{t_\mu} = g_\mu^{s_\mu + \alpha t_\mu} \quad ,$$
$$v_\mu = g^{s_\mu} h^{t_\mu} = g_\mu^{\beta s_\mu + \gamma t_\mu}$$

for random s_μ and t_μ, and the expressions $s_\mu + \alpha t_\mu$ and $\beta s_\mu + \gamma t_\mu$ are linearly independent combinations of s_μ and t_μ when $\alpha\beta \neq \gamma$. Thus, the distribution of c_μ is independent of m_μ. Therefore, the view of $\mathcal{Z}$ is identically distributed in the real world and the ideal world.

Next, we compute the probability that (g_0, h_0, g_1, h_1) is a DDH tuple in some subsessions. Below, we show that this probability is negligible in the real world. In this case, since simulator $\mathcal{S}$ interacts with internal $\mathcal{A}$ honestly (with uniformly chosen m_μ) and the computation of (g_0, h_0, g_1, h_1) is independent of the message m_μ, we conclude that this probability is also negligible in the ideal world.

Then, we show that (g_0, h_0, g_1, h_1) is a DDH tuple with negligible probability in the real world. Let us consider the following experiment $\mathsf{Exp}_i^{\mathcal{B}}(\lambda)$ for the hiding property of $\langle C, R\rangle$. First, adversary $\mathcal{B}$ sends $(a_{0,0}, a_{0,1}, a_{1,0}, a_{1,1})$ to the challenger. Then, the challenger commits to $a_{i,0}$ and $a_{i,1}$ for $\mathcal{B}$ by invoking $\langle C, R\rangle$ sequentially. Finally, $\mathcal{B}$ outputs bit i', which is the output of $\mathsf{Exp}_i^{\mathcal{B}}(\lambda)$. Advantage $\mathsf{Adv}_{\mathcal{B}}(\lambda)$ of $\mathcal{B}$ is

$$\mathsf{Adv}_{\mathcal{B}}(\lambda) := \left|\Pr\left[\mathsf{Exp}_0^{\mathcal{B}}(\lambda) = 1\right] - \Pr\left[\mathsf{Exp}_1^{\mathcal{B}}(\lambda) = 1\right]\right| .$$

Using the hiding property of $\langle C, R\rangle$, we can show that we have $\mathsf{Adv}_{\mathcal{B}}(\lambda) < \epsilon(\lambda)$ for any $\mathcal{B}$. Below, we show that, if in the real world (g_0, h_0, g_1, h_1) is a DDH tuple in some subsession j^* with probability $1/\lambda^c$ for some constant $c > 0$, we can construct adversary $\mathcal{B}^*$ such that $\mathsf{Adv}_{\mathcal{B}^*}(\lambda)$ is non-negligible, which contradicts the hiding property of $\langle C, R\rangle$.

Adversary $\mathcal{B}^*$ chooses $j \xleftarrow{\mathsf{U}} \{1, \ldots, \ell\}$ (here, ℓ is the upper bound of the number of subsessions), and internally executes the real-world execution until step (2.1) of subsession j. Let $(sid, ssid_j, \mathbb{G}_j, q_j, g_{j,0}, h_{j,0}, g_{j,1})$ be the message of step (1) in subsession j. Then, $\mathcal{B}^*$ chooses $a_{0,0}, a_{0,1}, a_{1,0}, a_{1,1} \xleftarrow{\mathsf{U}} \mathbb{G}_j$ and sends them to the challenger. When the challenger starts $\langle C, R\rangle$, adversary $\mathcal{B}^*$ forwards it to the internal execution as step (2) of subsession j. We call this internal execution exec_0. Let $(sid, ssid_j, b_0)$ be the message of step (3) in subsession j of exec_0. Next, $\mathcal{B}^*$ rewinds exec_0 to step (2) of subsession j. Then, $\mathcal{B}^*$ receives the next commitment of $\langle C, R\rangle$ from the challenger and forwards it to the rewound internal execution as step (2) of subsession j. We call this second execution exec_1. Let $(sid, ssid_j, b_1)$ be the message of step (3) in subsession j of exec_1. Then, $\mathcal{B}^*$ outputs 1 if and only if $b_0/b_1 = a_{0,0}^{-1}/a_{0,1}^{-1}$ holds.

Let ρ be a partial transcript such that step (2) of subsession j^* begins immediately after ρ in the real execution. Then, from the average argument, it holds that

$$\Pr\left[(g_0, h_0, g_1, h_1) \text{ is a DDH tuple in subsession } j^* \mid \rho \text{ occurs}\right] \geq \frac{1}{2\lambda^c}$$

with probability at least $1/2\lambda^c$ over the choice of ρ.

In $\mathcal{B}^*$, we have $j = j^*$ with probability $1/\ell$. In addition, in $\mathsf{Exp}_0^{\mathcal{B}^*}(\lambda)$, we have

$$\Pr\left[b_0 = a_{0,0}^{-1} g_{j,0}^{x_j y_j} \bigwedge b_1 = a_{0,1}^{-1} g_{j,0}^{x_j y_j} \middle| j = j^*\right] \geq \frac{1}{2\lambda^c} \cdot \left(\frac{1}{2\lambda^c}\right)^2 ,$$

where $x_j := \log_{g_{j,0}} h_{j,0}$ and $y_j := \log_{g_{j,0}} g_{j,1}$. Thus, we have $\Pr\left[\mathsf{Exp}_0^{\mathcal{B}^*}(\lambda) = 1\right] \geq 1/(8\ell\lambda^{3c})$. On the other hand, since no information about $a_{0,0}$ and $a_{0,1}$ is fed into exec_0 and exec_1 in $\mathsf{Exp}_1^{\mathcal{B}^*}(\lambda)$, we have $\Pr\left[\mathsf{Exp}_1^{\mathcal{B}^*}(\lambda) = 1\right] \leq 1/|\mathbb{G}| < \epsilon(\lambda)$. Therefore, we have $\mathsf{Adv}_{\mathcal{B}^*}(\lambda) \geq 1/\mathsf{poly}(\lambda)$. Since this contradicts the hiding property of $\langle C, R\rangle$, we conclude that the probability that (g_0, h_0, g_1, h_1) is a DDH tuple in some subsession is negligible in the real world.

Combining the above, we conclude that (1) holds in Case 2.

Analysis of Case 3. First, the outputs of the honest receiver in the real world are the same as in the ideal world. This is because, in each subsession of the real world, the receiver outputs 1 if and only if it holds that

$$g_0 = \frac{c_{\sigma,1}}{c_{\sigma,0}^r} = \frac{v_\sigma g_0^{m_\sigma}}{u_\sigma^r} = \frac{g^{s_\sigma} h^{t_\sigma} g_0^{m_\sigma}}{(g_\sigma^{s_\sigma} h_\sigma^{t_\sigma})^r} = g_0^{m_\sigma} .$$

Thus, to show the indistinguishability, it suffices to show that $\mathcal{Z}$ cannot tell whether it interacts with $\mathcal{A}$ in the real world or it interacts with the internal $\mathcal{A}$ (of $\mathcal{S}$) in the ideal world. Toward this, let us consider the following hybrid.

Hybrid H_0 is the same as the ideal world except that, in each subsession, simulator $\mathcal{S}$ uses honest parties' inputs m_0, m_1, and σ instead of 0. Note that the view of $\mathcal{Z}$ in H_0 is the same as in the real world.

Hybrid H_1 is the same as H_0 except that $\mathcal{S}$ sets $\sigma := 1$ in each subsession. The view of $\mathcal{Z}$ in H_1 is indistinguishable from the one in H_0 since, from the DDH assumption, $(g_0, h_0, g_1, h_1, g_0^r, h_0^r)$ and $(g_0, h_0, g_1, h_1, g_1^r, h_1^r)$ are indistinguishable for $\mathcal{Z}$.

Hybrid H_2 is the same as H_1 except that $\mathcal{S}$ sets $m_0 := 0$ in each subsession. The view of $\mathcal{Z}$ in H_2 is identical with the one in H_1 except with negligible probability since, from the same argument as in Case 2, the distribution of c_0 in each subsession is independent of the value of m_0 except with negligible probability.

Hybrid H_3 is the same as H_2 except that $\mathcal{S}$ sets $\sigma := 0$ in each subsession. From the same argument as in H_1, the view of $\mathcal{Z}$ in H_3 is indistinguishable from the one in H_2.

Hybrid H_4 is the same as H_3 except that $\mathcal{S}$ sets $m_1 := 0$ in each subsession. From the same argument as in H_2, the view of $\mathcal{Z}$ in H_4 is identical with the one in H_3 except with negligible probability.

Since H_4 is the same as the ideal world, it holds that the view of $\mathcal{Z}$ in the real world is indistinguishable from the one in the ideal world. We therefore conclude that (1) holds in Case 3.

Since we have (1) for all three cases, we conclude that protocol Π_{OT} UC-SPS-realizes $\mathcal{F}_{\mathsf{cOT}}$. □

4 Conclusion

This paper showed a concurrently-secure oblivious transfer protocol in the SPS security without any setup. Our protocol is efficient since it does not use any inefficient primitive such as general zero-knowledge proofs for all NP statements. Therefore, our protocol may be useful for practical purposes.

It should be noted that, unlike many previous studies on SPS security, we considered only concurrent security and do not considered other security notions such as *non-malleability* [10] and UC security. Thus, our protocol achieves somewhat restricted security. However, we believe that concurrent security is sufficient for various settings such as a network in which one party is a server and the others are clients. It would be interesting to improve our protocol so that non-malleability or the UC security is also guaranteed.

References

1. Aiello, W., Ishai, Y., Reingold, O.: Priced Oblivious Transfer: How to Sell Digital Goods. In: Pfitzmann, B. (ed.) EUROCRYPT 2001. LNCS, vol. 2045, pp. 119–135. Springer, Heidelberg (2001)
2. Barak, B., Sahai, A.: How to play almost any mental game over the net - concurrent composition via super-polynomial simulation. In: FOCS, pp. 543–552. IEEE Computer Society (2005)
3. Canetti, R.: Universally composable security: A new paradigm for cryptographic protocols. In: FOCS, pp. 136–145. IEEE Computer Society (2001)
4. Canetti, R.: Universally composable signature, certification, and authentication. In: CSFW, pp. 219–233. IEEE Computer Society (2004)
5. Canetti, R., Fischlin, M.: Universally Composable Commitments. In: Kilian, J. (ed.) CRYPTO 2001. LNCS, vol. 2139, pp. 19–40. Springer, Heidelberg (2001)
6. Canetti, R., Lin, H., Pass, R.: Adaptive hardness and composable security in the plain model from standard assumptions. In: FOCS, pp. 541–550. IEEE Computer Society (2010)
7. Canetti, R., Rabin, T.: Universal Composition with Joint State. In: Boneh, D. (ed.) CRYPTO 2003. LNCS, vol. 2729, pp. 265–281. Springer, Heidelberg (2003)
8. Choi, S.G., Dachman-Soled, D., Malkin, T., Wee, H.: Simple, Black-Box Constructions of Adaptively Secure Protocols. In: Reingold, O. (ed.) TCC 2009. LNCS, vol. 5444, pp. 387–402. Springer, Heidelberg (2009)
9. Damgård, I., Nielsen, J.B., Orlandi, C.: Essentially Optimal Universally Composable Oblivious Transfer. In: Lee, P.J., Cheon, J.H. (eds.) ICISC 2008. LNCS, vol. 5461, pp. 318–335. Springer, Heidelberg (2009)
10. Dolev, D., Dwork, C., Naor, M.: Nonmalleable cryptography. SIAM J. Comput. 30(2), 391–437 (2000)
11. Garay, J.A., MacKenzie, P.D.: Concurrent oblivious transfer. In: FOCS, pp. 314–324. IEEE Computer Society (2000)

12. Garg, S., Goyal, V., Jain, A., Sahai, A.: Concurrently Secure Computation in Constant Rounds. In: Pointcheval, D., Johansson, T. (eds.) EUROCRYPT 2012. LNCS, vol. 7237, pp. 99–116. Springer, Heidelberg (2012)
13. Goldreich, O., Micali, S., Wigderson, A.: How to play any mental game or a completeness theorem for protocols with honest majority. In: Aho, A.V. (ed.) STOC, pp. 218–229. ACM (1987)
14. Green, M., Hohenberger, S.: Universally Composable Adaptive Oblivious Transfer. In: Pieprzyk, J. (ed.) ASIACRYPT 2008. LNCS, vol. 5350, pp. 179–197. Springer, Heidelberg (2008)
15. Green, M., Hohenberger, S.: Practical Adaptive Oblivious Transfer from Simple Assumptions. In: Ishai, Y. (ed.) TCC 2011. LNCS, vol. 6597, pp. 347–363. Springer, Heidelberg (2011)
16. Haitner, I., Ishai, Y., Kushilevitz, E., Lindell, Y., Petrank, E.: Black-box constructions of protocols for secure computation. SIAM J. Comput. 40(2), 225–266 (2011)
17. Ishai, Y., Prabhakaran, M., Sahai, A.: Founding Cryptography on Oblivious Transfer – Efficiently. In: Wagner, D. (ed.) CRYPTO 2008. LNCS, vol. 5157, pp. 572–591. Springer, Heidelberg (2008)
18. Kilian, J.: Founding cryptography on oblivious transfer. In: Simon, J. (ed.) STOC, pp. 20–31. ACM (1988)
19. Kurosawa, K., Nojima, R., Phong, L.T.: Efficiency-Improved Fully Simulatable Adaptive OT under the DDH Assumption. In: Garay, J.A., De Prisco, R. (eds.) SCN 2010. LNCS, vol. 6280, pp. 172–181. Springer, Heidelberg (2010)
20. Lindell, A.Y.: Efficient Fully-Simulatable Oblivious Transfer. In: Malkin, T. (ed.) CT-RSA 2008. LNCS, vol. 4964, pp. 52–70. Springer, Heidelberg (2008)
21. Lindell, Y.: Lower bounds and impossibility results for concurrent self composition. J. Cryptology 21(2), 200–249 (2008)
22. Lindell, Y., Oxman, E., Pinkas, B.: The IPS Compiler: Optimizations, Variants and Concrete Efficiency. In: Rogaway, P. (ed.) CRYPTO 2011. LNCS, vol. 6841, pp. 259–276. Springer, Heidelberg (2011)
23. Naor, M., Pinkas, B.: Efficient oblivious transfer protocols. In: Kosaraju, S.R. (ed.) SODA, pp. 448–457. ACM/SIAM (2001)
24. Naor, M., Pinkas, B.: Computationally secure oblivious transfer. J. Cryptology 18(1), 1–35 (2005)
25. Pass, R.: Simulation in Quasi-Polynomial Time, and its Application to Protocol Composition. In: Biham, E. (ed.) EUROCRYPT 2003. LNCS, vol. 2656, pp. 160–176. Springer, Heidelberg (2003)
26. Pass, R., Venkitasubramaniam, M.: On Constant-Round Concurrent Zero-Knowledge. In: Canetti, R. (ed.) TCC 2008. LNCS, vol. 4948, pp. 553–570. Springer, Heidelberg (2008)
27. Pass, R., Wee, H.: Black-Box Constructions of Two-Party Protocols from One-Way Functions. In: Reingold, O. (ed.) TCC 2009. LNCS, vol. 5444, pp. 403–418. Springer, Heidelberg (2009)
28. Peikert, C., Vaikuntanathan, V., Waters, B.: A Framework for Efficient and Composable Oblivious Transfer. In: Wagner, D. (ed.) CRYPTO 2008. LNCS, vol. 5157, pp. 554–571. Springer, Heidelberg (2008)
29. Prabhakaran, M., Rosen, A., Sahai, A.: Concurrent zero knowledge with logarithmic round-complexity. In: FOCS, pp. 366–375. IEEE Computer Society (2002)
30. Prabhakaran, M., Sahai, A.: New notions of security: achieving universal composability without trusted setup. In: Babai, L. (ed.) STOC, pp. 242–251. ACM (2004)
31. Rabin, M.O.: How to exchange secrets by oblivious transfer. Tech. rep., TR-81, Harvard Aiken Computation Laboratory (1981)

Efficient Secure Primitive for Privacy Preserving Distributed Computations

Youwen Zhu[1,2,⋆], Tsuyoshi Takagi[1], and Liusheng Huang[2]

[1] Institute of Mathematics for Industry, Kyushu University, Fukuoka, 819-0395, Japan
[2] National High Performance Computing Center at Hefei, Department of Computer Science and Technology, University of Science and Technology of China, Hefei, 230026, China
zhuyw@mail.ustc.edu.cn

Abstract. Scalar product protocol aims at securely computing the dot product of two private vectors. As a basic tool, the protocol has been widely used in privacy preserving distributed collaborative computations. In this paper, at the expense of disclosing partial sum of some private data, we propose a linearly efficient Even-Dimension Scalar Product Protocol (EDSPP) without employing expensive homomorphic cryptosystem and third party. The correctness and security of EDSPP are confirmed by theoretical analysis. In comparison with six most frequently-used schemes of scalar product protocol (to the best of our knowledge), the new scheme is a much more efficient one, and it has well fairness. Simulated experiment results intuitively indicate the good performance of our novel scheme. Consequently, in the situations where divulging very limited information about private data is acceptable, EDSPP is an extremely competitive candidate secure primitive to achieve practical schemes of privacy preserving distributed cooperative computations. We also present a simple application case of EDSPP.

Keywords: privacy preserving, distributed computation, scalar product protocol.

1 Introduction

The advances of flexible and ubiquitous transmission mediums, such as wireless networks and Internet, have triggered tremendous opportunities for collaborative computations, where independent individuals and organizations could cooperate with each other to conduct computations on the union of data they each hold. Unfortunately, the collaborations have been obstructed by security and privacy concerns. For example, a single hospital might not have enough cases to analyze some special symptoms and several hospitals need to cooperate with each other to study their joint database of case samples for the comprehensive analysis

⋆ corresponding author.

G. Hanaoka and T. Yamauchi (Eds.): IWSEC 2012, LNCS 7631, pp. 233–243, 2012.

results. A simple way is that they share respective private database and bring the data together in one station for analysis. However, despite various shared benefits, the hospitals may be unwilling to compromise patients' privacy or violate any relevant law and regulation [1,2]. Consequently, some techniques [3,4] for privacy preserving distributed collaborative computations were introduced to address the concerns by privacy advocates. Nowadays, a large amount of attention [5–7] has been paid to dealing with the challenges of how to extract information from distributed data sets owned by independent parties while no privacy is breached.

Actually, many privacy preserving problems in distributed environments can essentially be reduced to securely computing the scalar product of two private vectors. Some recent examples are as follows. Murugesan *et al.* [8] proposed privacy preserving protocols to securely detect similar documents between two parties while documents cannot be publicly disclosed to each other, and the main process of their schemes, securely computing the cosine similarity between two private documents, is achieved by scalar product protocol. A privacy preserving hop-distance computation protocol in wireless sensor networks is introduced in [9] and secure scalar product protocol is used to privately compute the value of $\sum x_i y_i$, where x_i and y_i are the private coordinates. Then, the distance $S^2 = \sum(x_i - y_i)^2 = \sum x_i^2 - 2 * \sum x_i y_i + \sum y_i^2$ can be securely obtained. See [6,7,10,11] for more concrete applications of scalar product protocol.

As secure computation of private vectors is fundamental for many privacy preserving distributed computing tasks, several schemes [12–16] have been proposed to perform the secure computation. Du and Zhan presented two practical schemes in [12]: scalar product protocol employing commodity server (denoted as SPP-CS) and scalar product protocol using random invertible matrix (denoted as SPP-RIM). Through algebraic transformation, another scalar product protocol was introduced in [13] (denoted as ATSPP). Based on homomorphic encryption, two solutions for securely computing dot product of private vectors are given in [14] (denoted as GLLM-SPP) and [15] (denoted as AE-SPP) respectively. A polynomial-based scalar product protocol (denoted as PBSPP) was lately presented by Shaneck and Kim [16]. The computational complexity of SPP-RIM and ATSPP is $O(n^2)$ where n is the dimensionality of private vectors. SPP-CS and PBSPP have good linear complexity, but they employ one or more semi-trusted third parties, such as the commodity server in SPP-CS. GLLM-SPP and AE-SPP encrypt the private elements by using expensive homomorphic cryptosystem. As is well known, the public key cryptosystems are typically computationally expensive and they are far from efficient enough to be used in practice. The protocols will be vulnerable to unavoidable potential collusion attacks while employing the semi-trusted third parties. As a result, previous schemes of scalar product protocol are still far from being practical in most situations.

In this paper, we focus on the useful secure primitive, scalar product protocol [12], and propose a simple and linearly efficient protocol for securely computing the scalar product of two private vectors, even-dimension scalar product protocol (EDSPP). The novel scheme does not employ homomorphic encryption system and any auxiliary third party. Theoretical analysis confirms that the protocol is correct and no private raw data is revealed although it brings about some limited information disclosure. Simulated experiment results and comparison indicate that the new scheme has good fairness and it is much more efficient than the previous ones. As a result, our new scheme is a competitive secure candidate to achieve practical schemes of privacy preserving distributed cooperative computations while disclosing partial information is acceptable. Similar to the existing works [12–16], our protocol is also under semi-honest model [17], where each participant will correctly follow the protocols while trying to find out potentially confidential information from his legal medium records. It is remarkable that the semi-honest assumption is reasonable and practicable, as the participants in reality may strictly follow the protocols to exactly obtain the profitable outputs.

The rest of the paper is organized as follows. Section 2 proposes the new solution for scalar product protocol, and then presents the theoretical analysis of its correctness, security, communication overheads and computation complexity. The performance comparison and experiment results are displayed in section 3. At last, section 4 concludes the paper.

2 Even-Dimension Scalar Product Protocol

2.1 Problem Definition and Our Scheme

In scalar product protocol, there are two participants, denoted as Alice and Bob. Alice privately holds a vector $\boldsymbol{x} = (x_1, x_2, \cdots, x_n)$ and Bob has the other private vector $\boldsymbol{y} = (y_1, y_2, \cdots, y_n)$, where n is a positive integer. Their goal is that Alice receives a confidential number u and Bob obtains his private output v while the private vector is not disclosed to the other party or anyone else. Here, u and v meet $\boldsymbol{x} \cdot \boldsymbol{y} = u + v$. That is, scalar product protocol enables two participants to securely share the dot product of their confidential vectors in the form of addition.

As a secure primitive, scalar product protocol [12, 14] has extensive privacy preserving applications and an efficient scalar product protocol will boost the practical process of privacy preserving distributed cooperative computation. In this paper, we consider a special case where n is an even number (suppose $n = 2k$, k is a positive integer). Then, at the expense of disclosing partial sum of some private data, we propose an efficient Even-Dimension Scalar Product Protocol (EDSPP). In our scheme, the private data is hidden by stochastic transformation, and each participant obtains a private share of the scalar product of their private even-dimension vectors at last. The novel scheme has linear complexity and no third party is employed. Besides, it just needs a secure channel to securely transmit the data and does not use any public key cryptosystem. The detailed steps are displayed in protocol 1. In step 1.1 of the scheme, the participants protect their private numbers through randomization. Then, step 1.2 works out the secure share of the scalar product of each two dimensions. Finally, they privately obtain the expected outcomes in step 2. As can be seen from protocol 1, the private vectors are handled two by two dimensions, thus, our new scheme can only compute the dot product of even-dimension vectors.

Protocol 1. Even-Dimension Scalar Product Protocol (EDSPP)

Input: Alice has a private $2k$-dimension vector $\boldsymbol{x} = (x_1, x_2, \cdots, x_{2k})$ and Bob holds another confidential $2k$-dimension vector $\boldsymbol{y} = (y_1, y_2, \cdots, y_{2k})$. $(k \in \mathbb{Z}^+,\ x_i, y_i \in \mathbb{R}, i = 1, 2, \cdots, 2k)$

Output: Alice obtains private output u and Bob securely gets v which meet

$$u + v = \boldsymbol{x} \cdot \boldsymbol{y} = \sum_{i=1}^{2k} x_i y_i.$$

1: **Step 1**:
2: **for** $j = 1$ to k **do**
3: **Step 1.1**: Alice locally generates two random real numbers a_j and c_j such that $a_j + c_j \neq 0$. Then, she computes $p_j = a_j + c_j$, $x'_{2j-1} = x_{2j-1} + a_j$ and $x'_{2j} = x_{2j} + c_j$, and sends $\{p_j, x'_{2j-1}, x'_{2j}\}$ to Bob by a secure channel. Bob randomly generates two real numbers b_j and d_j which meet $b_j - d_j \neq 0$, and computes $q_j = b_j - d_j$, $y'_{2j-1} = b_j - y_{2j-1}$ and $y'_{2j} = d_j - y_{2j}$. Then, he securely sends $\{q_j, y'_{2j-1}, y'_{2j}\}$ to Alice.
4: **Step 1.2**: Alice locally calculates

$$u_j = y'_{2j-1}(x_{2j-1} + 2a_j) + y'_{2j}(x_{2j} + 2c_j) + q_j(a_j + 2c_j)$$

and Bob, by himself, computes

$$v_j = x'_{2j-1}(2y_{2j-1} - b_j) + x'_{2j}(2y_{2j} - d_j) + p_j(d_j - 2b_j).$$

5: **end for**
6: **Step 2**: Alice obtains $u = \sum_{j=1}^{k} u_j$ and Bob gets $v = \sum_{j=1}^{k} v_j$.

To visually illustrate how our novel scheme works, we give a concrete example as follows. Alice has a 4-dimension vector $\boldsymbol{x} = (2.3, -81.9, 96.7, -27.1)$, and Bob's private vector is $\boldsymbol{y} = (-19.5, -78.1, 39.2, 52.8)$. According to protocol 1, they, by the following procedures, can obtain the scalar product's private shares u and v, which meet $u + v = \boldsymbol{x} \cdot \boldsymbol{y}$, respectively.

- Alice generates random numbers: $a_1 = -53.0$ and $c_1 = 99.8$ for the first two dimensions of $\boldsymbol{x}$. Then, she computes
  $p_1 = a_1 + c_1 = 46.8$, $x'_1 = 2.3 + a_1 = -50.7$, $x'_2 = -81.9 + c_1 = 17.9$,
  and sends $\{p_1, x'_1, x'_2\}$ to Bob. At the same time, Bob randomly selects: $b_1 = 28.7$ and $d_1 = 11.3$ for the first two dimensions of $\boldsymbol{y}$. Then, he computes
  $q_1 = b_1 - d_1 = 17.4$, $y'_1 = b_1 - (-19.5) = 48.2$, $y'_2 = d_1 - (-78.1) = 89.4$,
  and sends $\{q_1, y'_1, y'_2\}$ to Alice.
- Analogously, for the latter two dimensions, Alice and Bob generates random numbers $\{a_2 = -81.1,\ c_2 = -17.5\}$ and $\{b_2 = -56.9,\ d_2 = -31.2\}$, respectively. Alice computes $p_2 = -98.6$, $x'_3 = 15.6$, $x'_4 = -44.6$, and Bob

computes $q_2 = -25.7$, $y'_3 = -96.1$, $y'_4 = -84.0$. Then, they send $\{p_2, x'_3, x'_4\}$ and $\{q_2, y'_3, y'_4\}$ to each other.

- Alice and Bob computes $\{u_1, u_2\}$ and $\{v_1, v_2\}$, respectively, by the following way.

$$u_1 = y'_1(x_1 + 2a_1) + y'_2(x_2 + 2c_1) + q_1(a_1 + 2c_1) = 8074.88$$
$$u_2 = y'_3(x_3 + 2a_2) + y'_4(x_4 + 2c_2) + q_2(a_2 + 2c_2) = 14494.72$$
$$v_1 = x'_1(2y_1 - b_1) + x'_2(2y_2 - d_1) + p_1(d_1 - 2b_1) = -1723.34$$
$$v_2 = x'_3(2y_3 - b_2) + x'_4(2y_4 - d_2) + p_2(d_2 - 2b_2) = -12134.96$$

- At last, Alice obtains the secure share $u = u_1 + u_2 = 22569.6$, and Bob gets his private output $v = u_1 + u_2 = -13858.3$.

If we directly calculates the dot product of $\boldsymbol{x}$ and $\boldsymbol{y}$, it is $2.3*(-19.5)+(-81.9)*(-78.1)+96.7*39.2+(-27.1)*52.8 = 8711.3$ which is exactly equal to the sum of $u = 22569.6$ and $v = -13858.3$. It shows the above steps are correct.

2.2 Correctness Analysis

To confirm the correctness of EDSPP, we need to consider,

Theorem 1. *After performing* EDSPP, *Alice's private output u and Bob's secret output v meet $u + v = \boldsymbol{x} \cdot \boldsymbol{y} = \sum_{i=1}^{2k} x_i y_i$. That is,* EDSPP *is correct.*

Proof. In step 1.1 of EDSPP, there are $x'_{2j-1} = x_{2j-1} + a_j$, $x'_{2j} = x_{2j} + c_j$, $p_j = a_j + c_j$, $y'_{2j-1} = b_j - y_{2j-1}$, $y'_{2j} = d_j - y_{2j}$ and $q_j = b_j - d_j$. Then,

$$x'_{2j-1}(2y_{2j-1} - b_j) = 2x_{2j-1}y_{2j-1} - b_j x_{2j-1} + 2a_j y_{2j-1} - a_j b_j,$$
$$x'_{2j}(2y_{2j} - d_j) = 2x_{2j}y_{2j} - d_j x_{2j} + 2c_j y_{2j} - c_j d_j,$$
$$p_j(d_j - 2b_j) = a_j d_j - 2a_j b_j + c_j d_j - 2b_j c_j,$$
$$y'_{2j-1}(x_{2j-1} + 2a_j) = b_j x_{2j-1} + 2a_j b_j - x_{2j-1}y_{2j-1} - 2a_j y_{2j-1},$$
$$y'_{2j}(x_{2j} + 2c_j) = d_j x_{2j-1} + 2c_j d_j - x_{2j}y_{2j} - 2c_j y_{2j},$$
$$q_j(a_j + 2c_j) = a_j b_j + 2b_j c_j - a_j d_j - 2c_j d_j.$$

According to step 1.2, we have $u_j = y'_{2j-1}(x_{2j-1}+2a_j)+y'_{2j}(x_{2j}+2c_j)+q_j(a_j+2c_j)$ and $v_j = x'_{2j-1}(2y_{2j-1} - b_j) + x'_{2j}(2y_{2j} - d_j) + p_j(d_j - 2b_j)$. Thus,

$$u_j + v_j = x_{2j-1}y_{2j-1} + x_{2j}y_{2j}. \tag{1}$$

There are $u = \sum_{j=1}^{k} u_j$ and $v = \sum_{j=1}^{k} v_j$ in step 2, then, $u+v = \sum_{j=1}^{k}(u_j+v_j) = \sum_{j=1}^{k}(x_{2j-1}y_{2j-1} + x_{2j}y_{2j})$. Therefore,

$$u + v = \sum_{i=1}^{2k} x_i y_i \tag{2}$$

That is, $u + v = \boldsymbol{x} \cdot \boldsymbol{y}$ holds at the end of EDSPP, which completes the proof.

2.3 Security Analysis

In this subsection, we will analysis the security of EDSPP under semi-honest model [17], where each participant correctly follow the protocol while trying to find out potentially confidential information from his legal medium records. Generally, we consider the view of each participant in this protocol and whether some privacy can be deduced from the view.

During the execution of EDSPP, Alice receives y'_{2j-1}, y'_{2j} and q_j, symmetrically, Bob learns x'_{2j-1}, x'_{2j} and p_j.

From y'_{2j-1} and y'_{2j}, Alice cannot learn any information about y_{2j-1} and y_{2j}. While q_j is known to her, the sum of $-y_{2j-1}$ and y_{2j} will be derived by $y_{2j} - y_{2j-1} = y'_{2j-1} - y'_{2j} - q_j$, however, Bob's private numbers y_{2j-1} and y_{2j} are still unrevealed. Analogously, Bob can figure out $x_{2j-1} + x_{2j} = x'_{2j-1} + x'_{2j} - p_j$, while he cannot obtain any more information about Alice's privacy x_{2j-1} and x_{2j}. Therefore, each real element of the private vectors of both participants is not disclosed in EDSPP. If the elements of the vectors are 0 or 1, EDSPP is not secure. GLLM-SPP [14] is more fit for securely computing the scalar product of binary vectors.

Quantification of Disclosure Level. Here, we give the quantification of disclosure level about Alice's private data x_{2j-1} and x_{2j}. While EDSPP has been applied, if $T = x'_{2j-1} + x'_{2j} - p_j$, then, Bob learns that (x_{2j-1}, x_{2j}) is randomly located at the line $T = x_{2j-1} + x_{2j}$, the slope of which is exactly equal to -1.

(1) While $x_{2j-1}, x_{2j} \in \mathbb{R}$, that is, before EDSPP being applied, according to Bob's view, (x_{2j-1}, x_{2j}) is randomly located at two-dimensional real space $\mathbb{R}^2$. After EDSPP, the distribution space of (x_{2j-1}, x_{2j}) is reduced to a line. However, as both x_{2j-1} and x_{2j} are random in Bob's view, then, he cannot extract the original private numbers x_{2j-1} and x_{2j} from their sum $T = x'_{2j-1} + x'_{2j} - p_j$.

(2) While $L \leqslant x_{2j-1}, x_{2j} \leqslant U$ $(L < U)$, then, before EDSPP, (x_{2j-1}, x_{2j}) is randomly located at a $(U - L) \times (U - L)$-square area in Bob's view. At the end of EDSPP, Bob can figure out $T = x'_{2j-1} + x'_{2j} - p_j$ which is equal to $x_{2j-1} + x_{2j}$. Furthermore, $x_{2j-1} = T - x_{2j}$ and $x_{2j} = T - x_{2j-1}$, thus, Bob knows $T - U \leqslant x_{2j-1}, x_{2j} \leqslant T - L$. Then, he obtains

$$\max\{L, T - U\} \leqslant x_{2j-1}, x_{2j} \leqslant \min\{U, T - L\}.$$

According to the range of x_{2j-1} and x_{2j}, it is easy to get $2L \leqslant T \leqslant 2U$.

If $2L \leqslant T < L + U$, then, $\max\{L, T - U\} = L$ and $\min\{U, T - L\} = T - L$. Therefore, Bob can find out $L \leqslant x_{2j-1}, x_{2j} \leqslant T - L$.

If $L + U \leqslant T \leqslant 2U$, then, $\max\{L, T - U\} = T - U$ and $\min\{U, T - L\} = U$. In Bob's view, there will be $T - U \leqslant x_{2j-1}, x_{2j} \leqslant U$.

In this situation, Bob can obtain a more narrow range about x_{2j-1} and x_{2j}, but he cannot exactly deduce the value of them except the following two extreme cases: $x_{2j-1} = x_{2j} = L, T = 2L$ and $x_{2j-1} = x_{2j} = U, T = 2U$.

In general, the new scheme sacrifices some security in a certain level, but the private raw data is still protected especially when the elements of the private vectors are real number. Alice and Bob disclose nothing but the sum $x_{2j-1}+x_{2j}$, $y_{2j-1}+y_{2j}$ to each other in EDSPP. Besides, two participants carry out symmetric computations, send and receive symmetrical data, consequently, EDSPP is quite fair.

2.4 Communication Overheads and Computational Complexity

The following contributes to the computational cost: (1) In step 1.1 of EDSPP, Alice and Bob respectively generate two random number and perform three additions. In step 1.2, each party performs three multiplications and two additions. All the above operations loop for k times. (2) In step 2, they each carry out $k-1$ additions.

Therefore, the computational complexity of EDSPP is $O(n)$ in total. Here, n is the dimension number of their private vectors and $n = 2k$ in the protocol.

The transmitting data contains $x'_{2j-1}, x'_{2j}, p_j, y'_{2j-1}, y'_{2j}$ and q_j $(j = 1, 2, \cdots, k)$ in EDSPP. Thus, the total communication overheads are $3nb_0$ bits $(n = 2k)$. Here, b_0 is the bit length of a message.

2.5 A Simple Application Case

In many privacy-preserving distributed computations [18, 19], a key step is to securely find out which one of the points holden by one party is nearest to another point of the other participant. For simplicity, we deal with the problem that Alice has two private points $\mathrm{P}_1(P_{11}, P_{12}, \cdots, P_{1d})$ and $\mathrm{P}_2(P_{21}, P_{22}, \cdots, P_{2d})$, and Bob privately holds another point $\mathrm{Q}(Q_1, Q_2, \cdots, Q_d)$. They want to find out which one of P_1 and P_2 is closer to Q without disclosing the private coordinates of each point to each other or anybody else. Here, we use the scalar product of the coordinates as the distance of two points, that is, $|\mathrm{P}_i\mathrm{Q}| = \sum_{j=1}^{d} P_{ij}Q_j \ \ (i = 1, 2)$. In fact, comparison of distances measured by other metrics, such as Euclidean distance and consine similarity, can be easily transferred into comparison of the dot products. Based on EDSPP, we present a simple but efficient solution for the above problem.

- Alice locally generates a random positive real numbers α and d random real numbers $r_1, r_2, \cdots, r_d$. Then, she sets the $2d$-dimensional vectors

$$\boldsymbol{P}'_i = (\alpha P_{i1}, r_1, \alpha P_{i2}, r_2, \cdots, \alpha P_{id}, r_d), \quad (i = 1, 2).$$

  Bob randomly generates a random positive real numbers β and d random real numbers $R_1, R_2, \cdots, R_d$, and computes his private $2d$-dimensional vector by the following way

$$\boldsymbol{Q}' = (\beta Q_1, R_1, \beta Q_2, R_2, \cdots, \beta Q_d, R_d).$$

- Alice and Bob collaboratively perform EDSPP such that Alice obtains U_1, U_2 and Bob gets his private outputs V_1, V_2 which meet $U_i + V_i = \boldsymbol{P}'_i \cdot \boldsymbol{Q}'$ $(i = 1, 2)$.

– At last, Alice sends $\delta = U_1 - U_2$ to Bob. Then Bob computes $\Delta = \delta + V_1 - V_2$ and finds out the closer one by comparing Δ with 0.

In the above scheme, we can obtain

$$\Delta = (U_1 + V_1) - (U_2 + V_2) = \boldsymbol{P}_1' \cdot \boldsymbol{Q}' - \boldsymbol{P}_2' \cdot \boldsymbol{Q}' = \alpha\beta(|\mathrm{P}_1\mathrm{Q}| - |\mathrm{P}_2\mathrm{Q}|).$$

Thus, if $\Delta > 0$, P_2 is closer to Q; otherwise, P_1 is closer to Q.

Table 1. Comparison between EDSPP and Existing Schemes

Protocols	**Computational Complexity**	**Employ Third Party?**	**Security**	**Fairness**
GLLM-SPP [14]	$O(n * \mathcal{H})^*$	No	CR-sec**	Very Bad
AE-SPP [15]	$O(n * \mathcal{H})^*$	No	CR-sec**	Good
SPP-RIM [12]	$O(n^2)$	No	L-dis**	Bad
ATSPP [13]	$O(n^2)$	No	L-dis**	Good
SPP-CS [12]	$O(n)$	Yes	IT-sec**	Good
PBSPP [16]	$O(n)$	Yes	IT-sec**	Good
EDSPP	$O(n)$	No	L-dis**	Good

* Suppose the computational complexity of an encryption by homomorphic cryptosystem is $O(\mathcal{H})$. n is the dimension of private vectors.
** Here, IT-sec denotes "information-theoretically secure", CR-sec denotes "the security based on the intractability of the composite residuosity class problem", and L-dis denotes that the scheme will result in limited disclosure about private information of participants. SPP-CS and PBSPP are vulnerable to collusion attacks, though the schemes have the security based on information theory.

3 Performance Comparison and Experiment Results

The communication overheads of EDSPP and each previous scheme are $O(n)$, to demonstrate the special features of EDSPP, we compare it with six most frequently-used schemes (to the best of our knowledge) in table 1. It indicates that EDSPP has the best performance in many aspects except for the security. SPP-CS [12] and PBSPP [16] have the same linear computational complexity as EDSPP, but SPP-CS and PBSPP employ one or more semi-trusted third parties, which results in that they are extremely vulnerable to unavoidable potential collusion attacks. While the third party colludes with one party, the other participant's privacy will be seriously breached. The computational complexity of SPP-RIM [12] and ATSPP [13] are $O(n^2)$ which is bigger than that of EDSPP. GLLM-SPP [14] and AE-SPP [15] use the expensive homomorphic cryptosystem. Additionally, participants execute very similar operations in EDSPP, thus, the scheme has good fairness. In GLLM-SPP [14] the participant, who generates the homomorphic encryption system and encrypts each element of his private vector, will load much more computation and communication than the other one, thus the fairness of GLLM-SPP is very bad.

We implement three most computationally efficient schemes, SPP-CS, PBSPP and EDSPP. In the experiments, each participant is performed on a computer with Intel Core2 Duo 2.93GHz CPU and 2.0GB memory, and the average **ping** time of them is shorter than 1 ms. Figure 1 exhibits the simulated results, which indicates that all the runtime linearly increase with dimension and EDSPP costs least time. While the vectors' dimension are 200 ($k = 100$), the total running time of EDSPP is only a little more than 100 ms which is less than one-third of that of PBSPP and is about one-sixth of the running time cost by SPP-CS.

In summary, the comparative advantages of EDSPP are its simpleness, linear efficiency, good fairness and it does not employ the expensive homomorphic cryptosystem and any auxiliary third party. As ideal security is too expensive to achieve, especially in large-scale systems, and it may be unnecessary in practice, if disclosing partial information about private data is still acceptable, EDSPP will be a competitive low-cost candidate secure primitive for privacy preserving distributed collaborative computations.

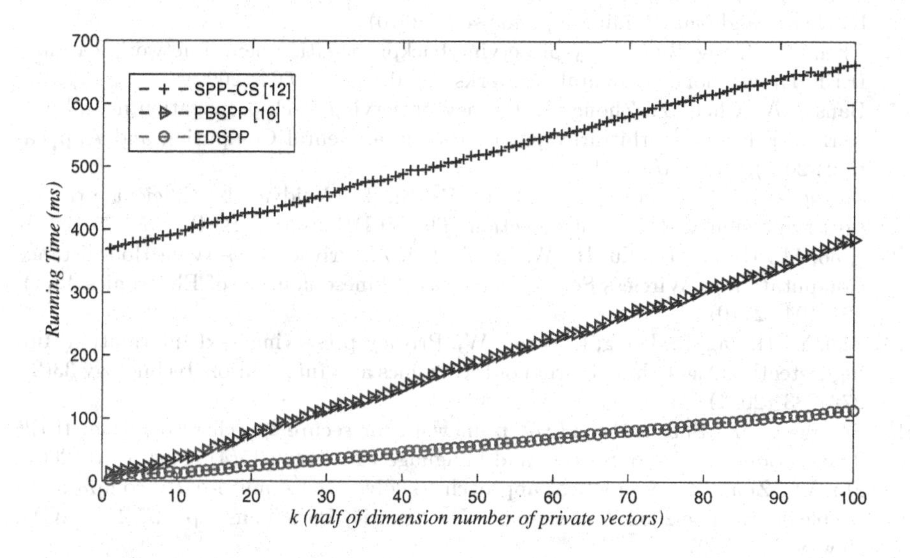

Fig. 1. Running Time of SPP-CS [12], PBSPP [16] and EDSPP ($ms = 10^{-3}s$, the private vectors' dimension $n = 2k$)

4 Conclusion

In this paper, a linearly efficient scheme for scalar product protocol, EDSPP, has been proposed. The protocol has no use of expensive homomorphic crypto-system and third party, which have been employed by existing solutions. Theoretical analysis and simulated experiment results confirm that the novel scheme is a competitive candidate for securely computing the scalar product of two private vectors.

Acknowledgement. This work was supported by the Japan Society for the Promotion of Science fellowship (No. P12045) , the National Natural Science Foundation of China (Nos. 60903217 and 60773032) and the Natural Science Foundation of Jiangsu Province of China (No. BK2010255).

References

1. HIPAA. The health insurance portability and accountability act of 1996 (October 1998), http://www.ocius.biz/hipaa.html
2. Cios, K.J., Moore, G.W.: Uniqueness of medical data mining. Artificial Intelligence in Medicine 26(1-2), 1–24 (2002)
3. Agrawal, R., Srikant, R.: Privacy-preserving data mining. ACM Sigmod Record 29, 439–450 (2000)
4. Lindell, Y., Pinkas, B.: Secure multiparty computation for privacy-preserving data mining. Journal of Privacy and Confidentiality 1(1), 59–98 (2009)
5. Yang, B., Nakagawa, H., Sato, I., Sakuma, J.: Collusion-resistant privacy-preserving data mining. In: The 16th ACM SIGKDD Conference on Knowledge Discovery and Data Mining, pp. 483–492 (2010)
6. Chen, T., Zhong, S.: Privacy-preserving backpropagation neural network learning. IEEE Transactions on Neural Networks 20(10), 1554–1564 (2009)
7. Bansal, A., Chen, T., Zhong, S.: Privacy preserving Back-propagation neural network learning over arbitrarily partitioned data. Neural Computing and Applications 20(1), 143–150 (2011)
8. Murugesan, M., Jiang, W., Clifton, C., Si, L., Vaidya, J.: Efficient privacy-preserving similar document detection. The VLDB Journal 19(4), 457–475 (2010)
9. Xiao, M., Huang, L., Xu, H., Wang, Y., Pei, Z.: Privacy Preserving Hop-distance Computation in Wireless Sensor Networks. Chinese Journal of Electronics 19(1), 191–194 (2010)
10. Zhu, Y., Huang, L., Dong, L., Yang, W.: Privacy-preserving Text Information Hiding Detecting Algorithm. Journal of Electronics and Information Technology 33(2), 278–283 (2011)
11. Smaragdis, P., Shashanka, M.: A framework for secure speech recognition. IEEE Transactions on Audio, Speech, and Language Processing 15(4), 1404–1413 (2007)
12. Du, W., Zhan, Z.: A practical approach to solve secure multi-party computation problems. In: The 2002 Workshop on New Security Paradigms, pp. 127–135. ACM, New York (2002)
13. Vaidya, J., Clifton, C.: Privacy preserving association rule mining in vertically partitioned data. In: The 8th ACM SIGKDD International Conference on Knowledge Discovery and Data Mining, pp. 639–644. ACM, New York (2002)
14. Goethals, B., Laur, S., Lipmaa, H., Mielikäinen, T.: On Private Scalar Product Computation for Privacy-Preserving Data Mining. In: Park, C.-s., Chee, S. (eds.) ICISC 2004. LNCS, vol. 3506, pp. 104–120. Springer, Heidelberg (2005)
15. Amirbekyan, A., Estivill-Castro, V.: A new efficient privacy-preserving scalar product protocol. In: The Sixth Australasian Conference on Data Mining and Analytics, vol. 70, pp. 209–214. Australian Computer Society (2007)

16. Shaneck, M., Kim, Y.: Efficient Cryptographic Primitives for Private Data Mining. In: The 2010 43rd Hawaii International Conference on System Sciences, pp. 1–9. IEEE Computer Society (2010)
17. Goldreich, O.: Foundations of Cryptography: Volume II, Basic Applications. Cambridge University Press, Cambridge (2004)
18. Qi, Y., Atallah, M.J.: Efficient privacy-preserving k-nearest neighbor search. In: The 28th International Conference on Distributed Computing Systems (ICDCS 2008), pp. 311–319. IEEE (2008)
19. Shaneck, M., Kim, Y., Kumar, V.: Privacy preserving nearest neighbor search. In: 6th IEEE International Conference on Data Mining Workshops, pp. 541–545. IEEE (2006)

Generic Construction of GUC Secure Commitment in the KRK Model

Itsuki Suzuki, Maki Yoshida, and Toru Fujiwara

Graduate School of Information Science and Technology, Osaka University
1-5 Yamadaoka, Suita, Osaka 565-0871, Japan

Abstract. This paper proposes a generic construction of GUC secure commitment against static corruptions in the KRK (Key Registration with Knowledge) model. The GUC security is a generalized version of universally composable security which deals with global setup used by arbitrary many protocols at the same time. The proposed construction is the first GUC secure protocol in which the commit phase is non-interactive (whereas the reveal phase is interactive). Thus, the proposed construction is suitable for applications where many values are committed to a few receivers within a short time period. The proposed construction uses simple tools, a public key encryption (PKE) scheme, a Sigma protocol, a non-interactive authenticated key exchange (NI-AKE) scheme, a message authentication code (MAC), for which efficient constructions have been presented. For the sake of simplicity of the proposed construction, which uses GUC secure authenticated communication (constructed from MAC and NI-AKE), we have not achieve full adaptive security because GUC secure authenticated communication in the KRK model is impossible.

Keywords: Commitment, GUC security, KRK, Static adversary.

1 Introduction

Commitment protocols are one of the most important components for cryptographic protocols. Thus, commitment protocols need providing universal composablity like Universally Composable (UC) security [3] and Generalized UC (GUC) security [6]. UC/GUC-secure protocols guarantee security even when the protocols are run concurrently with arbitrarily many other protocols. GUC-secure protocols further guarantee security even when the used setup is a global one accessed by arbitrary many protocols. For protocols with global setup, GUC is significantly stronger. In fact, while UC-secure commitment protocols in the Common Reference String (CRS) model are presented in [2,4,5,8,11,13,14,15], realizing GUC-secure commitment is provably impossible in the CRS model [6]. A globally available setup that can be used through the system is realistic and convenient. Thus, alternative and reasonable setup assumptions, called the key registration with knowledge (KRK) model and the augmented CRS (ACRS) model, are presented in [6], and on the ACRS model, GUC-secure commitment protocols are presented in [6,12].

G. Hanaoka and T. Yamauchi (Eds.): IWSEC 2012, LNCS 7631, pp. 244–260, 2012.

Table 1. Comparison of the previous GUC secure protocols, Lindell's UC secure protocols, and the proposed GUC secure protocol

	Security notion	Setup assum.	Adaptive security	Non-inter. commit	Non-inter. reveal
CDPW07 [6]	GUC	ACRS	√		√
DSW08 [12]	GUC	ACRS	√		√
Lin11 [14]	UC	CRS	√	√	
Proposed construction	GUC	KRK		√	

The common feature of the previous GUC-secure commitment protocols in [6,12] is that the commit phase is interactive (whereas the reveal phase is non-interactive). Thus, the previous constructions are suitable for applications where many values need revealing within a short time period. Considering a possibility of applications where commitments rush into a few receivers like e-voting, commitment protocols with non-interactive commit phase are also needed.

This paper presents the first GUC secure commitment protocol with non-interactive commit phase. Our approach is to extend an existing UC secure commitment protocol with non-interactive commit phase to GUC secure one. In [14], Lindell presented a generic construction of UC-secure commitment with non-interactive commit phase and its highly-efficient implementations under the standard DDH assumption. One of the main advantages of Lindell's construction is its conceptual simplicity (as mentioned by the author himself). Thus, we aim to extend Lindell's UC-secure generic construction while keeping its simplicity.

In both UC and GUC frameworks, proving the security of a protocol in some setup is to show how to simulate information that could be obtained via a real attack on the protocol. While in the UC framework a simulator can freely generate the setup information and use its trapdoor information, in the GUC framework a simulator cannot use the trapdoor since the setup information is given as global one. Thus, we need to establish a mechanism to simulate the protocol without the trapdoor. In the security proof of Lindell's construction, the commit phase is simulated only with the public parameter (i.e., CRS). However, in the reveal phase, the simulator uses the trapdoor of the CRS for attacks to impersonate the receiver without corrupting. Our idea to simulate the reveal phase without the trapdoor is to add an authentication mechanism in order to detect the impersonation and terminate the execution. Thus, the simulator does not need to use the trapdoor. As a GUC-secure authentication mechanism, we use the protocol in the KRK model presented by Dodis et al. in [10], which is constructed from simple tools, message authentication code (MAC) and non-interactive authenticated key exchange (NI-AKE) scheme. We note that GUC-secure authenticated communication against adaptive corruptions in the KRK model is impossible [10].

As a result, the proposed generic construction is GUC secure against static corruptions in the KRK model whereas Lindell's UC-secure constructions in [14] and the previous GUC-secure constructions in [6,10] are secure against adaptive corruptions. The advantage of the proposed construction is a conceptual simplicity same as Lindell's one: only MAC secure against one-time chosen message attack and NI-AKE are used in addition to the building blocks of Lindell's construction, a CCA2-secure public-key encryption scheme, a dual mode cryptosystem, and a Sigma protocol.

The rest of this paper is organized as follows. This paper uses a simplified variant of GUC security called Externalized UC (EUC) presented in [6] because its equivalence is proved. Sect. 2 shows the definitions of GUC/EUC framework and building blocks used in this paper. Sect. 3 overviews Lindell's construction and shows our idea. In Sect. 4, we present a generic construction of EUC secure commitment protocols and a security proof. Sect. 5 concludes this paper.

2 Definitions

When A is a random variable or distribution, $y \leftarrow A$ denotes that y is randomly selected from A according to its distribution. A function $f : \mathbb{N} \rightarrow \mathbb{R}$ is negligible in k if for all polynomial q, and all large k, $f(k) \leq \frac{1}{q(k)}$. If f is negligible in k, we write $f \leq neg(k)$.

2.1 Generalized Universally Comosable (GUC) Security

Generalized Universally Composable (GUC) security is an extension of UC security [6], that considers an execution of a protocol in a setting involving a global setup modeled by a shared functionality $\overline{\mathcal{G}}$, that is accessible by an environment $\mathcal{Z}$, in addition to the honest parties and adversary. As with the definition of UC security, ideal and real models are considered where the real protocol is run in the real model and a trusted party carries out the computation in the ideal model. Here, the trusted party is modeled by an ideal functionality $\mathcal{F}$. In the ideal model, an ideal protocol $\text{IDEAL}_{\mathcal{F}}$ is run. Parties running $\text{IDEAL}_{\mathcal{F}}$ simply forward their inputs to $\mathcal{F}$ and output any message received from $\mathcal{F}$. The essential difference of GUC security from UC security is that $\mathcal{Z}$ is allowed to invoke any party of multiple concurrent instances of the challenge protocol and other protocols. Such an environment is called unconstrained. The unconstrained environment $\mathcal{Z}$ invokes a shared functionality, chooses the inputs for the honest parties invoked by $\mathcal{Z}$, interacts with the adversary throughout the computation, and receives the honest parties' outputs. The adversary $\mathcal{A}$ can read all message sent by the parties and send arbitrary messages to any party. The simulator $\mathcal{S}$ whereas may not interact with the parties, but interacts with $\mathcal{F}$. The environment $\mathcal{Z}$ outputs a single bit when it halts. Let $\text{GEXEC}^{\overline{\mathcal{G}}}_{\pi,\mathcal{A},\mathcal{Z}}$ denote the output of the unconstrained environment $\mathcal{Z}$ when $\mathcal{Z}$ runs with π and $\mathcal{A}$. Security is formulated by requiring the existence of an ideal model simulator $\mathcal{S}$ so that no environment $\mathcal{Z}$ can distinguish between the case that it runs with $\mathcal{A}$ in the real model and the case that it runs with $\mathcal{S}$ in the ideal model.

Definition 1 (GUC-Emulation [6]). *Let k be security parameter. Let π and ϕ be PPT multi-party protocols. π is said to be GUC-emulating ϕ if, for any PPT adversary $\mathcal{A}$, there exists a PPT simulator $\mathcal{S}$, for any unconstrained PPT*

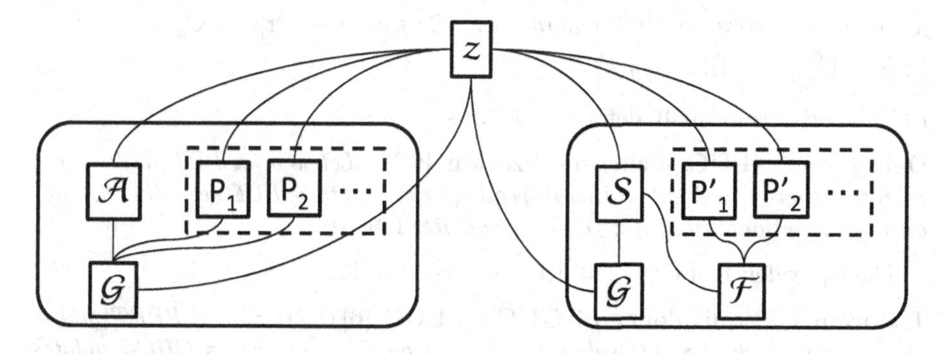

Fig. 1. Overview of EUC framework: The solid lines represent interactions by local input and output. The dashed box represent that $\mathcal{A}$ or $\mathcal{S}$ controls interaction between parties of the protocol.

environment $\mathcal{Z}$, *we have* $|\Pr[\text{GEXEC}_{\phi,\mathcal{S},\mathcal{Z}} = 1] - \Pr[\text{GEXEC}_{\pi,\mathcal{A},\mathcal{Z}} = 1]| \leq$ *neg*(k). □

In this paper, we restrict adversaries to *static* ones, which are not allowed to corrupt new parties during protocol execution.

GUC security is formally defined as follows.

Definition 2 (GUC secure realization [6]). *Let* π *be a PPT multi-party protocol.* π *is said to be GUC securely realizing an ideal functionality* $\mathcal{F}$ *if* π *GUC-emulates* $\text{IDEAL}_{\mathcal{F}}$. □

2.2 Externalized Universally Composable (EUC) Security

Externalized Universally Composable (EUC) security is a simplified variant of GUC security [6], that considers only single instance of the challenge protocol, that is, $\mathcal{Z}$ for EUC is only allowed to invoke parties of a single instance of the challenge protocol. Figure 1 depicts the overview of the EUC framework.

Before showing the definition of EUC security, we show some terminologies defined in [6]. A protocol instance M is said to be a subroutine of another instance M' if M either receives inputs from M', or outputs the message to M'. Recursively, M is said to be a sub-party of protocol π if M is a subroutine of a party running π or a sub-party of π. π is said to be $\overline{\mathcal{G}}$-subroutine respecting if none of the sub-parties of an instance of π provides outputs to or receives inputs from any instance that is not also party/sub-party of that instance of π, except for communicating with a single instance of $\overline{\mathcal{G}}$.

Definition 3 (EUC-Emulation [6]). *Let* k *be security parameter. Let* π *and* ϕ *be PPT multi-party protocols, where* π *is* $\overline{\mathcal{G}}$*-subroutine respecting. Let* $\text{EXEC}^{\overline{\mathcal{G}}}_{\pi,\mathcal{A},\mathcal{Z}}$ *denote the output of* $\mathcal{Z}$ *when* $\mathcal{Z}$ *runs with* π *and* $\mathcal{A}$. π *is said to be* $\overline{\mathcal{G}}$*-EUC-emulating* ϕ *if, for any PPT adversary* $\mathcal{A}$, *there exists a PPT simulator* $\mathcal{S}$, *for any unconstrained PPT environment $\mathcal{Z}$, we have* $|\Pr[\mathrm{EXEC}^{\overline{\mathcal{G}}}_{\pi,\mathcal{S},\mathcal{Z}} = 1] - \Pr[\mathrm{REAL}^{\overline{\mathcal{G}}}_{\phi,\mathcal{A},\mathcal{Z}} = 1]| \leq neg(k)$. □

EUC security is formally defined as follows.

Definition 4 (EUC secure realization [6]). *Let π be a PPT multi-party protocol, where π is $\overline{\mathcal{G}}$-subroutine respecting. π is said to be EUC securely realizing an ideal functionality $\mathcal{F}$ if π $\overline{\mathcal{G}}$-EUC-emulates* $\mathrm{IDEAL}_{\mathcal{F}}$. □

The equivalence of GUC and EUC is proved in [6].

Theorem 1 (Equivalence of GUC to EUC [6]). *Let π be a PPPT multi-party protocol, where π is $\overline{\mathcal{G}}$-subroutine respecting. Then protocol π GUC-emulates a protocol ϕ, if and only if π $\overline{\mathcal{G}}$-EUC-emulates ϕ.* □

In [6], the following generalized universally composition theorem is proved.

Theorem 2 (Generalized universally composition [6]). *Let ρ, π, ϕ be PPT multi-party protocols, and such that both ϕ and π are $\overline{\mathcal{G}}$-subroutine respecting, and π $\overline{\mathcal{G}}$-EUC-emulates ϕ. Let $\rho^{\pi/\phi}$ denote a modified version of ρ that invokes π instead of ϕ. Then $\rho^{\pi/\phi}$ GUC-emulates ρ.* □

2.3 Functionalities

In this paper, we present a protocol which GUC securely realizes the multi-commitment ideal functionality $\mathcal{F}_{mcom}$ in the key registration with knowledge (KRK) model. We recall the definitions of $\mathcal{F}_{mcom}$ and the KRK model.

Definition 5 (The ideal commitment functionality $\mathcal{F}_{mcom}$ [14]). *$\mathcal{F}_{mcom}$ proceeds as follows, running with parties $P_1, \ldots, P_m$, a parameter 1^n, and an adversary $\mathcal{S}$.*

- *Commit phase: Upon receiving a message* (commit, $sid, ssid, P_i, P_j, x$) *from P_i where $x \in \{0,1\}^{n \log^2 n}$, record the tuple ($ssid, P_i, P_j, x$) and send the messages* (receipt, $sid, ssid, P_i, P_j$) *to $\mathcal{S}$, and, after a delay, provide the same output to P_j. Ignore any future commit messages with the same ssid from P_i to P_j.*
- *Reveal phase: Upon receiving a message* (reveal, $sid, ssid$) *from P_i: If a tuple ($ssid, P_i, P_j, x$) was previously recorded, then send the message* (reveal, $sid, ssid, P_i, P_j, x$) *to $\mathcal{S}$ and, after a delay, provide the same output to P_j. Otherwise, ignore.* □

In the KRK model, the shared functionality $\overline{\mathcal{G}}^{\Pi}_{krk}$ chooses a private and public key pair for each registered party, and lets all parties know the public key. And parties can obtain their own secret keys. $\overline{\mathcal{G}}^{\Pi}_{krk}$ is defined as follows.

Definition 6 (The Π-key registration with knowledge shared functionality $\overline{\mathcal{G}}^{\Pi}_{krk}$ [6]). *$\overline{\mathcal{G}}^{\Pi}_{krk}$ proceeds as follows, given a (deterministic) key generation function* KRK.Gen *(with security parameter λ), running with parties $P_1, \ldots, P_n$ and an adversary $\mathcal{S}$.*

- *Registration: When receiving a message* (register) *from an honest party* P_i *that has not previously registered, sample* $r \leftarrow \{0,1\}^\lambda$*, then compute* $(PK_i, SK_i) \leftarrow \mathsf{KRK.Gen}^\lambda(r)$ *and record the tuple* (P_i, PK_i, SK_i)*.*
- *Corrupt Registration: When receiving a message* (register, r) *from a corrupt party* P_i *that has not registered, compute* $(PK_i, SK_i) \leftarrow \mathsf{KRK.Gen}^\lambda(r)$ *and record the tuple* (P_i, PK_i, SK_i)*.*
- *Public Key Retrieval: When receiving a message* (retrieve, P_i) *from any party* P_j *(where* $i = j$ *is allowed), if there is a previously recorded tuple of the form* (P_i, PK_i, SK_i)*, then return* (P_i, PK_i) *to* P_j*. Otherwise, return* $(P_i, \bot)$ *to* P_j*.*
- *Secret Key Retrieval: When receiving a message* (retrievesecret, P_i) *from a party* P_i *that is either corrupt or honestly running the protocol code for* Π*, if there is a previously recorded tuple of the form* (P_i, PK_i, SK_i) *then return* (P_i, PK_i, SK_i) *to* P_i*. In all other cases, return* $(P_i, \bot)$*.* □

2.4 Building Blocks

We show the definitions of the building blocks used in this paper.

Definition 7 (Message authentication functionality $\mathcal{F}_{auth}$ [10]). $\mathcal{F}_{auth}$ *proceeds as follows, running with a sender* S*, a receiver* R*, and an adversary* $\mathcal{S}$*.*

1. *Upon receiving an input* (send, sid, m) *from* S*, do: If* $sid = (S, R, sid_0)$ *for* R*, then output* (sent, sid, m) *to the adversary, and, after a delay, provide the same output to* R *and halt. Otherwise, ignore the input.*
2. *Upon receiving* (corruptsend, sid, m_0) *from the adversary, if* S *is corrupt and* (sent, sid, m) *was not yet delivered to* R*, then output* (sent, sid, m_0) *to* R *and halt.* □

In [10], Dodis et al. have presented a protocol which EUC securely realizes $\mathcal{F}_{auth}$ with the use of $\overline{\mathcal{G}}_{krk}^{\Pi}$. In the protocol, the sender S simply computes a message authentication code (MAC) tag for the tuple (sid, S, R, m) using the key that he non-interactively shares with R by non-interactive authenticated key exchange (NI-AKE), while R verifies the MAC tag using the same key. The definitions of MAC and NI-AKE and the protocol in [10] are given below.

Definition 8 (Message authentication code (MAC) [1]). *A message authentication code (MAC)* MAC *is a pair of algorithms* (MAC.Sign, MAC.Ver)*.*

- $\sigma \leftarrow \mathsf{MAC.Sign}_{mk}(\tau)$
  An algorithm that on input a mac key $mk \in \mathcal{K}_{MAC}$ *and a message* $\tau \in \{0,1\}^*$*, outputs a string* σ*, where* $\mathcal{K}_{MAC}$ *denotes the mac key space.*
- $b \leftarrow \mathsf{MAC.Ver}_{mk}(\sigma, \tau)$
  An algorithm that verifies that σ *is the signature for* τ *using the mac key* mk*, and outputs a boolean* $b \in \{0,1\}$*.* □

We say that (σ, τ) is valid with regard to a mac key mk if $\sigma = \mathsf{MAC.Sign}_{mk}(\tau)$. The correctness of MAC requires that for any $mk \leftarrow \mathcal{K}_{MAC}$ and any $\tau \in \{0,1\}^*$, $\mathsf{MAC.Ver}_{mk}(\mathsf{MAC.Sign}_{mk}(\tau), \tau) = 1$.

Let A_{MAC} be a polynomial-time machine that plays the following game.
[GAME.MAC]
Step 1. $mk \leftarrow \mathcal{K}_{MAC}$ and $\tau \leftarrow \{0,1\}^*$.
Step 2. $\sigma \leftarrow \mathsf{MAC.Sign}_{mk}(\tau)$.
Step 3. $(\sigma_0, \tau_0) \leftarrow A_{MAC}(\sigma, \tau)$.
In Step 3, A_{MAC} is restricted not to output $(\sigma_0, \tau_0) = (\sigma, \tau)$. We define $\epsilon_{\mathsf{mac},A_{MAC}} = \Pr[\mathsf{MAC.Ver}_{mk}(mk, \sigma_0, \tau_0) = 1]$ and $\epsilon_{\mathsf{mac}} = \max_{A_{MAC}}(\epsilon_{\mathsf{mac},A_{MAC}})$ where maximum is taken over all PPT machines. We say that a MAC is secure against one-time chosen message attack (OT-CMA secure) if ϵ_{mac} is negligible in λ.

Definition 9 (Non-interactive authenticated key exchange (NI-AKE) [10]). *A non-interactive authenticated key exchange (NI-AKE) scheme* NI-AKE *consists of two algorithms* AKE.Gen *and* SymExt.

- $(pk, sk) \leftarrow \mathsf{AKE.Gen}(1^\lambda)$
  A probabilistic algorithm that on input the security parameter λ, generates a pair of public and private keys (pk, sk).
- $k = \mathsf{SymExt}(sk_i, pk_j)$
  A deterministic algorithm that computes a shared key $k \in \mathcal{K}_{AKE}$, where $\mathcal{K}_{AKE}$ denotes the shared key space. □

The correctness of NI-AKE requires that for any $(pk_i, sk_i) \leftarrow \mathsf{AKE.Gen}(1^\lambda)$, and any $(pk_j, sk_j) \leftarrow \mathsf{AKE.Gen}(1^\lambda)$, $\mathsf{SymExt}(sk_i, pk_j) = \mathsf{SymExt}(sk_j, pk_i)$.

Let A_{AKE} be a polynomial-time machine that plays the following game.
[GAME.AKE]
Step 1. $(pk_0, sk_0) \leftarrow \mathsf{AKE.Gen}(1^\lambda)$,
and $(pk_1, sk_1) \leftarrow \mathsf{AKE.Gen}(1^\lambda)$.
Step 2. $b \leftarrow \{0,1\}$.
Step 3. If $b = 0$, $k = \mathsf{SymExt}(sk_0, pk_1)$.
Otherwise, $k \leftarrow \mathcal{K}_{AKE}$.
Step 4. $\tilde{b} \leftarrow A_{AKE}(pk_0, pk_1, k)$.
We define $\epsilon_{\mathsf{ake},A_{AKE}} = |\Pr[\tilde{b} = b] - \frac{1}{2}|$ and $\epsilon_{\mathsf{ake}} = \max_{A_{AKE}}(\epsilon_{\mathsf{ake},A_{AKE}})$ where maximum is taken over all PPT machines. We say that an NI-AKE is secure if ϵ_{ake} is negligible in λ. The NI-AKE can be implement by using the Diffie Hellman key exchange scheme[9].

Definition 10 (EUC secure message authentication protocol in [10]). $\mathsf{KRK.Gen}^\lambda(r)$

1. $(AKE.pk, AKE.sk) \leftarrow \mathsf{AKE.Gen}(1^\lambda; r)$.
2. *Outputs* $(AKE.pk, AKE.sk)$

The message authentication protocol Φ proceeds as follows.

1. *Upon input* (send, sid, m), *S computes* $k = \mathsf{SymExt}(AKE.sk_S, AKE.pk_R)$ *and* $\sigma = \mathsf{MAC.Sign}_k(m)$. *$S$ sends* (sid, S, R, m, σ).
2. *Upon receiving* (sid, S, R, m', σ') *from S, R computes* $k' = \mathsf{SymExt}(AKE.sk_S, AKE.pk_R)$. *If* $\mathsf{MAC.Ver}_k(\sigma', m') = 1$ *then R outputs* (sent, sid, m'), *else R aborts.* □

Definition 11 (Public key encryption (PKE) [1]). *A public key encryption (PKE) scheme* PKE *consists of three algorithms,* PKE.Gen, PKE.Enc, *and* PKE.Dec.

- $(pk, sk) \leftarrow \mathsf{PKE.Gen}(1^\lambda)$
  A probabilistic algorithm that on input the security parameter λ, generates public and private keys (pk, sk). The public key defines the message space $\mathcal{M}$.
- $c \leftarrow \mathsf{PKE.Enc}_{pk}(m)$
  A probabilistic algorithm that encrypts a message $m \in \mathcal{M}$ into a ciphertext c.
- $m \leftarrow \mathsf{PKE.Dec}_{sk}(c)$
  An algorithm that decrypts c. It outputs either $m \in \mathcal{M}$ or a special symbol $\perp \notin \mathcal{M}$. □

The correctness of PKE requires that for any $(pk, sk) \leftarrow \mathsf{PKE.Gen}(1^\lambda)$ and any $m \in \mathcal{M}$, $\mathsf{PKE.Dec}_{sk}(\mathsf{PKE.Enc}_{pk}(m)) = m$.

Let A_E be a polynomial-time oracle machine that plays the following game. By $\mathcal{O}$, we denote the decryption oracle, $\mathsf{PKE.Dec}_{sk}(\cdot)$

[GAME.PKE]
Step 1. $(pk, sk) \leftarrow \mathsf{PKE.Gen}(1^\lambda)$.
Step 2. $(m_0, m_1, \rho) \leftarrow A_E^{\mathcal{O}}(pk)$.
Step 3. $b \leftarrow \{0, 1\}, c \leftarrow \mathsf{PKE.Enc}_{pk}(m_b)$.
Step 4. $\tilde{b} \leftarrow A_E^{\mathcal{O}}(\rho, c)$.

In Step 4, A_E is restricted not to ask c to $\mathcal{O}$. In addition, m_0 and m_1 must be of the same length. We define $\epsilon_{\mathsf{pke},A_E} = |\Pr[\tilde{b} = b] - \frac{1}{2}|$ and $\epsilon_{\mathsf{pke}} = \max_{A_E}(\epsilon_{\mathsf{pke},A_E})$ where maximum is taken over all PPT machines. We say that a PKE is CCA-secure if ϵ_{pke} is negligible in λ. Cramer and Shoup in [7] presented the first truly practical CCA-secure encryption scheme.

A Sigma protocol is a 3-round honest-verifier zero-knowledge protocol. We denote the messages sent by a prover P and a verifier V by (a, e, z). We say that a transcript (a, e, z) is an accepting transcript for x if the protocol instructs V to accept based on the values (x, a, e, z). A Sigma protocol is formally defined as follows.

Definition 12 (Sigma protocol [14]). *Let k be security parameter. A protocol is a Σ-protocol for relation R if it is a three-round public-coin protocol and the following requirements hold.*

- *Completeness: If P and V follow the protocol on input x and private input w to P where $(x, w) \in R$, then V always accepts.*
- *Special soundness: There exists a polynomial-time algorithm A that given any x and any pair of accepting transcripts (a, e, z), (a, e', z) for x where $e \neq e'$, outputs w such that $(x, w) \in R$.*
- *Special honest verifier zero knowledge: There exists a probabilistic polynomial-time simulator M, which on input x and e outputs a transcript of the form*

(a, e, z) with the same probability distribution as transcripts between the honest P and V on common input x. Formally, for every x and w such that $(x, w) \in R$ *and every* $e \in \{0,1\}^t$ *it holds that* $|\Pr[M(x,e) = 1] - \Pr[\langle P(x,w), V(x,e)\rangle]| \leq neg(k)$ *where* $M(x,e)$ *denotes the output of simulator M upon input x and e, and* $\langle P(x,w), V(x,e)\rangle$ *denotes the output transcript of an execution between P and V, where P has input* (x,w)*, V has input x, and V's challenge is e.* □

A dual mode cryptosystem DUAL has a regular key generation algorithm and an alternative one. When a regular key is used, it behaves as a standard public key encryption scheme. On the other hand, when an alternative key is used, it perfectly hides the encrypted value. The regular and alternative keys are indistinguishable. A simple version of a dual mode cryptosystem is formally defined as follows.

Definition 13 (Dual mode cryptosystem [16]). *A dual mode cryptosystem* DUAL *is four algorithms* (DUAL.RegGen, DUAL.AlterGen, DUAL.Enc, DUAL.Dec).

- $(pk, sk) \leftarrow \mathsf{DUAL.RegGen}(1^\lambda)$
  A probabilistic algorithm that on input the security parameter λ*, generates regular public and private keys* (pk, sk)*. The public key defines the message space* $\mathcal{M}$*.*
- $(pk, sk) \leftarrow \mathsf{DUAL.AlterGen}(1^\lambda)$
  A probabilistic algorithm that on input the security parameter λ*, generates alternative public and private keys* (pk, sk)*. The public key defines the message space* $\mathcal{M}$*.*
- $c \leftarrow \mathsf{DUAL.Enc}_{pk}(m)$
  A probabilistic algorithm that encrypts a message $m \in \mathcal{M}$ *into a ciphertext c.*
- $m \leftarrow \mathsf{DUAL.Dec}_{sk}(c)$
  An algorithm that decrypts c. It outputs either $m \in \mathcal{M}$ *or a special symbol* $\bot \notin \mathcal{M}$*.* □

The correctness of DUAL requires that for any $(pk, sk) \leftarrow \mathsf{DUAL.RegGen}(1^\lambda)$ and any $m \in \mathcal{M}$, $\mathsf{DUAL.Dec}_{sk}(\mathsf{DUAL.Enc}_{pk}(m)) = m$ and, for any $(pk, sk) \leftarrow \mathsf{DUAL.AlterGen}(1^\lambda)$ and any $m \in \mathcal{M}$, $\mathsf{DUAL.Dec}_{sk}(\mathsf{DUAL.Enc}_{pk}(m)) = m$. For any $(pk, sk) \leftarrow \mathsf{DUAL.RegGen}(1^\lambda)$ and any $m \in \mathcal{M}$, $c_0 = \mathsf{DUAL.Enc}_{pk}(m)$ and $c_1 = \mathsf{DUAL.Enc}_{pk}(m)$ are indistinguishable without negligible probability. On the other hand, when $(pk, sk) \leftarrow \mathsf{DUAL.AlterGen}(1^\lambda)$, $c = \mathsf{DUAL.Enc}_{pk}(m)$ is perfectly hiding m.

Let A_{DUAL} be a polynomial-time machine that plays the following game.

[GAME.DUAL]
Step 1. $b \leftarrow \{0,1\}$.
Step 2. If $b = 0$, $(pk, sk) \leftarrow \mathsf{DUAL.RegGen}(1^\lambda)$.
Otherwise, $(pk, sk) \leftarrow \mathsf{DUAL.AlterGen}(1^\lambda)$.
Step 3. $\tilde{b} \leftarrow A_{DUAL}(pk)$.

We define $\epsilon_{\mathsf{dual},A_{DUAL}} = |\Pr[\tilde{b} = b] - \frac{1}{2}|$ and $\epsilon_{\mathsf{dual}} = \max_{A_{DUAL}}(\epsilon_{\mathsf{dual},A_{DUAL}})$ where maximum is taken over all PPT machines. We say that public keys and the alternative keys of a DUAL are indistinguishable if ϵ_{dual} is negligible in λ.

3 Overview

As described in [14], a UC-secure commitment protocol must be both extractable and equivocal without a simulator rewinding the adversary. The extractable property here means that the simulator can extract the value that a corrupted party commits to. The equivocal property means that the simulator can generate commitments that can be opened to any value. Since EUC security is stronger than UC one [6], a EUC secure protocol must be also both extractable and equivocal.

In the following, we first overview Lindell's construction in [14] and then show our extension for EUC secure commitment. Lindell's construction assumes the CRS that consists of a public key of a CCA2-secure PKE scheme PKE and a public key of a dual mode cryptosystem DUAL. The committer C encrypts a string x with PKE and sends the ciphertext to the receiver R as a commitment. In the reveal phase, C sends x with zero-knowledge proof that the commitment is a ciphertext of x. The proof is based on a Sigma protocol and DUAL: Have R first commit to its challenge by encrypting it with DUAL; run the Sigma protocol with R decommitting. We note that the "dual mode" (meaning that DUAL behaves as a regular public-key encryption scheme when a regular key is generated, but perfectly hides the encrypted value when an alternative key is generated) plays an important role to guarantee soundness of this transformation from a Sigma protocol to a zero-knowledge proof. The soundness can only be proven if the commitment of challenges is perfect hiding.

In the UC framework, a simulator freely generates the CRS, and knows its trapdoor (in this case, the secret keys). Thus, Lindell's construction obviously satisfies the extractable property since the simulator can decrypt any commitments of x with the secret key of PKE. The equivocal property is satisfied since the simulator can obtain the challenge before running the Sigma protocol by decrypting its commitment with the secret key of DUAL. In contrast, in the EUC framework, a simulator cannot use the trapdoor (except for personalized ones of corrupted parties) because the setup is given as global one. To satisfy the extractable property, we use the KRK (i.e., each party's keys) instead of the CRS (i.e., the common keys). Specifically, in the commit phase (resp. the reveal phase), the committer C (resp. the receiver R) uses his own public key of PKE (resp. DUAL) obtained from the KRK. From the definition of the KRK, when C is corrupted, the simulator is able to obtain C's secret key and decrypt any commitments. Thus, the extractable property is satisfied.

On the other hand, the equivocal property is satisfied if the simulator can obtain a challenge before running the Sigma protocol. In the KRK model of EUC, if R is corrupted, then the simulator can obtain the challenge by decrypting the commitment with R's secret key of DUAL. Even for honest R, if the commitment of challenge is not tampered by the adversary, then in the simulation, a challenge is chosen by the simulator simulating R and can be used without any change. However, in the case that R is honest and a commitment of challenge is tampered, the simulator cannot obtain the corresponding tampered value from the security of DUAL. Thus, satisfying the equivocal property in this case is essentially difficult.

We solve this by avoiding the need to satisfy the equivocal property for the case that R is honest and R's commitment is tampered by the adversary. Specifically, we use EUC secure authenticated communication for detecting tampering and terminate the execution. Thus, the simulator need not to open the commitment. We use the ideal message authentication functionality $\mathcal{F}_{auth}$. In [10], Dodis et al. prove that realizing EUC secure authentication against adaptive corruptions in the KRK model is impossible, and present a construction against non-adaptive corruptions from MAC and NI-AKE. Thus, we prove the EUC security of the proposed commitment protocol in the presence of static adversaries.

4 Proposed Generic Construction

In this section, we propose an GUC secure commitment protocol Π. Let λ be security parameter. The proposed protocol Π uses a PKE, a Σ-protocol, DUAL, and the ideal message authentication functionality $\mathcal{F}_{auth}$.
$\mathsf{KRK.Gen}^\lambda(r)$

1. $(PKE.pk, SKE.sk) \leftarrow \mathsf{PKE.Gen}(1^\lambda; r)$.
2. $(DUAL.pk,\ DUAL.sk) \leftarrow \mathsf{DUAL.RegGen}(1^\lambda; r)$.
3. $PK = (PKE.pk, DUAL.pk)$, $SK = (PKE.sk, DUAl.sk)$.
4. Outputs (PK, SK).

Commit phase: Upon input (commit, $sid, ssid, P_i, P_j, x$) where $x \in \{0,1\}^{poly(\lambda)}$.

1. The committer P_i sets $m = sid\|ssid\|i\|j\|x$, computes a ciphertext $c = \mathsf{PKE.Enc}_{PKE.pk_i}(m; r_C)$ as a commitment of x, and sends $(sid, ssid, c)$.
2. Upon receiving a message (sid, $ssid$, c) from P_i, the receiver P_j outputs (receipt, $sid, ssid, P_i$, P_j).

Reveal phase:

1. Upon input (reveal, $sid, ssid$), P_i reveals the committed value by sending $(sid, ssid, x)$ to P_j.
2. Let (α, ε, z) be the message of a Σ-protocol for proving that c is an ciphertext of $sid\|ssid\|i\|j\|x$ using witness r_C.
   (a) P_j chooses a random challenge ε for the Σ-protocol and a random r_R, sets $m' = sid\|ssid\|\varepsilon$, and computes $c' = \mathsf{DUAL.Enc}_{DUAL.pk_j}(m'; r_R)$.
   (b) P_j inputs (send, $(P_j, P_i, (sid, ssid)), m'$) into $\mathcal{F}_{auth}$.
   (c) If P_i receives (sent, $(P_j, P_i, (sid, ssid)), m'$) from $\mathcal{F}_{auth}$, proceeds the next step.
   (d) P_i sends $(sid, ssid, \alpha)$.
   (e) P_j sends $(sid, ssid, \varepsilon, r_R)$.

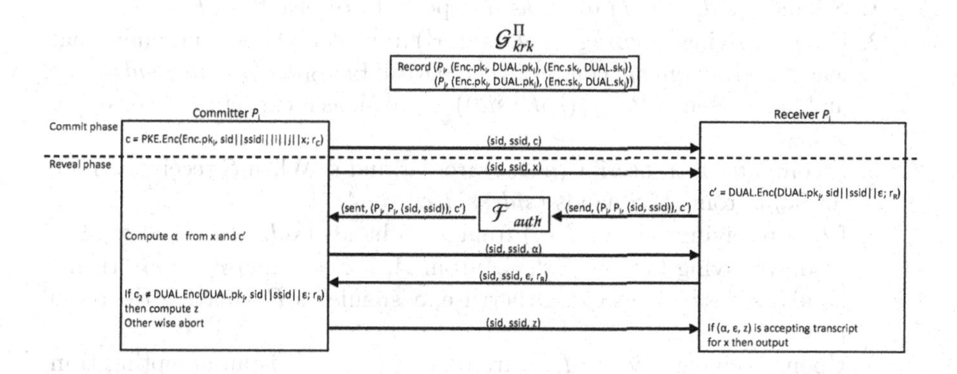

Fig. 2. Overview of the proposed construction

(f) P_i checks that $\mathsf{DUAL.Enc}_{DUAL.pk_j}(sid\|ssid\|\varepsilon; r_R) = c'$ and if yes, sends the reply $(sid, ssid, z)$. Otherwise, P_i aborts.

(g) P_j outputs (reveal, $sid, ssid, P_i, P_j, x$) if and only if $(\alpha,\ \varepsilon,\ z)$ is an accepting transcript.

Theorem 3. *Assuming the existence of a CCA2-secure* PKE, *a Σ-protocol for relation $R = \{(c, m) | c = \text{PKE.Enc}_{pk}(m)\}$, and a* DUAL *of which a regular key and an alternative one are indistinguishable, the proposed protocol Π GUC securely realizes $\mathcal{F}_{mcom}$ in the presence of static adversaries.* □

Proof: We show that the proposed protocol Π $\overline{\mathcal{G}}_{krk}^{\Pi}$-EUC-emulates $\text{IDEAL}_{\mathcal{F}}$, because the proposed protocol Π is $\overline{\mathcal{G}}_{krk}^{\Pi}$-subroutine respecting. We first show a simulator $\mathcal{S}$ for an adversary $\mathcal{A}$, and prove $|\Pr[\text{EXEC}^{\overline{\mathcal{G}}}_{\text{IDEAL}_{\mathcal{F}},\mathcal{S},\mathcal{Z}} = 1] - \Pr[\text{EXEC}^{\overline{\mathcal{G}}}_{\Pi,\mathcal{A},\mathcal{Z}} = 1]| \le neg(\lambda)$ for any environment $\mathcal{Z}$.

In the commit phase, $\mathcal{S}$ encrypts 0 as a commitment instead of x. In the reveal phase, $\mathcal{S}$ reveals x and proves that the commitment is a ciphertext of x by simulating the Σ-protocol. Let P_i and P_j be committer and receiver, respectively. For any PPT static adversary $\mathcal{A}$, the simulator $\mathcal{S}$ behaves as follows.

- Simulating the communication with $\mathcal{Z}$: $\mathcal{S}$ inputs every input value that $\mathcal{S}$ receives from $\mathcal{Z}$ to $\mathcal{A}$. $\mathcal{S}$ outputs every output value of $\mathcal{A}$.
- $\mathcal{S}$ obtains the public keys of P_i and P_j, and the secret key for any corrupted party from $\overline{\mathcal{G}}_{krk}^{\Pi}$.
- Simulating the commit phase when both P_i and P_j are honest: Upon receiving (receipt, sid, $ssid$, P_i, P_j) from $\mathcal{F}_{mcom}$, $\mathcal{S}$ chooses a random r_C, sets $m = sid\|ssid\|i\|j\|0$, computes a commitment $c = \mathsf{PKE.Enc}_{PKE.pk_i}(m; r_C)$ as the committed value $x = 0$, and hands $(sid, ssid, c)$ to $\mathcal{A}$, as it expects to receive from P_i. Upon receiving $(sid, ssid, c'')$ from $\mathcal{A}$, $\mathcal{S}$ sends (receipt, $sid, ssid, P_i, P_j$) to $\mathcal{F}_{mcom}$.
- Simulating the reveal phase when both P_i and P_j are honest: Upon receiving (reveal, sid, $ssid$, P_i, P_j, x) from $\mathcal{F}_{mcom}$, $\mathcal{S}$ behaves as follows.

1. $\mathcal{S}$ hands $(sid, ssid, x)$ to $\mathcal{A}$, as it expects to receive from P_i.
2. Upon receiving a message $(sid, ssid, x')$ from $\mathcal{A}$, $\mathcal{S}$ chooses a random challenge ε, a random r_R, computes $c' = \mathsf{DUAL.Enc}_{DUAL.pk_j}(sid\|ssid\|\varepsilon; r_R)$, and hands $(\text{sent}, (P_j, P_i, (sid, ssid)), c')$ to $\mathcal{A}$, as it expects to receive from $\mathcal{F}_{auth}$.
3. $\mathcal{S}$ computes a transcript (α, ε, z) from x and ε. When $\mathcal{S}$ receives a reply to $\mathcal{F}_{auth}$ from $\mathcal{A}$, $\mathcal{S}$ hands $(sid, ssid, \alpha)$ to $\mathcal{A}$.
4. Upon receiving $(sid, ssid, \alpha')$ from $\mathcal{A}$, $\mathcal{S}$ hands $(sid, ssid, \varepsilon, r_R)$ to $\mathcal{A}$.
5. Upon receiving $(sid, ssid, \varepsilon', r'_R)$ from $\mathcal{A}$, if $\varepsilon' = \varepsilon$ and $r'_R = r_R$, then $\mathcal{S}$ hands $(sid, ssid, z)$ to $\mathcal{A}$. Otherwise, $\mathcal{S}$ simulates P_i aborting the reveal phase.
6. Upon receiving $(sid, ssid, z')$ from $\mathcal{A}$, if $(\alpha', \varepsilon, z')$ is an accepting transcript, then $\mathcal{S}$ sends $(\text{reveal}, sid, ssid, P_i, P_j, x)$ to $\mathcal{F}_{mcom}$. Otherwise, it does nothing.

- Simulating the commit phase when P_i is corrupted and P_j is honest: Upon receiving $(sid, ssid, c)$ from $\mathcal{A}$ as it intends to send from P_i to P_j, $\mathcal{S}$ decrypts c to obtain x. If the result is $\perp$, then $\mathcal{S}$ sends a dummy commitment $(\text{commit}, sid, ssid, P_i, P_j, 0)$ to $\mathcal{F}_{mcom}$. Otherwise, $\mathcal{S}$ sends $(\text{commit}, sid, ssid, P_i, P_j, x)$ to $\mathcal{F}_{mcom}$. Upon receiving a message $(\text{receipt}, sid, ssid, P_i, P_j)$ from $\mathcal{F}_{mcom}$, $\mathcal{S}$ sends $(\text{receipt}, sid, ssid, P_i, P_j)$ to $\mathcal{F}_{mcom}$.
- Simulating the reveal phase when P_i is corrupted and P_j is honest: Upon receiving $(\text{reveal}, sid, ssid, P_i, P_j, x)$ from $\mathcal{F}_{mcom}$, $\mathcal{S}$ behaves as follows.
  1. Upon receiving $(sid, ssid, x)$ from $\mathcal{A}$, $\mathcal{S}$ chooses a random challenge ε and a random r_R. $\mathcal{S}$ computes $c' = \mathsf{DUAL.Enc}_{DUAL.pk_j}(sid\|ssid\|\varepsilon; r_R)$, hands $(\text{sent}, (P_j, P_i, (sid, ssid)), c')$ to $\mathcal{A}$, as it expects to receive from $\mathcal{F}_{auth}$.
  2. Upon receiving a reply to $\mathcal{F}_{auth}$ from $\mathcal{A}$, $\mathcal{S}$ inputs $(\text{sent}, (P_j, P_i, (sid, ssid)), c')$ to the corrupted committer P_i.
  3. Upon receiving $(sid, ssid, \alpha)$ from $\mathcal{A}$, $\mathcal{S}$ hands $(sid, ssid, \varepsilon, r_R)$ to $\mathcal{A}$.
  4. Upon receiving $(sid, ssid, z)$ from $\mathcal{A}$, if (α, ε, z) is an accepting transcript, then $\mathcal{S}$ sends $(\text{reveal}, sid, ssid, P_i, P_j)$ to $\mathcal{F}_{mcom}$. Otherwise, it does nothing.
- Simulating the commit phase when P_i is honest and P_j is corrupted: Upon receiving $(\text{receipt}, sid, ssid, P_i, P_j)$ from $\mathcal{F}_{mcom}$, $\mathcal{S}$ chooses a random r_C, computes a commitment $c = \mathsf{PKE.Enc}_{PKE.pk_i}(0; r_C)$ as $x = 0$, and hands $(sid, ssid, c)$ to $\mathcal{A}$, as it expects to receive from P_i. $\mathcal{S}$ sends $(\text{receipt}, sid, ssid, P_i, P_j)$ to $\mathcal{F}_{mcom}$.
- Simulating the reveal phase when P_i is honest and P_j is corrupted: Upon receiving $(\text{reveal}, sid, ssid, P_i, P_j, x)$ from $\mathcal{F}_{mcom}$, $\mathcal{S}$ works as follows.
  1. $\mathcal{S}$ hands $(sid, ssid, x)$ to $\mathcal{A}$, as it expects to receive from P_i.
  2. When the corrupted receiver P_j produces an input $(\text{sent}, (P_j, P_i, (sid, ssid)), c')$ to $\mathcal{F}_{auth}$, hands $(\text{sent}, (P_j, P_i, (sid, ssid)), c')$ to $\mathcal{A}$, as it expects to receive from $\mathcal{F}_{auth}$.
  3. Upon receiving a reply to $\mathcal{F}_{auth}$ from $\mathcal{A}$, $\mathcal{S}$ decrypts c' to obtain ε' and proceeds the step 5.

4. When $\mathcal{S}$ receives (corruptsend, $(P_j, P_i, (sid, ssid)), c'$) from $\mathcal{A}$, if $\mathcal{A}$ did not send reply to $\mathcal{F}_{auth}$, $\mathcal{S}$ decrypts c' to obtain ε' and proceeds the next step.
5. Let c be as computed by $\mathcal{S}$ in the commit phase. $\mathcal{S}$ computes a transcript $(\alpha, \varepsilon', z)$ from x and ε', and hands $(sid, ssid, \alpha)$ to $\mathcal{A}$.
6. Upon receiving $(sid, ssid, \varepsilon, r_R)$ from $\mathcal{A}$, if $\mathsf{PKE.Enc}_{PKE.pk_j}(\varepsilon; r_R) = c'$, then $\mathcal{S}$ hands $(sid, ssid, z)$ to $\mathcal{A}$. Otherwise, $\mathcal{S}$ simulates P_i aborting the reveal phase.

For the above $\mathcal{S}$, we now show that for any $\mathcal{Z}$, $|\Pr[\mathrm{EXEC}^{\overline{\mathcal{G}}}_{\mathrm{IDEAL}_{\mathcal{F}},\mathcal{S},\mathcal{Z}} = 1] - \Pr[\mathrm{EXEC}^{\overline{\mathcal{G}}}_{\Pi,\mathcal{A},\mathcal{Z}} = 1]| \leq neg(\lambda)$ by a series of hybrid games, HYB-GAME^1, HYB-GAME^2, HYB-GAME^3. Let HYB-GAME$^i_{\mathcal{S},\mathcal{Z}}$ be the output of $\mathcal{Z}$ which runs with $\mathcal{S}$ in HYB-GAME^i.

Hybrid game HYB-GAME^1: In this game, the ideal functionality $\mathcal{F}_{mcom}$ gives the simulator $\mathcal{S}_1$ the value x committed to by an honest P_i, in addition to the message (receipt, $sid, ssid, P_i, P_j$). $\mathcal{S}_1$ behaves in exactly the same way as $\mathcal{S}$ except that when simulating the commit phase when P_i is honest, it computes c as an encryption of x as an honest P_i would. We show that HYB-GAME$^1_{\mathcal{S}_1,\mathcal{Z}}$ is indistinguishable from the output of $\mathcal{Z}$ in the ideal model by reduction to CCA2-security of the PKE scheme.

We construct an adversary A_E for GAME.PKE as follows. Let pk_{pke} be the public key given to A_E. Then A_E simulates an execution of $\mathrm{EXEC}^{\overline{\mathcal{G}}}_{\mathrm{IDEAL}_{\mathcal{F}},\mathcal{S},\mathcal{Z}}$ with the following differences.

1. Whenever the shared functionality $\overline{\mathcal{G}}^{\Pi}_{krk}$ outputs the public key for an honest P_i, A_E hands $(pk_{pke}, DUAL.pk_i)$ instead of $(PKE.pk_i, DUAL.pk_i)$.
2. Whenever an honest P_i commits to a value x, instead of $\mathcal{S}$ encrypting 0 (or $\mathcal{S}_1$ encrypting x), A_E generates the encryption in the ciphertext by asking for an encryption challenge of the pair $(0, x)$. The ciphertext c received back is sent as the commitment.
3. Whenever a corrupted P_i sends a commitment value c and the simulator needs to decrypt c, A_E queries its decryption oracle with c. If c was received as a ciphertext challenge then A_E has the simulator send a dummy commitment (commit, $sid, ssid, P_i, P_j, 0$) to $\mathcal{F}_{mcom}$.

Finally, A_E outputs whatever $\mathcal{Z}$ outputs.

If $b = 0$ in the GAME.PKE, then the commitments c are ciphertexts of 0 when the committer P_i is honest. Thus, the simulation is exactly like $\mathcal{S}$ and the output of A_E is exactly that of $\mathrm{IDEAL}^{\overline{\mathcal{G}}}_{\mathcal{F},\mathcal{S},\mathcal{Z}}$. In contrast, if $b = 1$, then the commitments c are ciphertexts of x and the simulation is exactly like $\mathcal{S}_1$. Thus, the output of A_E is exactly that of HYB-GAME$^1_{\mathcal{S}_1,\mathcal{Z}}$. We conclude that $|\Pr[\text{HYB-GAME}^1_{\mathcal{S}_1,\mathcal{Z}} = 1] - \Pr[\mathrm{EXEC}^{\overline{\mathcal{G}}}_{\mathrm{IDEAL}_{\mathcal{F}},\mathcal{S},\mathcal{Z}} = 1]| \leq neg(\lambda)$, by the assumption that PKE is CCA2-secure.

Hybrid game HYB-GAME^2: In this game, the simulator $\mathcal{S}_2$ behaves in exactly the same way as $\mathcal{S}_1$, except that when simulating the reveal phase in the case that P_i is honest, it computes the messages α and z from x and c as same as the honest committer. $\mathcal{S}_2$ can do this because the commitment c sent in the commitment phase is the correct value x and so it can play the honest prover. Therefore, $\mathcal{S}_2$ perfectly simulates the proof of the reveal phase. We can show that HYB-GAME$^2_{\mathcal{S}_2,\mathcal{Z}}$ is exactly the same as HYB-GAME$^1_{\mathcal{S}_1,\mathcal{Z}}$ by the property of the Σ-protocol, special honest verifier zero knowledge, because $\mathcal{S}_1$ can obtain the challenge of P_j and simulate the Σ-protocol. We therefore have that $|\Pr[\text{HYB-GAME}^2_{\mathcal{S}_2,\mathcal{Z}} = 1] - \Pr[\text{HYB-GAME}^1_{\mathcal{S}_1,\mathcal{Z}} = 1]| \leq neg(\lambda)$.

Hybrid game HYB-GAME^3: In this game, the simulator $\mathcal{S}_3$ behaves in exactly the same way as $\mathcal{S}_2$, except that when simulating the reveal phase in the case that P_j is honest, it encrypts ε by the alternative key $DUAL.pk'$ instead of the regular key $DUAL.pk_j$. We show that the output of $\mathcal{Z}$ in HYB-GAME^3 is indistinguishable from the output of $\mathcal{Z}$ in HYB-GAME^2 by reduction to the indistinguishability of a regular key and an alternative one of DUAL.

We construct an adversary A_{DUAL} for GAME.DUAL as follows. Let pk be the input given to A_{DUAL}. Then A_{DUAL} simulates an execution of HYB-GAME$^2_{\mathcal{S}_2,\mathcal{Z}}$ with the following differences.

1. Whenever the shared functionality $\overline{\mathcal{G}}^{\Pi}_{krk}$ outputs the public key for an honest P_j, A_{DUAL} hands $(PKE.pk_j, pk)$ instead of $(PKE.pk_j, DUAL.pk_j)$.
2. When P_j is honest, A_{DUAL} uses pk to encrypt ε in the reveal phase.

Finally, A_{DUAL} outputs whatever $\mathcal{Z}$ outputs.

Now, if $b = 0$ in GAME.DUAL, pk is the regular key. Thus, the simulation is exactly like $\mathcal{S}_2$ and the output of A_{DUAL} is exactly that of HYB-GAME$^2_{\mathcal{S}_2,\mathcal{Z}}$. In contrast, if $b = 1$, pk is the alternative key. Thus, the output of A_{DUAL} is exactly that of HYB-GAME$^3_{\mathcal{S}_3,\mathcal{Z}}$. We conclude that $|\Pr[\text{HYB-GAME}^2_{\mathcal{S}_2,\mathcal{Z}} = 1] - \Pr[\text{HYB-GAME}^3_{\mathcal{S}_3,\mathcal{Z}} = 1]| \leq neg(\lambda)$, by the assumption that a regular key and an alternative one are indistinguishable.

Completing the proof: It remains to show that the output of $\mathcal{Z}$ after an execution in the real model is indistinguishable from the output of $\mathcal{Z}$ in HYB-GAME^3. We show that the outputs of P_j in the reveal phase are identical in both cases. We observe that the case that the outputs of P_j are different only occurs when P_i is corrupted and P_j is honest. Specifically, in the real model, even though P_i committed x in the commit phase, P_j outputs that x_0 is committed. In contrast, P_j in HYB-GAME^3 always outputs that x is committed. However, the zero-knowledge proof in HYB-GAME^3 is sound because $\mathcal{S}_3$ uses an alternative key of DUAL. If P_j outputs that x_0 is committed with non-negligible probability, then we can construct an adversary for GAME.DUAL which can distinguish regular and alternative keys with non-negligible probability. Therefore, the outputs of P_j in the reveal phase are identical in both cases, that is, we conclude that $|\Pr[\text{HYB-GAME}^3_{\mathcal{S}_3,\mathcal{Z}} = 1] - \Pr[\text{REAL}^{\mathcal{G}}_{\Pi,\mathcal{A},\mathcal{Z}} = 1]| \leq neg(\lambda)$.

Therefore, for every $\mathcal{A}$ and $\mathcal{Z}$, $|\Pr[\text{EXEC}^{\mathcal{G}}_{\text{IDEAL}_{\mathcal{F}},\mathcal{S},\mathcal{Z}} = 1] - \Pr[\text{EXEC}^{\mathcal{G}}_{\Pi,\mathcal{A},\mathcal{Z}} = 1]| \leq neg(\lambda)$. Thus, the proposed protocol Π $\overline{\mathcal{G}}^{\Pi}_{krk}$-EUC-emulates $\text{IDEAL}_{\mathcal{F}_{mcom}}$.

By Theorem 1, Π GUC-emulates $\text{IDEAL}_{\mathcal{F}_{mcom}}$, that is, Π GUC securely realizes $\mathcal{F}_{mcom}$. □

Corollary 1. *Let Π be the proposed protocol and Φ be the EUC secure message authentication protocol proposed in [10]. Let $\Pi^{\Phi/\text{IDEAL}_{\mathcal{F}_{auth}}}$ denote a modified version of Π that invokes Φ instead of the ideal protocol for $\mathcal{F}_{auth}$. Specifically, $\Pi^{\Phi/\text{IDEAL}_{\mathcal{F}_{auth}}}$ is as follows.*

- $\mathsf{KRK.Gen}^{\lambda}(r)$ *outputs* $PK = (PKE.pk, DUAL.pk, AKE.pk)$ *and* $SK = (PKE.sk, DUAL.sk, AKE.sk)$, *where* $(PKE.pk, SKE.sk) \leftarrow \mathsf{PKE.Gen}(1^{\lambda}; r)$, $(DUAL.pk, DUAL.sk) \leftarrow \mathsf{DUAL.RegGen}(1^{\lambda}; r)$, *and* $(AKE.pk, AKE.sk) \leftarrow \mathsf{AKE.Gen}(1^{\lambda}; r)$.
- *Steps (b) and (c) in the reveal phase are replaced with following (b') and (c').*
  - *(b')* P_j *computes* $k = \mathsf{SymExt}(AKE.sk_j, AKE.pk_i)$, *and* $\sigma = \mathsf{MAC.Sign}_k(c')$. P_j *sends* $(sid, ssid, c', \sigma)$.
  - *(c') Upon receiving a message* $(sid, ssid, c', \sigma)$ *from* P_j, P_i *computes* $k = \mathsf{SymExt}(AKE.sk_i, AKE.pk_j)$. *If* $\mathsf{MAC.Ver}_k(\sigma, c') = 1$, *proceeds the next step. Otherwise* P_i *aborts.*

Assuming the existence of a CCA2-secure PKE, *a Σ-protocol for relation* $R = \{(c, m) | c = \text{PKE.Enc}_{pk}(m)\}$, *a* DUAL *of which a regular key and an alternative one are indistinguishable, a secure* NI-AKE, *and an OT-CMA-secure* MAC, $\Pi^{\Phi/\text{IDEAL}_{\mathcal{F}_{auth}}}$ *GUC-emulates* Π. □

This corollary is proved by Theorem 2.

5 Conclusion

In this paper, we have proposed a generic construction of GUC secure commitment in the KRK model which uses a GUC secure authentication protocol, a CCA2-secure PKE scheme, a dual mode cryptosystem, and a Σ-protocol. We have showed that the proposed construction is GUC secure in the presence of static adversaries. The proposed construction is the first GUC secure one in which the commit phase is non-interactive. A possible future work is to present a construction with non-interactive commit phase which is GUC secure even in the presence of adaptive adversaries.

References

1. Abe, M., Gennaro, R., Kurosawa, K.: Tag-KEM/DEM: A New Framework for Hybrid Encryption. J. Cryptology 21(1), 97–130 (2008)
2. Camenisch, J.L., Shoup, V.: Practical Verifiable Encryption and Decryption of Discrete Logarithms. In: Boneh, D. (ed.) CRYPTO 2003. LNCS, vol. 2729, pp. 126–144. Springer, Heidelberg (2003)
3. Canetti, R.: Universally Composable Security: A New Paradigm for Cryptographic Protocols. IACR Cryptology ePrint Archive, Report 2000/067 (2000)

4. Canetti, R., Fischlin, M.: Universally Composable Commitments. In: Kilian, J. (ed.) CRYPTO 2001. LNCS, vol. 2139, pp. 19–40. Springer, Heidelberg (2001)
5. Canetti, R., Lindell, Y., Ostrovsky, R., Sahai, A.: Universally Composable Two-party and Multi-party Secure Computation. In: STOC 2002, pp. 494–503 (2002)
6. Canetti, R., Dodis, Y., Pass, R., Walfish, S.: Universally Composable Security with Global Setup. In: Vadhan, S.P. (ed.) TCC 2007. LNCS, vol. 4392, pp. 61–85. Springer, Heidelberg (2007)
7. Cramer, R., Shoup, V.: A Practical Public Key Cryptosystem Provably Secure against Adaptive Chosen Ciphertext Attack. In: Krawczyk, H. (ed.) CRYPTO 1998. LNCS, vol. 1462, pp. 13–25. Springer, Heidelberg (1998)
8. Damgard, I., Groth, J.: Non-interactive and Reusable Non-Malleable Commitment Schemes. In: STOC 2003, pp. 426–437 (2003)
9. Diffie, W., Hellman, M.E.: New Directions in Cryptography. IEEE Trans. on Information Theory IT-22(6), 644–654 (1976)
10. Dodis, Y., Katz, J., Smith, A., Walfish, S.: Composability and On-Line Deniability of Authentication. In: Reingold, O. (ed.) TCC 2009. LNCS, vol. 5444, pp. 146–162. Springer, Heidelberg (2009); The full version is [17]
11. Damgård, I.B., Nielsen, J.B.: Perfect Hiding and Perfect Binding Universally Composable Commitment Schemes with Constant Expansion Factor. In: Yung, M. (ed.) CRYPTO 2002. LNCS, vol. 2442, pp. 581–596. Springer, Heidelberg (2002)
12. Dodis, Y., Shoup, V., Walfish, S.: Efficient Constructions of Composable Commitments and Zero-Knowledge Proofs. In: Wagner, D. (ed.) CRYPTO 2008. LNCS, vol. 5157, pp. 515–535. Springer, Heidelberg (2008)
13. Fischlin, M., Libert, B., Manulis, M.: Non-interactive and Re-usable Universally Composable String Commitments with Adaptive Security. In: Lee, D.H., Wang, X. (eds.) ASIACRYPT 2011. LNCS, vol. 7073, pp. 468–485. Springer, Heidelberg (2011)
14. Lindell, Y.: Highly-Efficient Universally-Composable Commitments Based on the DDH Assumption. In: Paterson, K.G. (ed.) EUROCRYPT 2011. LNCS, vol. 6632, pp. 446–466. Springer, Heidelberg (2011)
15. Nishimaki, R., Fujisaki, E., Tanaka, K.: Efficient Non-interactive Universally Composable String-Commitment Schemes. In: Pieprzyk, J., Zhang, F. (eds.) ProvSec 2009. LNCS, vol. 5848, pp. 3–18. Springer, Heidelberg (2009)
16. Peikert, C., Vaikuntanathan, V., Waters, B.: A Framework for Efficient and Composable Oblivious Transfer. In: Wagner, D. (ed.) CRYPTO 2008. LNCS, vol. 5157, pp. 554–571. Springer, Heidelberg (2008)
17. Walfish, S.: Enhanced Security Models for Network Protocols. Ph.D. thesis, New York University (2007)

Author Index

Ahn, Gail-Joon 1

Chen, Chien-Ning 37
Chen, Kefei 182

Fouque, Pierre-Alain 197
Fujiwara, Toru 244

Hu, Hongxin 1
Huang, Liusheng 233

Iokibe, Kengo 51
Isobe, Takanori 138

Jacob, Nisha 37
Jing, Yiming 1

Kawachi, Akinori 123
Kiyoshima, Susumu 216
Kutzner, Sebastian 37

Ling, San 37
Lu, Jiqiang 197

Manabe, Yoshifumi 216
Mohamad, Moesfa Soeheila 105
Morii, Masakatu 138

Nekado, Kenta 51
Nogami, Yasuyuki 51

Ohigashi, Toshihiro 138
Ohta, Kazuo 156, 170
Okamoto, Tatsuaki 216

Pasalic, Enes 197
Poh, Geong Sen 105
Poschmann, Axel 37

Rangan, C. Pandu 87

Saetang, Sirote 37
Sakiyama, Kazuo 156, 170
Sakurai, Kouichi 19
Sasaki, Yu 156, 170
Selvi, S. Sharmila Deva 87
Suzuki, Itsuki 244

Takagi, Tsuyoshi 19, 233
Takasaki, Yasuhiro 156
Takebe, Hirotoshi 123
Tanaka, Keisuke 123

Vivek, S. Sree 87

Wang, Jingjing 182
Wang, Lei 156, 170
Wei, Yongzhuang 197

Yasuda, Takanori 19
Yoneyama, Kazuki 69
Yoshida, Maki 244

Z'aba, Muhammad Reza 105
Zhu, Shixiong 182
Zhu, Youwen 233